Fisheries Biology and Assessment

Fisheries Biology and Assessment

Edited by Roger Creed

SYRAWOOD
PUBLISHING HOUSE

New York

Published by Syrawood Publishing House,
750 Third Avenue, 9th Floor,
New York, NY 10017, USA
www.syrawoodpublishinghouse.com

Fisheries Biology and Assessment
Edited by Roger Creed

International Standard Book Number: 978-1-68286-378-7 (Hardback)

Cataloging-in-publication Data

Fisheries biology and assessment / edited by Roger Creed.
    p. cm.
Includes bibliographical references and index.
ISBN 978-1-68286-378-7
1. Fish populations. 2. Fishes--Ecology. 3. Fisheries--Environmental aspects. 4. Fish culture. 5. Aquaculture. I. Creed, Roger.
QL618.3 .F57 2017
639.3--dc23

Printed in the United States of America.

TABLE OF CONTENTS

PREFACE

Fisheries are an organization that farm and harvest fish for commercial purposes. Fisheries as an academic discipline deal with fisheries management, conservation and ethical practices of food harvesting and consumption. This book strives to provide a fair idea of fisheries biology and to help develop a better understanding of the latest advances within this field. It includes some of the vital pieces of work being conducted across the world, on various topics related to fisheries biology and assessment. It elucidates the concepts and innovative models around prospective developments with respect to this discipline. For all those who are interested in management and regulation of fisheries such as oceanographers, marine biologists and aquatic conservationists, this book can prove to be an essential guide.

The main aim of this book is to educate learners and enhance their research focus by presenting diverse topics covering this vast field. This is an advanced book which compiles significant studies by distinguished experts. This book addresses successive solutions to the challenges arising in the area of application, along with it; the book provides scope for future developments.

It was a great honour to edit this book, though there were challenges, as it involved a lot of communication and networking between me and the editorial team. However, the end result was this all-inclusive book covering diverse themes in the field.

Finally, it is important to acknowledge the efforts of the contributors for their excellent chapters, through which a wide variety of issues have been addressed. I would also like to thank my colleagues for their valuable feedback during the making of this book.

Editor

Persistent Oxytetracycline Exposure Induces an Inflammatory Process That Improves Regenerative Capacity in Zebrafish Larvae

Francisco Barros-Becker[1], Jaime Romero[2], Alvaro Pulgar[1], Carmen G. Feijóo[1]*

1 Departamento de Ciencias Biologicas, Facultad de Ciencias Biologicas, Universidad Andres Bello, Santiago, Chile, 2 Instituto de Nutrición y Tecnología de los Alimentos, Universidad de Chile, Santiago, Chile

Abstract

Background: The excessive use of antibiotics in aquaculture can adversely affect not only the environment, but also fish themselves. In this regard, there is evidence that some antibiotics can activate the immune system and reduce their effectiveness. None of those studies consider in detail the adverse inflammatory effect that the antibiotic remaining in the water may cause to the fish. In this work, we use the zebrafish to analyze quantitatively the effects of persistent exposure to oxytetracycline, the most common antibiotic used in fish farming.

Methodology: We developed a quantitative assay in which we exposed zebrafish larvae to oxytetracycline for a period of 24 to 96 hrs. In order to determinate if the exposure causes any inflammation reaction, we evaluated neutrophils infiltration and quantified their total number analyzing the $Tg(mpx:GFP)^{i114}$ transgenic line by fluorescence stereoscope, microscope and flow cytometry respectively. On the other hand, we characterized the process at a molecular level by analyzing several immune markers (*il-1β*, *il-10*, *lysC*, *mpx*, *cyp1a*) at different time points by qPCR. Finally, we evaluated the influence of the inflammation triggered by oxytetracycline on the regeneration capacity in the lateral line.

Conclusions: Our results suggest that after 48 hours of exposure, the oxytetracycline triggered a widespread inflammation process that persisted until 96 hours of exposure. Interestingly, larvae that developed an inflammation process showed an improved regeneration capacity in the mechanosensory system lateral line.

Editor: Bernhard Ryffel, French National Centre for Scientific Research, France

Funding: CGF was supported by Fondo Nacional de Desarrolo Cientifico y Tecnologico (Fondecyt) 11090102, Universidad Andres Bello (UNAB) DI39-11/R, and Comision Nacional de Investigacion Cientifica y Tecnologica (Conicyt) 79090006. JR was supported by Fondecyt 1110253. The funders had no role in study design, data collection and analysis, decision to publish, or preparation of the manuscript.

Competing Interests: The authors have declared that no competing interests exist.

* E-mail: cfeijoo@unab.cl

Introduction

The aquaculture of fish, constitutes a rapidly world wide growing industry, especially for salmon and trout business. However this growth has been also accompanied by an increasing number of infectious diseases affecting fish, and therefore it has become very common to dose animals with antibiotics in the food to protect against illness [1]. As fish pens are typically located in rivers or lakes, feces, uneaten food pellets as well as antibiotic residues, are distributed over the entire ecosystem. Due to approximately 70 to 80% of the antibiotics administered are released into the aquatic environment, these chemicals must be considered potential environmental micropollutants [2]. Unfortunately, there are no reports about this problem and only limited information is available regarding the presence of antibiotics in the sediment surrounding the fish farms in a few countries [3–6].

The effects of antibiotics on the immune system in fish are numerous and can be different for each drug used. There is evidence suggesting that they can suppress immune functions in carp, rainbow trout, turbot and Atlantic cod [7–16]. However, the results reported from experimental investigations are contradictory

as they depend on the type of assay developed and the fish species studied. Importantly, only the effect of antibiotics ingested or present in the bloodstream has been analyzed, but not the effect on the immune system triggered by antibiotics remaining in the water. As an alternative to analyzing the repercussions on a specific fish species and for unifying criteria, we decided to use the zebrafish (*Danio rerio*) as a model. This teleost fish has an especially well known biology, rapid development, is very easy to handle and allows us to make all the analysis *in vivo* and with a high number of specimens per data point [17–18]. Furthermore it has become of widespread use in ecotoxicology and toxicology research [19–22] and a particularly attractive and powerful new model for immunity research as it has number of strengths, including genetic tractability and transparency in embryonic and larval stage, which facilitates monitoring of infection processes [23–27].

In the zebrafish, the innate immune system becomes active early during somitogenesis, with fully functional macrophages appearing by 16 hpf and neutrophils at approximately 26 hpf [28]. In the case of an injury, both macrophages and neutrophils are able to migrate from the intermediate cell mass (ICM) (it will become the caudal hematopoietic tissue later during development) to the affected territories indicating that they are mature enough to fulfill a role as first defenders against an aggressor. Early during life of the fish, this innate immune system exists in isolation of an adaptive system, which only develops later in larval stages requiring 4–6 weeks to achieve a fully functional sate [29]. However, this adaptive immune system is very rudimentary in the zebrafish and consists mainly of B lymphocytes [30]. Therefore the gap between both systems allows considering infection of zebrafish larvae as a way to study exclusively the innate immune response without any adaptive immune contribution.

The hallmark of innate immunity response is inflammation. This process is triggered in response to injury, irritants, or pathogens [31]. If inflammation occurs there are influx, accumulation, and activation of leukocytes (predominantly neutrophils) at the site of injury during the early stages of the response [32]. These cells destroy the *injury agent* through the production of non-specific toxins, such as superoxide radicals, hypochlorite, and hydroxyl radicals [33]. Neutrophils also release several cytokines, including interleukin Il-1β, which is essential for inflammation progression. If the *injury agent* is removed, macrophage will secrete anti-inflammatory cytokines, such Il-10, which marks the end of the inflammatory process [34].

In the present study we investigate the effect of a prolonged exposure to oxytetracycline, one of the most commonly used antibiotics in finfish farming, on the innate immune response and regeneration capacity of zebrafish larvae. The goal is to establish an experimental model that allows us to analyze *in vivo* the effect on fish immunity of a persistent exposure to antibiotics, in many cases in high doses, present in the aquatic environment of fish farms.

Results and Discussion

Determination of LC$_{50}$

Oxytetracycline is a broad-spectrum antibiotic with considerable activity against Gram-negative bacteria with worldwide use in fish farming [5]. The drug is administered to fish mixed in food at a dose rate of 50–100 mg per kg fish per day for 3 to 21 days. As during infection, fish usually show reduced feed intake and considering the low oxytetracycline bioavailability, it can be assumed that a considerable part of the medicated feed pass the treated fish uneaten and unabsorbed to the environment of the fish farm. Although, there is no current information about the concentrations of oxytetracycline detected in the water column or sediment in fish farms, only a few reports from the nineties are available and indicate that oxytetracycline is very persistent in fish farm sediments. In the environment surrounding pens up to 4.4 ug/g (4,4 ppm) of antibiotic has been detected even after 308 days of administration [35–37]. These levels rise dramatically in a freshwater recirculating system were after administrating the medicated feed for 10 days, the concentration in the sediment can reach 2150 µg/g (2150 ppm) [38]. This suggests that commercially relevant fish could be exposed to important doses of this compound after controlled administration.

Another important point related to the environmental oxytetracycline concentration is the amount of drug used, which differs greatly between countries. In the salmon industry, Norway and Chile are two world leader producers. During 2007 and 2008, to produce a tone of salmon Norway used 0,02 and 0,07 gr of antibiotic respectively. For the same period and production, Chile used 732 and 560 gr of antibiotic. This means that Chile used 36.600 and 8.000 times more antibiotics than in Norway [39] and therefore the fish farm environment is considerably more contaminated with these drugs.

Due to the lack of information about the concentration and effect of oxytetracycline in fish farms we decided to determine the LC$_{50}$ at 24 hours post incubation (hpi), during embryonic development. The oxytetracycline used is a chloride form, which allows solubilization in aqueous media. Embryos at blastula stage were incubated in the antibiotic in 125 ppm, 250 ppm, 500 ppm, 750 ppm, 1000 ppm, and 1500 ppm and monitored every 6 hrs. At 6 hpi all the embryos at 1500 ppm and 1000 ppm were dead; later at 12 hpi, only a single embryo survived at 500 ppm. Finally at 24 hpi no more than two embryos survived at 125 ppm (Figure S1 A). Since it was not possible to determine the LC$_{50}$ during embryonic stages, we decided to determine the highest sub-lethal oxytetracycline concentration (Oxy sub-lethal) in larval stages. To perform this experiment, we incubated 48hpf larvae in the same oxytetracycline dilutions used above for 4 days (Figure S1 B). On day 2 all larvae exposed to a concentration of 1500 ppm were dead; later at day 4, we find two (2/45) dead larvae at 1000 ppm. Due to we did not find larvae mortality with 750 ppm oxytetracycline in any of the 4 days studied, we decided to perform the next experiments using this concentration. To be sure that the amount of drug chosen remains sub-lethal after 4 days of incubation and that larvae will not die within a few hours after analyzed, we decided to incubate them 2 more days at 750 ppm. At day six of incubation all larvae were alive and we did not find any larvae with pericardial edema, abnormal heart function, delay in development or any other obvious detrimental phenotypic effect (Figure S2). These results indicate that the selected oxytetracycline working concentration is indeed sub-lethal, during at least 6 days of incubation.

Persistent exposure to oxytetracycline induce neutrophils migration to superficial tissues of larval tail

As there is evidence suggesting that antibiotics can affect immune functions in other fish [7–16], we addresses the question if oxytetracycline exposure could modulate the innate immune function by triggering an inflammatory process. With this aim we decide to analyze neutrophils behavior in live zebrafish larvae. These types of leukocytes are the first cells to be mobilized in response to an injury and the first to infiltrate the damaged territory. To develop the assay we took advantage of the *Tg(mpx:GFP)^{i114}* transgenic line, from now on *Tg(mpx:GFP)*, that expresses GFP under the control of the *myeloperoxidase* entire regulatory region and allows tracking individual immune cells in live animals [40]. We incubated *Tg(mpx:GFP)* larvae in 750 ppm oxytetracycline and monitored neutrophils migration at 10 hpi, 24 hpi, 48 hpi 72 hpi and 96 hpi utilizing a fluorescence stereoscope. To establish if indeed oxytetracycline induces neutrophils migration, we focused on the larval tail. In this region neutrophils are normally restricted to the caudal hematopoietic tissue (CHT) and only a few are circulating (Figure 1 A). It has been shown previously that tissue damage caused by injury, irritants, or pathogens, promotes neutrophils migration from the CHT to the site of damage [41] (Figure 1 A). Our results show that there is no significant recruitment of neutrophils in the tail at 10hpi nor at 24hpi (Figure 1 B–F). However at 48hpi we find a few neutrophils distributed throughout the experimental larvae tail, but by 72hpi an increasing number of neutrophils was evident in the analyzed territory (Figure 1 B, G–J). Finally, at 96hpi the migration becomes even more intense, indicating that we are clearly in presence of an inflammation reaction (Figure 1 K, L).

To gain detailed insights into the inflammation process, we investigate neutrophils localization across the larval tail by using confocal microscopy. We thought that if the inflammation process is indeed triggered by the presence of oxytetracycline in the water

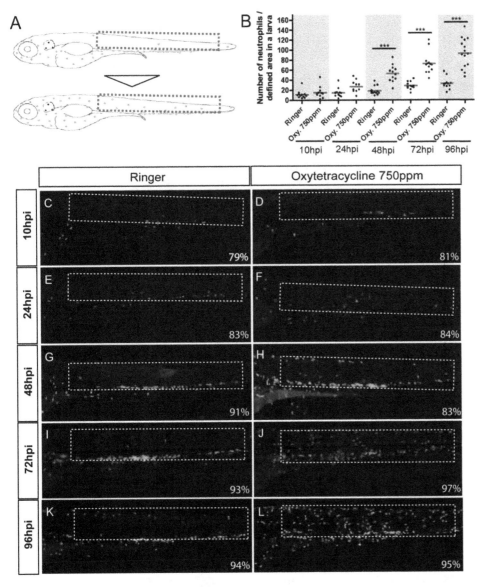

Figure 1. Exposure to oxytetracycline triggered an inflammation process. Incubation of *Tg(mpx:GFP)* transgenic larvae (that express GFP exclusively in neutrophils) in oxytetracycline 750 ppm induce a progressive migration of neutrophils from to the caudal hematopoietic tissue (CHT) to the entire tail. (A) Scheme of neutrophils localization in control and experimental larvae. (B) Quantification of neutrophils migration into the selected area. (C–L) Close up of the tail of control and experimental embryos at 10 hour post incubation (hpi), 24hpi, 48hpi, 72hpi and 96hpi. At 10hpi (C, D) and 24hpi (E, F) we did not detect any significant neutrophils migration. Later, at 48hpi (G, H) a discrete number of neutrophils migrate from de CHT to the tail. At 72hpi (I, J) and 96hpi (K, L), the presence of the GFP positive cells in the tail becomes clearly evident. The numbers of larvae that presented the phenotype shown is expressed as a percentage. For all experiments, at least 13 larvae were used for each condition. *** p<0.001.

and is at not a consequence of a general toxic effect, the inflammation must be more intense at superficial tissue (those that are in contact with the water such the mucosal epithelia) then internal organs, at least at early stages. We determined that in experimental larvae neutrophils were localized at superficial tissues both at the beginning of the inflammatory process (48hpi) and later (96hpi) (Figure 2 B′, D′), while in control larvae these polymorphonuclear cells remained throughout the ventral region, at the CHT (Figure 2 A′, C).

Together, our results suggest that waterborne exposure of larvae to oxytetracycline triggered a specific inflammatory response in external territories at larval tail. However at this point we cannot rule out that these observations could be due to massive cell death in treated larvae.

Early stages of oxytetracycline induced inflammation are independent of cell death

Our previous results lead us to hypothesize that oxytetracycline triggers the inflammation process by inducing cell death in larvae. If we are right, we should observe leukocytes migrating to the areas where cells are dying. To explore this, we determined cell death levels by using acridine orange (AO), in transgenic larvae that express RFP in leukocytes *Tg(lyzC:DsRED2)nz50* [42,43], herein named *Tg(LyzC:DsRED2)*. We found that there is no enhanced cell death in the whole embryo at 24hpi or 48hpi (data not shown) suggesting that the inflammation is not started due to massive apoptosis. Only at 72hpi, after neutrophils migration is initiated, a clear increase in the AO stain in the tail of larvae exposed to oxytetracycline was detected (Figure 3 A–D); this was further

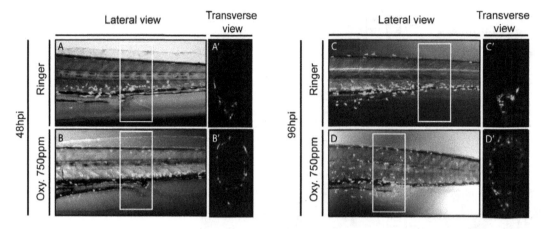

Figure 2. Oxytetracycline exposure induces neutrophils migration to superficial tissues. We analyzed the *Tg(mpx:GFP)* transgenic line, by confocal microscopy, detecting that in oxytetracycline exposure larvae neutrophils migrated to superficial tissue. (A–D) lateral view, (A'–D') transverse view. Transverse views show a section of a larval tail of approximately 250 μm (white rectangle in lateral view).

intensified at 96hpi (Figure 3 E, G). We analyzed if there was a correlation of this data and the leukocyte migration, but we found that these immune cells migrated in all directions, independently of the physical location of cell death (Figure 2 F, H).

All together these results suggest that cell death was not the initial step of this inflammatory process. Although we cannot exclude that the observed cell death is a belatedly consequence of the direct action of oxytetracycline, a possible explanation to the late apoptosis, is that it could be a consequence of neutrophils action. These cells eliminate the invading agents by releasing toxic contents of their granules, but they do not discriminate between external and host targets. In this way, undesired damage is produced to host tissues [31].

The ever-increasing number of neutrophils that infiltrate the larval tail under oxytetracycline exposure indicates that the inflammatory process becomes more intense and resolution is far from occurring, insinuating an increase in the total number of this type of granulocytes. To quantify the amount of neutrophils at every point analyzed before, we performed flow cytometry analysis (Figure 4 A, B) using the *Tg(mpx:GFP)* line. The results obtained confirmed our previous observations, and a clear increase in the number of GFP positive cells was seen since 72hpi (Figure 4 B).

All together our results show that oxytetracycline promotes inflammation independently of cell death, at least at early stages of exposure, and it does induce an increase in the total amount of neutrophils in zebrafish larvae.

Prolonged exposure to oxytetracycline induces expression of innate immune molecular markers

An early pivotal event in an inflammatory process is the increase in the transcription of pro-inflammatory cytokines, such Il-1β, enabling organisms to respond to different kinds of insults [31]. On the opposite side are the anti-inflammatory cytokines, such as Il-

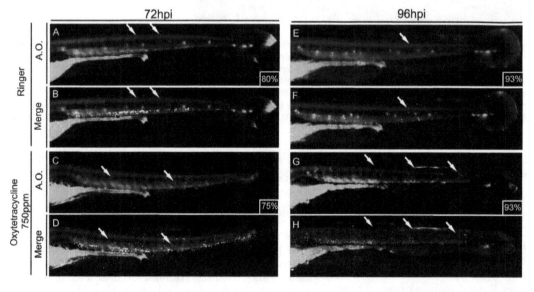

Figure 3. Cell death is detected in advanced steps of oxytetracycline induced inflammation. 48hpf *Tg(LyzC:DsRED2)* transgenic larvae were incubated for 72 hrs (A–D) or 96 hrs (E–H) in oxytetracycline 750 ppm followed by an acridine orange stain to address cell death. Experimental larvae showed higher level of cell death (white arrows) compared to control at both time analyzed. The numbers of larvae that presented the phenotype shown is expressed as a percentage.

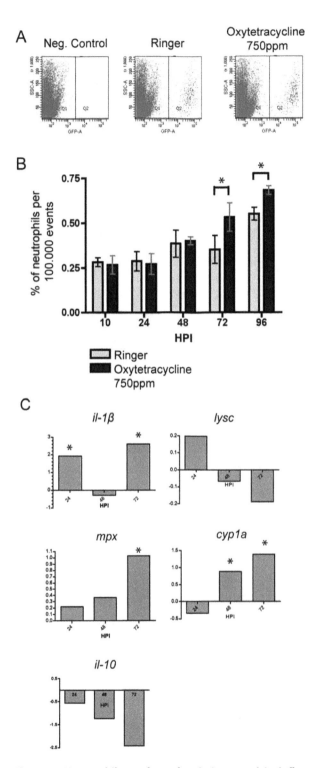

Figure 4. Neutrophils total number is increased in inflamed larvae. (A) To determine the total number of neutrophils a flow cytometry analysis was done. Three representative SSC v/s GFP graphics are showed. Cells from wild-type larvae were used to set the lowest level of GFP expression. (B) A significant increase in the number of total neutrophils is induced after 72 hrs of exposure to oxytetracycline. (C) Transcription levels of interleukin 1β (*il-1β*), interleukin 10 (*il-10*), mielloperoxidase (*mpx*), Lysozime C (*lysc*) and the cytochrome p450 1a (*cyp1a*) were quantified by qPCR. Transcript data were normalized to β-actin1 and to the corresponding control. * p<0.05.

10, which are secreted mainly by macrophages when the inflammatory agent is removed, promoting the end of the inflammatory process [34].

As we mention before, other fundamental player of inflammation are neutrophils. These cells express myeloperoxidase (*mpx*), which is an enzyme stored in large amount in azurophilic granules [44]. Thus when new neutrophils differentiate, transcriptional level of *mpx* increase. On the other hand, when the inflammation occurs due to the presence of organic compounds, the transcription of the cytochrome *P*-450 monooxygenase enzymes is also induced, in particular the *cyp1a* subfamily. These enzymes are pivotal for the metabolism and biotransformation of many drugs [45]. In this circumstance, the nitric oxide generated by Mpx activity, can react with superoxide, produced by Cytochrome p450, generating the highly reactive peroxynitrite molecule, causing tissue damage [46].

In order to complement our previous results, we decided to monitor at molecular level the progression of the inflammatory process by evaluating several immune markers by qPCR. We evaluated well known immune markers at 24hpi, 48hpi and 72hpi (Figure 4 C). At 24hpi we detected a significant increase in the transcript levels of *il-1β*, indicating that the inflammation process is already triggered at a molecular level. Later at 72hpi we identified a new peak of *il-1β*, event that support the increasing number of neutrophils outside the CHT. Likewise, a marked increase in the *mpx* transcription levels at the same time point was detected, result that also correlates with the enhanced total neutrophils number described before (Figure 4 B). Finally, *il-10* levels were indistinguishable from control, at all the analyzed time points, which is consistent with our data showing an unresolved inflammation process (Figure 4 B). Therefore, these data provides a quantitative support for our assay.

Oxytetracycline induced inflammation increases regeneration capability

Due to an inflammatory process may alter the course of other physiological properties, we decided to evaluate changes in regeneration capacity in oxytetracycline treated animals. To assay regeneration capacity, we took advantage of a well-established protocol for monitoring effects of contaminants on the regeneration of mechanosensory hair cells in the fish posterior lateral line organ [41,47]. The posterior lateral-line organ is a sensory system comprising a number of discrete sense organs, the neuromasts, distributed over fish tail in specific patterns. In turn, neuromasts comprise a core of mechanosensory hair cells, surrounded by support cells. Hair cells are a very good sensor of harmful compounds present in the water due to their localization on the surface of the fish [48,49].

To perform the assay, we used the SqET4 transgenic line that specifically labels the mechanosensory hair cells (HC), allowing to monitor the presence of this cell type in live fish by detecting GFP expression [50]. With this aim we exposed 48hpf SqET4 larvae during 48 hrs to 750 ppm oxytetracycline (Figure 5 A). Then, we incubated control and experimental larvae for 2 hrs with 10 μM CuSO₄ to eliminate HC and washed with E3 medium to remove remaining traces of CuSO₄ as reported before [41,47]. At this point, we divided the experimental larvae into to groups, one incubated in ringer and other in oxytetracycline. This strategy allowed us to distinguish whether a possible effect in the regeneration is due to the antibiotic itself or to the inflammation triggered by it beforehand. After the copper treatment we monitored the appearance of HC in 3 specific neuromast, L1, L6 and L9, of the posterior lateral line in control larvae and the two experimental groups every 12 hrs. We choose those three

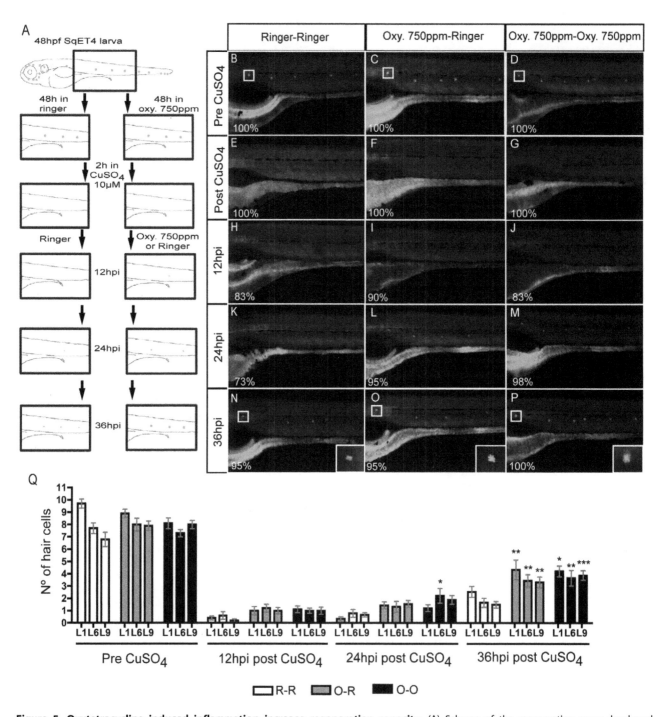

Figure 5. Oxytetracycline induced inflammation increase regeneration capacity. (A) Scheme of the regeneration assay developed. Incubation of SqET4 transgenic larvae (that express GFP in neuromast hair cells, HC) in oxytetracycline 750 ppm increases the rate of GFP/hair cells appearance. (B–D) HC before copper treatment. (E–G) CuSO$_4$ treatment eliminates all GFP positive cells. (H–M, Q) Although we observe a tendency, there is no significant difference in the number of regenerated HC. (N–Q) At 36hpi the regeneration capacity of HC is significantly increased on oxytetracycline exposed larvae. (Q) The quantification of hair cells was performed in three neuromast; L1, L6 and L9. White box indicate L1 neuromast hair cells. The numbers of larvae that presented the phenotype shown is expressed as a percentage. * p<0.05, ** p<0.01, ***p<0.001.

neuromast since the position in the larval tail in which the new HC shall appear is easily identifiable. The obtained results show an evident acceleration in the onset of appearance of GFP positive cells in larvae treated with oxytetracycline after copper (Figure 5). Immediately after CuSO$_4$ treatment, both experimental and control larvae loose all their hair cells (Figure 5 E–G) and although we observed a tendency, there is no significant difference

in the number of HC at 12 and 24hpi (Figure 5 H–M). Later at 36hpt, neuromast in control larvae have a very low number of hair cells, between 1 and 3 (Figure 5 N, Q). In turn, in both experimental assays L1, L6 and L9 have between 4 and 5 HC (Figure 5 O–Q). This result indicates that it is not the antibiotic itself, but the inflammation that promotes the regeneration process. A similar effect of an inflammation process on regener-

Table 1. Primer sequences used for amplification of specific gene production with the RT-qPCR technique.

Gene	Forward Primer	Reverse Primer	Amplicon Size
il-1β	TGGACTTCGCAGCACAAAATG	GTTCACTTCACGCTCTTGGATG	150 bp
il-10	CACTGAACGAAAGTTTGCCTTAAC	TGGAAATGCATCTGGCTTTG	120 pb
mpx	TCCAAAGCTATGTGGGATGTGA	GTCGTCCGGCAAAACTGAA	90 bp
lysc	TGGGAGGCAATAGGCATGA	GCGTAGGATCCATCTGGTTTG	100 bp
cyp1a	GCATTACGATACGTTCGATAAGGAC	GCTCCGAATAGGTCATTGACGAT	120 bp
β-actin1	GCCAACAGAGAGAAGATGACACAG	CAGGAAGGAAGGCTGGAAGAG	110 pb

ation was proposed by D'Alençon and collaborators [41]. A possible explanation is that in the inflammation process triggered by the exposure to oxytetracycline, a higher number of M2 macrophages are recruited to the damage zone compared to control larvae, thus increasing the regeneration capacity. Evidence in mammals indicates that when M2 macrophages are depleted after an injury, diminishing muscle repair, differentiation, and regeneration occur [51].

Conclusion

In this paper we developed a battery of assays that allow, simply and rapidly, to evaluate aquatic environment impact on fish immune system. Our strategy considere the use of a number of biological markers of innate immune response that enables to determine if agents normally used or present in fish farms can induce these markers during or after exposure. Using a fish model increases the relevance of the study, as we are directly assaying bioavailable compounds with the target organism. Finally, it is of importance to establish which are the conditions that favor improvements in fish healthy and survival under conditions of stress, such as those found in salmon farms.

Materials and Methods

Ethics Statement

All animals subjected to experimentation were anesthetized and procedures complied with the guidelines of the Animal Ethics Committees of the Universidad Andres Bello, which approved this study.

Zebrafish strains and maintenance

Zebrafish were maintained and raised in our facility according to standard protocols [52]. The following strains of fish were used in this study: Tab5 (wild type), Tg(mpx:GFP)^{i114} [40], Tg(lyzC:Ds-RED2)nz50 [43] and SqEt4 [50]. All embryos were collected by natural spawning, staged according to Kimmel et al. [18] and raised at 28,5°C in E3 medium (5 mM NaCl, 0.17 mM KCl, 0.33 mM CaCl2, 0.33 mM MgSO4, without methylene blue, equilibrated to pH 7.0) in Petri dishes, as described previously [53]. Embryonic and larval ages are expressed in hours post fertilization (hpf).

Preparation of oxytetracycline solution and larvae Incubation

Six oxytetracycline (Zanil HCL 80%, Centrovet, Santiago, Chile) solutions at 125 ppm, 250 ppm, 500 ppm, 750 ppm, 1000 ppm and 1500 ppm were prepared in Ringer pH 7.0 medium (116 mM NaCl, 2,9 mM KCl, 1,8 mM CaCl2, 5 mM HEPES pH 7,2). The different solutions were stored in bottles protected from light and sealed with parafilm at 4°C for a maximum of 6 days. The use of a commercial oxytetracycline was validated by carrying out a set of incubations with pure oxytetracycline (Sigma). No differences between both antibiotics form at any concentration tested were identified. All embryos/larvae incubations were carried out basically in the same way. Briefly, 15 embryos, in the case of the LC$_{50}$, or 15 larvae were transferred to six-well plate, in a volume of 5 ml of Ringer pH 7 until the experiments begun. At 3hpf or 48hpf respectively, the Ringer solution was replaced by the appropriated oxytetracycline solution. In the case of the LC$_{50}$ assay, mortality was monitored every 6 hrs and for the determination of the heights sub-lethal oxytetracycline concentration (Oxy sub-lethal) every 24 hrs. All incubations were done in triplicate at a temperature of 28°C and at least three times. The time of incubation in oxytetracycline are expressed as hours post incubation (hpi).

Acridine Orange Staining

For cell death characterization, zebrafish larvae were stained according to Williams et al. [42] with minor modifications. Embryos were incubated for 20 minutes in 5 μg/ml acridine orange (Sigma) in Ringer pH 7 medium, washed five times for 5 minutes in Ringer medium and observed under fluorescence stereoscope.

Neutrophils migration and flow cytometry analysis

Neutrophils migration was addressed by analyzing, with a fluorescence stereoscope, the displacement of GFP positive cells in Tg(mpx:GFP) from the caudal hematopoietic tissue to the tail at 10hpi, 24hpi, 48hpi, 72hpi and 96hpi. For the flow cytometry, Tg(mpx:GFP) larvae at 10hpi, 24hpi, 48hpi, 72hpi and 96hpi (15 larvae per point in at least four independent experiment) were washed twice with PBS, followed by larvae disaggregation with 0.5% Trypsin-EDTA (Gibco) in PBS and resuspended in RPMI (Gibco) +1% FBS (Gibco) and passed through a filter with a 75 μm pore size and twice trough a 35 μm pore size. Cells from wild-type embryos were used to set the lower limit of GFP expression for each experiment. Flow cytometric analysis was performed on a FACScan (BD Biosciences).

qPCR quantification

Larvae exposed to oxytetracycline 750 ppm since 48 hpf were sampled at 24, 48 and 72hpi for total RNA extraction. Control fish were sampled at identical time points. Samples included 25 larvae per time point per treatment. Total RNA extraction was obtained using Trizol Reagent (Invitrogen) according to the manufacturer's instructions. cDNAs were synthesized from RNA samples in a reverse transcription reaction using Super Script II RT (Invitrogen) according to the manufacturer's instructions and using oligo-dt primers.

Real time PCR was performed following descriptions of Rawls et. al. [54]. Each gene was tested in octuplicate and verified for non-specificity using melting curves (primers sequence in table 1). The mean Ct values from each sample were normalized against the mean Ct value of a reference gene (β-actin1, housekeeping gene). Relative quantification of each gene was obtained with the Pfaffl method and the REST 2009 software (Qiagen). This software includes a statistical test to determine accuracy of relative expression, which is complex because ratio distributions do not have a standard deviation. REST 2009 software overcomes this limitation by using simple statistical randomization test, the permuted expression data rather than the raw Ct values input by the user [55].

Regeneration assay

Regeneration assay was carried out as described previously [47] with minor modifications. Briefly, fifteen 48hpf SqET4 larvae were incubated in Ringer pH 7,0 or oxytetracycline 750 ppm during 48 hrs. Then both control and experimental embryos were incubated in $CuSO_4$ 10 μM (Merck) for 2 hrs to eliminate the hair cells, followed by 5 washes with E3 pH 7,0 to eliminate traces of $CuSO_4$. Finally, control larvae were maintained in Ringer pH 7,0 and experimental larvae in oxytetracycline 750 ppm or Ringer during 36 hrs for monitoring hair cells regeneration and analyzed every 12 hrs.

Imaging and statistics

Photographs were taken in an Olympus SZX16 stereoscope with a QImaging MicroPublisher 5.0 RVT camera and an Olympus BX61 microscope with a Leica DC300F camera. Confocal images were acquired with Olympus FluoView FV1000 Spectral Confocal Microscope (software version 2.1). Images were processed with Photoshop CS4 or Image J 1.44o. For all the experiments described, the images shown are representative of the effects observed in at least 70% of the individuals.

For statistical analysis Prism 4 (GraphPad Software) was used. The data was analyzed using one-way ANOVA for the neutrophils migration assay and two-way ANOVA for flow cytometry and hair cell regeneration assays.

Supporting Information

Figure S1 Determination of the oxytetracycline LC_{50} and maximum sub-lethal concentration. Both, embryos (A) and larvae (B) were incubated in six different oxytetracycline concentrations ranging from 125 ppm to 1500 ppm and monitored every 6 hrs and 24 hrs respectively. The maximum sub-lethal concentration was the one where no mortality or any apparent phenotypic effect was detected. Results indicate that oxytetracycline was lethal to embryos in all the concentration analyzed. The highest sub-lethal concentration determined for larvae was 750 ppm.

Figure S2 6 days of oxytetracycline exposure does not produce mortality nor adverse effects on larvae. Larvae were incubated during 120 hrs and 144 hrs to ensure that the treatment with oxytetracycline does not produce any detrimental effects on larvae. We did not found phenotypic effects such cerebral edema, bending of the tail (A, D), pericardial edema, abnormal heart function (B, E), or any somite malformation (C, F). No change in survival rate was detected at the time analyzed (G).

Acknowledgments

The authors are thankful to Emilio Vergara for expert fish care and Dr. Leonardo Valdivia for manuscript discussion. Zebrafish strains were kindly provided by Dr. Vladimir Korhz (SqET4), Dr. Steve Renshaw (*Tg(mpx:GFP)*) and Dr. Phil Crosier (*Tg(LyzC:DsRED2)*).

Author Contributions

Conceived and designed the experiments: CGF FBB. Performed the experiments: FBB JR AP. Analyzed the data: CGF FBB JR. Contributed reagents/materials/analysis tools: CGF JR. Wrote the paper: CGF FBB.

References

1. Cabello FC (2004) Antibiotics and aquaculture in Chile: Implications for human and animal health. Rev Méd Chile 132: 1001–1006.

2. Christensen AM, Ingerslev F, Baun A (2006) Ecotoxicity of mixtures of antibiotics used in aquacultures. Environ Toxicol Chem 25: 2208–2215.

3. Carson MC, Bullock G, Bebak-Williams J (2002) Determination of oxytetracycline residues in matrixes from a freshwater recirculating aquaculture system. J AOAC Int. 85: 341–348.

4. Lalumera GM, Calamari D, Galli P, Castiglioni S, Crosa G, et al. (2004) Preliminary investigation on the environmental occurrence and effects of antibiotics used in aquaculture in Italy. Chemosphere 54: 661–668.

5. Rose PE, Pedersen JA (2005) Fate of oxytetracycline in streams receiving aquaculture discharges: model simulations. Environ Toxicol Chem 24: 40–50.

6. Pouliquen H, Delépée R, Thorin C, Haury J, Larhantec-Verdier M, et al. (2009) Comparison of water, sediment, and plants for the monitoring of antibiotics: a case study on a river dedicated to fish farming. Environ Toxicol Chem 28: 496–502.

7. Rijkers GT, van Oosterom R, van Muiswinkel WB (1981) The immune system of cyprinid fish. Oxytetracycline and the regulation of humoral immunity in carp (Cyprinus carpio L). Vet Immunol Immunopathol 2: 281–290.

8. Grondel JL, Nouws JFM, van Muiswinkel WB (1987a) The influence of antibiotics on the immunesystem: immuno-pharmacokinetic investigations on the primary anti-SRBC response in carp, Cyprinus carpio L, after oxytetracycline injection. J Fish Dis 10: 35–43.

9. Thorburn MA, Carpenter TE, Ljungberg O (1987) Effects of immersion in live Vibrio anguillarum and simultaneous oxytetracycline treatment on protection of vaccinated and non-vaccinated rainbow trout Salmo gairdneri against vibriosis. Dis Aquat Org 2: 167–171.

10. Siwicki AK, Anderson DP, Dixon OW (1989) Comparisons of nonspecific and specific immunomodulation by oxolinic acid, oxytetracycline and levamisole in salmonids. Vet Immunol Immunopathol 23: 195–200.

11. Kajite Y, Sakai M, Atsuta S, Kobayashi M (1990) The immunomodulatory effect of Levamisole on rainbow trout, Oncorhynchus mykiss. Fish Pathology 25: 93–98.

12. Van der Heijden MH, Helders GM, Booms GH, Huisman EA, Rombout JH, et al. (1996) Influence of flumequine and oxytetracycline on the resistance of the European eel against the parasitic swimbladder nematode Anguillicola crassus. Veterinary Immunology and Immunopathology 52: 127–134.

13. Lundén T, Miettinen S, Lönnström LG, Lilius EM, Bylund G (1998) Influence of oxytetracycline and oxolinic acid on the immune response of rainbow trout (Oncorhynchus mykiss). Fish Shellfish Immunol 8: 217–230.

14. Tafalla C, Novoa B, Alvarez JM, Figueras A (1999) *In vivo* and *in vitro* effect of oxytetracycline treatment on the immune response of turbot, *Scophthalmus maximus* (L.). J Fish Dis 22: 271–276.

15. Lundén T, Bylund G (2002) Effect of sulphadiazine and trimethoprim on the immune response of rainbow trout (*Oncorhynchus mykiss*). Vet Immunol Immunopathol. Feb; 85: 99–108.

16. Caipang CM, Lazado CC, Brinchmann MF, Berg I, Kiron V (2009) In vivo modulation of immune response and antioxidant defense in Atlantic cod, *Gadus morhua* following oral administration of oxolinic acid and florfenicol. Comp Biochem Physiol C Toxicol Pharmacol 150: 459–464.

17. Laale HW (1977) Culture and preliminary observations of follicular isolates from adult zebra fish, Brachydanio rerio. Can J Zool 55: 304–309.

18. Kimmel CB, Ballard WW, Kimmel SR, Ullmann B, Schilling TF (1995) Stages of embryonic development of the zebrafish. Dev Dyn 203: 253–310.

19. Roex EW, Giovannangelo M, van Gestel CA (2001) Reproductive impairment in the zebrafish, Danio rerio, upon chronic exposure to 1,2,3-trichlorobenzene. Ecotoxicol Environ Saf 48: 196–201.

20. Hinton DE, Kullman SW, Hardman RC, Volz DC, Chen PJ, et al. (2005) Resolving mechanisms of toxicity while pursuing ecotoxicological relevance? Mar Pollut Bull 51: 635–648.

21. Scholz S, Fischer S, Gündel U, Küster E, Luckenbach T, et al. (2008) The zebrafish embryo model in environmental risk assessment–applications beyond acute toxicity testing. Environ Sci Pollut Res Int 15: 394–404.

22. Froehlicher M, Liedtke A, Groh KJ, Neuhauss SC, Segner H, et al. (2009) Zebrafish (Danio rerio) neuromast: promising biological endpoint linking developmental and toxicological Studies. Aquat Toxicol 95: 307–319.

23. Traver D, Herbomel P, Patton EE, Murphey RD, Yoder JA, et al. (2003) The zebrafish as a model organism to study development of the immune system. Adv Immunol 81: 253–330.

24. Zapata A, Diez B, Cejalvo T, Gutiérrez-de Frías C, Cortés A (2006) Ontogeny of the immune system of fish. Fish Shellfish Immunol 20: 126–136.

25. Sullivan C, Kim CH (2008) Zebrafish as a model for infectious disease and immune function. Fish Shellfish Immunol 25: 341–350.

26. Lieschke GJ, Trede NS (2009) Fish immunology. Curr Biol 19: 678–682.

27. Allen JP, Neely MN (2010) Trolling for the ideal model host: zebrafish take the bait. Future Microbiol 5: 563–569.

28. Ellett F, Lieschke GJ (2010) Zebrafish as a model for vertebrate hematopoiesis. Curr Opin Pharmacol 10: 563–570.

29. Lam SH, Chua HL, Gong Z, Lam TJ, Sin YM (2004) Development and maturation of the immune system in zebrafish, Danio rerio: a gene expression profiling, in situ hybridization and immunological study, Dev Comp Immunol 28: 9–28.

30. Trede NS, Langenau DM, Traver D, Look AT, Zon LI (2004) The use of zebrafish to understand immunity. Immunity 20: 367–379.

31. Chen GY, Nuñez G Sterile inflammation: sensing and reacting to damage. Nature Rev Immunol 10: 826–837.

32. Witko-Sarsat V, Rieu P, Descamps-Latscha B, Lesavre P, Halbwachs-Mecarelli L (2000) Neutrophils: Molecules, Functions and Pathophysiological Aspects. Lab Invest 80: 617–653.

33. Fialkow L, Wang Y, Downey GP (2007) Reactive oxygen and nitrogen species as signaling molecules regulating neutrophil function. Free Radic Biol Med 42: 153–164.

34. Ouyang W, Rutz S, Crellin NK, Valdez PA, Hymowitz SG (2011) Regulation and Functions of the IL-10 Family of Cytokines in Inflammation and Disease. Annu Rev Immunol 29: 71–109.

35. Jacobsen P, Berglind L (1988) Persistence of oxytetracycline in sediments from fish farms. Aquaculture 70: 365–370.

36. Björklund H, Räbergh CMI, Bylund G (1991) Residues of oxolinic acid and oxytetracycline in fish sediments from fish farms. Aquaculture 97: 85–96.

37. Samuelsen OB, Torsvik V, Ervik A (1992) Long-range changes in oxytetracycline concentration and bacterial resistance toward oxytetracycline in a fish farm sediment after medication. Sci Total Environ 114: 25–36.

38. Bebak-Williams J, Bullock G, Carson MC (2002) Oxytetracycline residues in a freshwater recirculating system. Aquaculture 205: 221–223.

39. Millano B, Barrientos M, Gomez C, Tomova A, Buschmann A, et al. (2011) Injudicious and excessive use of antibiotics: Public health and salmon aquaculture in Chile. Rev Med Chile 137: 107–118.

40. Renshaw SA, Loynes CA, Trushell DM, Elworthy S, Ingham PW, et al. (2006) A transgenic zebrafish model of neutrophilic inflammation. Blood 108: 3976–3978.

41. D'Alençon CA, Peña OA, Wittmann C, Gallardo VE, Jones RA, et al. (2010) A high-throughput chemically induced inflammation assay in zebrafish. BMC Biology 8: 151–167.

42. Williams JA, Barrios A, Gatchalian C, Rubin L, Wilson SW, et al. (2000) Programmed cell death in zebrafish rohon beard neurons is influenced by TrkC1/NT-3 signaling. Dev Biol 226: 220–230.

43. Hall C, Flores MV, Storm T, Crosier K, Crosier P (2007) The zebrafish lysozyme C promoter drives myeloid-specific expression in transgenic fish. BMC Dev Biol 7: 42.

44. Arnhold J, Flemmig J (2010) Human myeloperoxidase in innate and acquired immunity. Arch Biochem Biophys 500: 92–106.

45. Goksoyr A, Forlin L (1992) The cytochrome-P-450 system in fish, aquatic toxicology and environmental monitoring. Aquat Toxicol 22: 287–311.

46. Morgan ET (2001) Regulation of cytochrome p450 by inflammatory mediators: why and how? Drug Metab Dispos 29: 207–212.

47. Hernandez PP, Moreno V, Olivari FA, Allende ML (2006) Sub-lethal concentrations of waterborne copper are toxic to lateral line neuromasts in zebrafish (Danio rerio). Hear Res. 213: 1–10.

48. Metcalfe WK, Kimmel CB, Schabtach E (1985) Anatomy of the posterior lateral line system in young larvae of the zebrafish. J Comp Neurol 233: 377–389.

49. Ghysen A, Dambly-Chaudière C (2004) Development of the zebrafish lateral line. Curr Opin Neurobiol 14: 67–73.

50. Parinov S, Kondrichin I, Korzh V, Emelyanov A (2004) Tol2 transposon-mediated enhancer trap to identify developmentally regulated zebrafish genes in vivo. Dev Dyn 231: 449–459.

51. Tidball JG, Villalta SA (2010) Regulatory interactions between muscle and the immune system during muscle regeneration. Am J Physiol Regul Integr Comp Physiol 298: 1173–1187.

52. Westerfield M (1994) The Zebrafish Book: A Guide for the Laboratory Use Of the Zebrafish (Danio rerio), 2.1 ed. Eugene: University of Oregon Press.

53. Haffter P, Granato M, Brand M, Mullins MC, Hammerschmidt M, et al. (1996) The identification of genes with unique and essential functions in the development of the zebrafish, Danio rerio. Development 123: 1–36.

54. Rawls JF, Samuel BS, Gordon JI (2004) Gnotobiotic zebrafish reveal evolutionarily conserved responses to the gut microbiota. P Natl Acad Sci USA 101: 4596–4601.

55. Pfaffl MW, Horgan GW, Dempfle L (2002) Relative expression software tool (REST) for group-wise comparison and statistical analysis of relative expression results in real-time PCR. Nucleic Acids Res 30: e36.

Denitrification and Anammox in Tropical Aquaculture Settlement Ponds: An Isotope Tracer Approach for Evaluating N₂ Production

Sarah A. Castine[1]*, Dirk V. Erler[2], Lindsay A. Trott[3], Nicholas A. Paul[4], Rocky de Nys[4], Bradley D. Eyre[2]

1 AIMS@JCU, School of Marine and Tropical Biology, Australian Institute of Marine Science, Centre for Sustainable Tropical Fisheries and Aquaculture, James Cook University, Townsville, Queensland, Australia, 2 School of Environmental Science and Management, Centre for Coastal Biogeochemistry, Southern Cross University, Lismore, New South Wales, Australia, 3 Australian Institute of Marine Science, Townsville, Queensland, Australia, 4 School of Marine, Tropical Biology and Centre for Sustainable Tropical Fisheries and Aquaculture, James Cook University, Townsville, Queensland, Australia

Abstract

Settlement ponds are used to treat aquaculture discharge water by removing nutrients through physical (settling) and biological (microbial transformation) processes. Nutrient removal through settling has been quantified, however, the occurrence of, and potential for microbial nitrogen (N) removal is largely unknown in these systems. Therefore, isotope tracer techniques were used to measure potential rates of denitrification and anaerobic ammonium oxidation (anammox) in the sediment of settlement ponds in tropical aquaculture systems. Dinitrogen gas (N₂) was produced in all ponds, although potential rates were low (0–7.07 nmol N cm^{-3} h^{-1}) relative to other aquatic systems. Denitrification was the main driver of N₂ production, with anammox only detected in two of the four ponds. No correlations were detected between the measured sediment variables (total organic carbon, total nitrogen, iron, manganese, sulphur and phosphorous) and denitrification or anammox. Furthermore, denitrification was not carbon limited as the addition of particulate organic matter (paired t-Test; $P = 0.350$, $n = 3$) or methanol (paired t-Test; $P = 0.744$, $n = 3$) did not stimulate production of N₂. A simple mass balance model showed that only 2.5% of added fixed N was removed in the studied settlement ponds through the denitrification and anammox processes. It is recommended that settlement ponds be used in conjunction with additional technologies (i.e. constructed wetlands or biological reactors) to enhance N₂ production and N removal from aquaculture wastewater.

2

Editor: Jacqueline Mohan, Odum School of Ecology, University of Georgia, United States of America

Funding: This research was funded by an AIMS@JCU research scholarship awarded to Sarah Castine at the beginning of her PhD. The funders had no role in study design, data collection and analysis, decision to publish, or preparation of the manuscript.

Competing Interests: The authors have declared that no competing interests exist.

* E-mail: sarah.castine@my.jcu.edu.au

Introduction

The release of anthropogenic N to the coastal zone poses a threat to many shallow marine ecosystems [1]. Discharge of aquaculture wastewaters has contributed to N enrichment of some coastal regions [2] and settlement ponds have been established as a remediation strategy from aquaculture wastewater prior to release to the environment [3,4]. Settlement pond technologies are widely implemented as a low cost option for treating municipal [5], fish farm [6] and dairy farm wastewater [7]. However, the nutrient removal efficiency of settlement ponds associated with land-based tropical aquaculture systems is unclear. Generally, newly established (<1 yr old) settlement ponds, with a basic design, provide significant reductions in total suspended solids, but are less efficient in the remediation of dissolved nutrients [3,8]. Furthermore, given that the efficiency of wetland wastewater treatment systems can decrease with age [9], it is likely that the performance of settlement ponds, which act as brackish water constructed wetlands, will decrease over time unless they are actively managed. Methods to improve the long term performance of tropical aquaculture settlement ponds include the use of extractive organisms such as algae, which can be cultured and subsequently harvested [10], and

also the removal of settled organic rich particulates (sludge) which prevents remineralization of dissolved N back into the water column [3,11]. Microbial nutrient transformation, which is largely un-quantified, also presents a potentially significant mechanism to reduce dissolved inorganic nitrogen (DIN) in aquaculture wastewater.

Denitrification and anammox are the major microbial processes removing fixed N from wastewater through the production of dinitrogen gas (N₂). During denitrification, nitrate (NO₃⁻) is reduced to nitrite (NO₂⁻), nitric oxide (NO) and nitrous oxide (N₂O), before eventually being converted to N₂. Anammox also directly removes fixed N and couples NO₂⁻ reduction with ammonium (NH₄⁺) oxidation to produce N₂ [12,13]. Denitrification and anammox are also important for the removal of N from natural system such as intertidal flats [14], marsh sediments [15], deep anoxic waters [16] and sediments from the continental shelf (50 m) and slope (2000 m) [17]. Denitrification and anammox in natural systems can remove up to 266 mmol m^{-2} d^{-1} and 61 mmol m^{-2} d^{-1} of N, respectively [16]. These processes may be active in the treatment of aquaculture effluent water and could be exploited to enhance treatment. However, to date there has been no published quantification of denitrification and anammox in settlement pond systems treating waste from tropical aquaculture farms.

The first step in optimizing the removal of fixed N through the denitrification and anammox pathways is to quantify their activity in settlement ponds and relate this to the environment of the ponds. Accordingly, the aim of this study was to determine if denitrification and anammox occur in sediments collected from tropical settlement ponds that are used to treat effluent from commercial production of prawns (shrimp) and fish. We used sediment slurry assays to investigate potential N_2 production in multiple zones of four settlement ponds on three farms (two prawn farms and one fish farm). We also investigated the relationship between the potential rates of N_2 production with the geochemical characteristics of the ponds. Additionally, the effect of carbon on N_2 production was tested since intensive aquaculture systems have N rich wastewaters where microbial N removal is typically limited by the supply of carbon as an electron donor [18]. Together these data provide new insight into N cycling processes in shallow tropical eutrophic marine systems in the context of N management.

Methods

Study site

The presence of denitrification and anammox and their potential rates were measured in sediment collected from four settlement ponds across two operational prawn (*Penaeus monodon*) farms and one barramundi (*Lates calcarifer* Bloch) farm. At Farm 1 sediment was collected from the two functional settlement ponds, this allowed comparison of N_2 production over small spatial scales (A and B; Figure 1). Additionally, sediment was collected from the only settlement pond at Farm 2 (Pond C) and the only settlement pond at Farm 3 (Pond D) (Figure 1). The three farms spanned the wet and dry tropics allowing comparison of N_2 production in different environments. Each pond was split into 3 zones (Z1, Z2 and Z3) (Figure 1). In all ponds Z1 was near the inlet, Z2 was near the middle of the settlement pond, and Z3 was near the outlet of the settlement pond. Ponds have diurnal fluctuations in dissolved oxygen (DO) concentration; from <31.2 μM at night to supersaturation (>312.5 μM) during the day, indicating rapid water column productivity. Similarly, there are diurnal pH fluctuations (1–1.5 pH). According to farm records, salinity fluctuates seasonally, with dramatic decreases from 35‰ to 5‰ caused by heavy precipitation over the summer wet season. During the wet season access to the farms by road is limited. All assays were, therefore run within the same dry season, although salinity at Farm 2 was still reasonably low due to particularly heavy rainfall over the 2009/2010 wet season (see results section).

Geochemical characteristics

To investigate the spatial variation of sediment characteristics within and between settlement ponds, and their role in driving N_2 production, sediments were collected at Z1, Z2 and Z3 in each of the four settlement ponds (total of 12 zones) (Figure 1). Sampling was conducted in March 2010 for Ponds B and C and August 2010 for Ponds A and D. Directly before taking sediment samples, surface water salinity, temperature and pH were also measured at each zone within each pond using specific probes (YSI-Instruments). Probes were calibrated 24 h before use. They were submerged directly below the surface and left to stabilize for 5 min before recording data. A known volume of sediment (30–60 mL) was subsequently collected in intact sediment cores ($n = 3$ per zone). The sediments were extruded, weighed and subsequently

oven dried (60°C) and reweighed for porosity (ϕ) determination ($n = 3$). Dried sediment was then milled (Rocklabs Ring Mill) for total N determination (LECO Truspec CN Analyzer). TOC was determined on a Shimadzu TOC-V Analyzer with a SSM-5000A Solid Sample Module. Solid phase S, P, Fe and Mn were also analyzed from milled sediment samples subjected to strong acid digestion. A THERMO Iris INTREPID II XSP ICP_AES was used to determine element content in triplicate sediment samples from each zone [19].

Denitrification and anammox potential

Slurry assays were conducted to test for the presence of N_2 (inclusive of both N_2 and N_2O) production through denitrification and anammox in March 2010 (Ponds B and C) or August 2010 (Ponds A and D). At the time of abiotic sample collection (see above), approximately 500 g of the most reactive sediments were collected from each zone in the four settlement ponds ($n = 1$ from each zone within each pond) (Figure 1) with a 30 mm i.d. corer [20]. The top 0–3 cm was collected because this includes the oxic and suboxic layers where NO_x is present or being reduced (denitrification) [21] and the anoxic layer below the interface, where NO_x penetrates but O_2 does not, making conditions favorable for anammox [22]. Each sediment sample was placed into sterile plastic bags with minimal air and subsequently homogenized by hand and doubled bagged before transportation to the laboratory. Sediments remained in initial plastic bags at room temperature for up to five days until the start of the experiment. Standard anammox assays were run according to Trimmer et al. [23] and Thamdrup and Dalsgaard [22] with modifications (artificial seawater of the same salinity as site water) according to Erler et al. [20]. Artificial seawater was used to preclude the potential interference of ambient NO_3^- in the isotope assay. A known volume of sediment (3–6 g) was loaded into Exetainers (Labco Ltd, High Wycombe, UK) and ~5 mL of degassed (flushed for 1 hr with ultra pure He), artificial seawater was added to form a slurry. Sediments were pre-incubated (overnight) under anoxic conditions to ensure all residual NO_3, NO_2^- and O_2 were consumed. Three different enrichment treatments (100 μM ^{15}N-NH_4^+, 100 μM ^{15}N-NH_4^+ plus 100 μM ^{14}N-NO_3^- or 100 μM ^{15}N-NO_3^-) were added to the slurries. After the isotope amendment, the Exetainers were filled with the degassed seawater, capped without headspace and homogenized by inverting 2–3 times. Triplicate samples were sacrificed from each treatment at 0, 0.5, 17 h and 24 h by introducing 200 μL 50% w/v $ZnCl_2$ through a rubber septum ($n = 3$). The 0 and 0.5 h time periods were chosen based on rapid turnover rates determined by Trimmer et al. [23] and 17 and 24 h were modified from Erler et al. [20]. Sacrifice of the slurry samples involved the addition of 2 mL He headspace to the samples through the septum. Samples were stored inverted and submerged in water at 4°C until analysis to ensure there was no diffusion of N_2 into or out of the Exetainers. A gas chromatograph (Thermo Trace Ultra GC) interfaced to an isotope ratio mass spectrometer (IRMS, Thermo Delta V Plus IRMS) was used to determine $^{29}N_2$ and $^{30}N_2$ content of dissolved nitrogenous gas (includes ^{15}N-N_2 and ^{15}N-N_2O, collectively referred to as N_2). Varying volumes (3–10 μL) of air were used as calibration standards.

The rate of N_2 production in the 24 h incubation trials (above) was calculated from the slope of the regression over the incubation period (0, 0.5, 17, 24 h) based on Dalsgaard and Thamdrup [24]. However, in some cases the production of $^{29}N_2$ and $^{30}N_2$ was nonlinear and rates were calculated based on the first two production points. Therefore a subsequent slurry assay was run to investigate N_2 production rates over short, regular time intervals (15 min) to

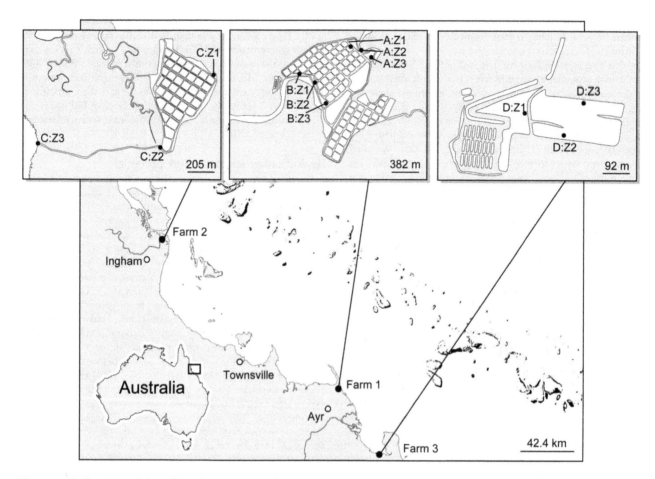

Figure 1. The location of three flow-through aquaculture farms along the North Australian coastline. The inset figures show the layout of each farm, the location of the settlement ponds and the 3 zones within each pond.

gain a more accurate insight into potential process rates. Sediment for the additional assays was collected from settlement Pond D, Zones 1 ($n=1$) and 3 ($n=1$) in October 2010. These zones were chosen because the production of N_2 was non-linear during the 24 h incubation assay (see results section). Assays were run as described above, following the same sediment collection, pre-incubation, amendment and analysis techniques. However, samples were sacrificed at 0, 15, 30, 45 and 90 min.

Slurry assay with carbon manipulation

The effect of an additional carbon source on the occurrence of denitrification and anammox was tested with a separate set of slurry assays because organic carbon limits N_2 production in some aquaculture systems [25]. Extra sediment was collected in March and August (2010) in the sampling described above. Sediments from Ponds A (August) and C (March) were assayed with and without addition of a carbon source because organic carbon has stimulated or correlated with N_2 production in some systems previously [17,26,27]. Concentrated particulate organic matter (POM) was used to test the effect of an *in situ* carbon source collected from Pond A. POM was collected by transporting settlement pond-influent water to the laboratory at the same time that sediments were collected. Suspended solids in influent water were concentrated by centrifugation (10 min at 3000 rpm). 400 µL aliquots of concentrated (~100 mg L^{-1}) POM were added to Exetainer vials prior to the addition of amendments. However, in the absence of a high total suspended solid load at

Pond C, methanol (MeOH) was used as the carbon source as it stimulates denitrification but inhibits anammox in some circumstances [28,29]. MeOH additions were carried out by adding MeOH at a concentration of 3 mM (based on Jensen et al. [29]) to a parallel set of samples from Pond C prior to amendments.

Modeling N removal

A simplistic model was constructed to estimate the mean dry season N removal (NR) capacity (%) of the four settlement ponds. NR was estimated using the potential N_2 production rates calculated in the present study, and N inputs into the pond through the wastewater. Given the substantial contribution of N remineralized from sludge in shrimp grow-out ponds (often exceeding inputs of N originating from feeds [30]), a variable to account for remineralization inputs was also added (N_{imin}). The following equation was used to calculate N removal and the parameters are further defined in Table 1:

Equation 1.

$$NR = \frac{N_2 \times A \times t \times A_r}{(N_{iww} + N_{i\,min})} \times 100$$

where N_2 = the mean total (inclusive of anammox) N_2 production rate measured during the 24 h incubation (nmol N cm^{-3} h^{-1}; Table 1). We adopted a conservative approach and assumed that N_2 production, driven by denitrification, only occurs in the top 1 cm of the sediment. Denitrification occurs at the oxic-anoxic

Table 1. An estimate of nitrogen inputs and microbial removal from settlement ponds, note TN = total nitrogen, WW = wastewater, min = mineralization.

Parameter	Value	Unit	Reference
Pond area	6000	m^2	Farm proprietors Pers. comm.
Mean TN WW input	14.8	kg N d^{-1}	EPA monitoring data
Mean net NH_4^+ min	27.8	mmol m^{-2} h^{-1}	[30]
Mean net DON min	0.6	mmol m^{-2} h^{-1}	[30]
Mean N_2 production	2.9	nmol N cm^{-3} h^{-1}	Slurry assay
Net N removal	2.5	%	Model

interface so the depth at which it occurs is dependent on O_2 penetration into the sediments. O_2 penetration is estimated at <0.5 mm in fish farm wastewater treatment ponds [4], 1.5–4 mm in sediments below fish cages and associated reference sites and up to 20 mm in a muddy macrotidal estuary [31]. This active zone is subsequently extrapolated to estimate rates for the entire area of the settlement pond. The remaining parameters are defined as follows: A = mean area of the settlement pond (m^2); t = 24 (h d^{-1}); A_r = atomic weight of N; N_{iww} = mean rate of TN input (inclusive of particulates and dissolved) via the wastewater (environmental protection agency (EPA) monitoring data, quantified monthly by Farm 1; kg N d^{-1}); N_{imin} = mean rate of N input via mineralisation (deduced from NH_4^+ and DON fluxes in Burford and Longmore [32]; Table 1; kg N d^{-1}).

Calculations and statistical analysis

The sediment characteristics data was analysed as a 2-factor nested design, pond and zone(pond) using permutational multivariate analysis of variance (PERMANOVA) [33]. PERMANOVA calculated p-values from 9999 permutations based on Bray-Curtis distances. A 1-factor PERMANOVA was subsequently used to compare differences in N_2 production rate data (three variables; denitrification, anammox and total N_2 production) between ponds with zones as replicates (n=3). 9999 permutations were again used to calculate p-values based on Bray-Curtis distance. PRIMER version 6 and PERMANOVA+ version 1.0.4 were used to conduct both analyses.

The relationship between N_2 production rate (three variables: denitrification, anammox and total N_2 production) and sediment characteristics was subsequently investigated using the BIOENV procedure in PRIMER. This procedure performs a rank correlation of the two similarity matrices (described above) and tests every combination of sediment characteristics to determine which set of variables best explains the observed N_2 production rates [34]. A Bray-Curtis similarity matrix comprised of both N_2 production rate data and sediment variable data was also used to conduct a hierarchical agglomerative cluster analysis which was superimposed on a multidimensional scaling (nMDS) plot. The nMDS plot provided a 2-D visualization of the relationship between sediment characteristics and N_2 production rates.

The effect of carbon addition on potential N_2 production rate in sediments was analyzed with paired t-Tests for each carbon source (POM and MeOH).

Results

Pond characteristics and abiotic factors

Surface water temperature (25.8°C±1.0) and pH (7.6±0.2) varied little similar across all ponds and zones. Surface water salinity in Pond C (Farm 2) was lower (17–18‰) than the other three ponds (31–35‰; Table 2) due to its location in the wet tropics where precipitation is high (Figure 1).

Sediment at all zones was uniformly dark black with minor color variation shown in a narrow lighter band (~3 mm oxic zone) at the surface of the sediment. The porosity ranged between 41–72% (Table 2) and sediments produced a rich hydrogen-sulfide smell and gaseous bubbles (presumably consisting of a mix of biogases) at the water surface when the sediment was disturbed. Very little bioturbation by burrowing organisms or flora was evident. There was significant variability between ponds (Table 3; PERMANOVA; pond; $Pseudo$ F=2.06, P=0.028) and between zones within ponds (Table 3; PERMANOVA; zone (pond); $Pseudo$ F=33.83, P<0.001). The variance in sediment characteristics at the finer scale (i.e. meters) between zones within ponds (52.4%) was greater than the variance between settlement ponds located kilometers apart (31.6%).

Denitrification and anammox potential

There was also a significant difference in the potential rate of N_2 production between ponds (Table 3; PERMANOVA; pond; $Pseudo$ F=3.91, P=0.001). The potential rate was highest in sediments collected from pond A, with denitrification the sole producer of N_2 (7.07±2.99 nmol N cm^{-3} h^{-1}; Table 4) and lowest in sediments collected from pond C, where again denitrification was the responsible for 100% of the N_2 produced (0.004±0.003 nmol N cm^{-3} h^{-1}; Table 4). However, there was no correlation between the potential production of N_2 in zones within ponds and different sediment characteristics that defined each pond (nMDS, Figure 2a & b). For example, pond B zone 3 had the highest anammox rates and low denitrification, whereas pond A, zones 2 and 3 had the opposite trend (Figure 2a). This is highlighted in the vector loadings for which the vectors for anammox and denitrification are clearly negatively correlated (Figure 2b).

Highly positive or negative loadings of the sediment characteristics appeared to have little influence on total N_2 production or denitrification (Figure 2b) as these are perpendicular to the positive

Table 2. Mean surface water salinity (n = 3 ± 1 SE) and abiotic sediment characteristics (n = 9 ± 1 SE) in the four settlement ponds (A, B, C and D) used to treat aquaculture wastewater (μmol g^{-1} unless stated).

	Pond A	Pond B	Pond C	Pond D
Salinity (‰)	31±0	34±0	18±0	35±0
Porosity (%)	0.5±0.0	0.5±0.0	0.5±0.0	0.6±0.0
TOC	61±13	62±6	43±5	63±4
TOC (%)	0.7±0.9	0.8±0.1	0.5±0.1	0.8±0.1
TN	5±1	6±1	4±1	8±1
TN (%)	0.1±1.0	0.1±0.4	0.1±0.8	0.1±0.6
TP	18±4	14±2	5±1	14±3
S	9±1	9±2	12±2	9±0
Fe	43±6	52±5	18±2	25±1
Mn	8±2	6±1	1±0	2±0

Table 3. A summary of statistical analyses; PERMANOVAs based on the Bray-Curtis similarities of transformed (4th root) sediment characteristic data and potential N_2 production rate data.

Sediment characteristics

Test				PERMANOVA
Factors	df	MS	Pseudo-F	P
Pond	3	39	2.06	0.028
Zone (Pond)	8	19	33.83	0.000

N_2 production rate

Test				PERMANOVA
Factors	df	MS	Pseudo-F	P
Pond	3	4029	3.91	0.001

loadings of all the sediment characteristics. Anammox did cluster near sediment variables (Figure 2b), however there was no correlation between the N_2 production matrix (inclusive of total N_2, denitrification and anammox) or the sediment variable matrix (BIOENV analysis; $\rho = 0.134$, $P = 0.730$).

In incubations with ^{15}N labeling of nitrate only, the majority of $^{15}N\text{-}NO_3^-$ converted to N_2 was found in $^{30}N_2$ (Figure 3). Only in pond B was more of $^{15}N\text{-}NO_3^-$ that was converted to N_2 found in $^{29}N_2$ than in $^{30}N_2$ (Figure 3). Anammox was detected in pond B sediments as indicated by the higher percent recovery ($0.67\pm0.28\%$) of $^{15}N\text{-}N_2$ in treatments where $^{15}N\text{-}NH_4^+$ and unlabelled $^{14}N\text{-}NO_3^-$ were added compared to treatments where $^{15}N\text{-}NH_4^+$ was added ($0.28\pm0.09\%$; Table 5). However, in this pond total recovery of $^{15}N\text{-}NO_3^-$ as $^{15}N\text{-}N_2$ was extremely low (0.20 ± 0.07; Table 5).

Slurry assay with carbon manipulation

There was no significant difference in the rate of N_2 production when either POM (Table 3; Pond A; paired t-Test; $P = 0.350$, $n = 3$) or methanol (Table 3; Pond C, paired t-Test; $P = 0.744$, $n = 3$) was added to the experimental sediment slurries (Table 4; 24 h incubation compared to carbon incubation).

Nitrogen removal capacity

We estimate that 2.5% of the total N inputs to the settlement pond are removed through denitrification and anammox (Table 1).

Table 4. The rate (nmol N cm^{-3} h^{-1}) of N_2 production in three incubations (i.e. 24 h, 1.5 h and in the incubation with carbon additions).

Pond	24 h incubation		1.5 h incubation		Carbon Incubation	
	DNT	ANA	DNT	ANA	DNT	ANA
A	7.07±2.99	ND			7.97±3.35	ND
B	0.06±0.06	0.22±0.12				
C	0.004±0.003	ND			0.004±0.003	0.03±0.02
D	4.36±2.01	ND	6.32±4.16	0.48±0.48		

DNT = denitrification; ANA = anammox.

Discussion

Total N_2 production and controlling mechanisms

Isotope tracer techniques confirmed the production of N_2 in sediment collected at all three zones within each of the four settlement ponds used to treat wastewater from commercial prawn and barramundi farms. The potential rates (0–7.07 nmol N cm^{-3} h^{-1}) were within the range of those reported for a subtropical constructed wetland (1.1 ± 0.2 to 13.1 ± 2.6 nmol N cm^{-3} h^{-1}) [20], but lower than those reported for subtropical mangrove and shrimp grow out pond sediments (21.5–78.5 nmol N cm^{-3} h^{-1}) [35]. Nevertheless, it can be assumed that both denitrifying bacteria and *Planctomycetes* (anammox bacteria) are present in the ponds and that there is potential to stimulate N_2 production rates and enhance N removal. To achieve this, an understanding of the mechanisms controlling N_2 production is required. We therefore investigated the effect of carbon additions on N_2 production rate and the relationship between the concentration of sediment elements and N_2 production rates. However, there was no significant change in the rate of N_2 production under carbon loading and there was no correlation between any of the measured sediment variables and N_2 production rate via denitrification or anammox.

Denitrification is often limited by carbon in aquaculture ponds, as carnivorous marine species require high inputs of protein rich feeds. N removal can be enhanced through the addition of an exogenous carbon source, for example glucose and cassava meal [26] or molasses [25] have been added to shrimp farm wastewater treatment processes, resulting in up to 99% removal of NH_4^+, NO_3^- and NO_2^-. Similarly, methanol is a common additive to enhance denitrification for municipal wastewater treatment, increasing degradation of NO_2^- in activated sludge from 0.27 mg NO_2^- g^{-1} volatile suspended solids (VSS) h^{-1} to 1.20 mg NO_2^- g^{-1} VSS h^{-1} [27]. However, in the present study N_2 production was not enhanced through the addition of carbon, suggesting that there are additional controlling mechanisms driving N_2 production. This concurs with the lack of significant correlation between measured sedimentary TOC and N_2 production. The lack of stimulation of N_2 production after the addition of carbon has also been demonstrated in the oxygen minimum zone of the Arabian Sea, where denitrification (and anammox) was only enhanced at one out of 11 depths [36]. Instead, Bulow et al. [36] highlighted a correlation between denitrification and NO_2^- concentration, a factor which likely also plays a role in controlling denitrification in settlement pond systems but was not measured in

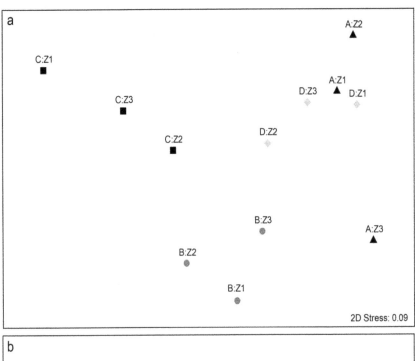

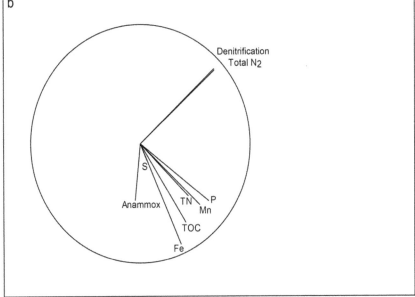

Figure 2. Similarity between N_2 production rates and sediment characteristics in the four settlement ponds. a) nMDS ordination; 2-D stress = 0.09. b) The same nMDS as a), with vectors superimposed, the length and direction of which indicated the strength of the correlation and direction of change between the two nMDS axes.

the present study. NO_3^- concentration also regulates anammox activity in estuarine sediments [37], so future work should aim to correlate extractable NO_3^-, NO_2^- and NH_4^+ with denitrification and anammox potentials to determine if these are driving process rates in settlement ponds.

It is also possible that the exogenous carbon source is instead stimulating nitrate ammonifiers (DNRA) and therefore competition for NO_x as a substrate [38]. Of the added $^{15}NO_3^-$ only $7.9\pm2.7\%$ was recovered as $^{15}N_2$, so a large portion (i.e. ~90%) of added $^{15}NO_3^-$ could be rapidly consumed by competing pathways such as DNRA or assimilation. The prevalence of DNRA or assimilation over denitrification will determine the balance between N being removed from the system through

gaseous N_2 production, or conserved within the system [39–41]. Furthermore, although dominance of DNRA over denitrification and anammox has been demonstrated in tropical estuaries [42] and under fish cages [43], DNRA has never been quantified in tropical settlement ponds and warrants further investigation.

Another potential controlling factor may be the presence of free sulfides. Sulfur is cycled rapidly in tropical sediments [44], and is the most important anaerobic decomposition pathway in tropical benthic systems, occurring at rates of $0.2–13$ mmol S m^{-2} d^{-1} and releasing free sulfides [45,46]. Free sulfides inhibit nitrification and therefore may be reducing N_2 production in the present study by reducing the amount of NO_3^- available to denitrifiers [47]. Additionally, DNRA may be stimulated in the presence of sulfur, increasing competition with denitrifiers for NO_3^- [48]. Again, the effect of sulfur on N_2 production in tropical settlement ponds is largely unknown and further studies are needed to elucidate the potential of this factor on stifling N removal in settlement ponds.

Denitrification verses anammox

In our study denitrification was the dominant N_2 production pathway. In coastal, hyper-nutrified sediments, low N_2 production through anammox has been attributed to the limitation of NO_2^- [49,50]. Further controlling factors for anammox are NH_4^+, total kilojoule nitrogen, TN, TP, salinity, redox state, and an inverse relationship with TOC [51]. Given these controlling factors anammox potential varies seasonally [51] and reported anammox contribution to N_2 production is highly variable with values of 1–8% [23], ≤3% [15], 10–15% [52], 19–35% [16], up to 65% [17], 2–67% [22] and 4–79% [53].

Anammox was detected in sediment collected in ponds B (24 h incubation), C (carbon incubation) and D (1.5 h incubation), notably, where overall N_2 production was exceptionally low. For example, during the 24 h incubation with sediment collected in pond B, N_2 production was lower than in sediment collected from all other ponds, but anammox contributed 95% to N_2 production. Low carbon oxidation rates and correspondingly low denitrification (and thus competition for substrate) have been proposed as the reason anammox contribution is high in environments where denitrification is low [17]. Bulow et al. [36] demonstrated that

Table 5. The percent recovery of added ^{15}N as labelled N_2 in three treatments.

	$^{15}N\text{-}NO_3^-$	$^{15}N\text{-}NH_4^+$ & $^{14}N\text{-}NO_3^-$	$^{15}NH_4^+$
A	11.8±1.17	0.00±0.00	0.00±0.00
B	0.20±0.07	0.67±0.28	0.28±0.09
C	10.92±1.99	0.26±0.08	0.43±0.07
D	8.79±0.61	0.01±0.32	0.00±0.00

high anammox rates corresponded with low denitrification rates at one site in the oxygen minimum zone in the Arabian Sea. At this site both anammox and denitrification were stimulated by the addition of organic carbon. This suggests that N_2 production was carbon limited giving anammox the competitive advantage. In tropical estuary systems where high temperatures, low sediment organic content and low water column NO_3^- concentrations prevail, the order of NO_x reduction pathways is proposed to be DNRA>denitrification>anammox [42].

The apparent detection of anammox in the presence of MeOH in sediments collected from Pond C is unusual given that anammox is inhibited by MeOH [28]. It is possible that during the 24 h incubation $^{15}NH_4^+$ was transformed through anoxic

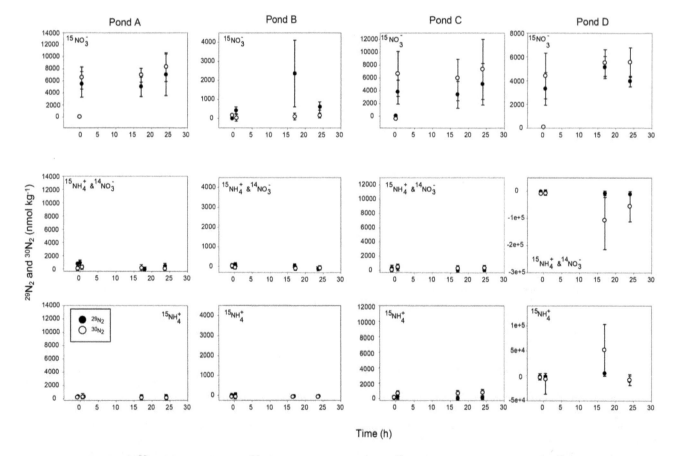

Figure 3. Production of $^{29}N_2$ (black circles) and $^{30}N_2$ (white circles) over 24 h. $^{15}N\text{-}N_2$ production in the presence of $^{15}N\text{-}NO_3^-$ is represented in row 1, $^{15}N\text{-}N_2$ production in the presence of $^{15}N\text{-}NH_4^+$ and $^{14}N\text{-}NO_3^-$ is represented in row 2 and $^{15}N\text{-}N_2$ production in the presence of $^{15}N\text{-}NH_4^+$ is represented in row 3. Column 1 represents $^{15}N\text{-}N_2$ production in sediments collected from pond A, column 2 represents $^{15}N\text{-}N_2$ production in sediments collected from pond B, column 3 represents $^{15}N\text{-}N_2$ production in sediments collected from pond C and column 4 represents $^{15}N\text{-}N_2$ production sediments collected from pond D.

nitrification [54], producing $^{15}NO_3^-$ and the resulting $^{15}N_2$ was produced as the result of denitrification.

Settlement pond functioning and implications

Microbial N_2 production has the potential to play a major role in removing N from aquaculture wastewater. However, we estimated that only 2.5% of total N added to the settlement pond via wastewater inputs and mineralization is removed through N_2 production. It is likely that the noxious compounds of H_2S and NH_4^+ are produced in settlement ponds when they are left unmanaged with no removal of settled particulate organic matter (sludge). These compounds have significant consequences for the inhibition of microbial processes that remove N from wastewater. In addition, H_2S accumulation causes a shift in the species of gaseous N produced from N_2 to N_2O due to the inhibition of the last step of denitrification [41]. This has detrimental consequences for global warming as N_2O is ~300 times more potent than CO_2 as a greenhouse gas whereas N_2 is relatively inert [55]. Future research should determine the concentration of H_2S at which the last reductive step of denitrification is inhibited and relate this to the amount of sludge that has built up in the settlement pond. We recommend that sludge be extracted at this point to prevent H_2S release and to prevent the recycling of soluble N through mineralization, DNRA or assimilation and subsequent senescence, as has been recommended for grow out ponds previously [30]. Innovative technology, such as anaerobic digesters and biogas capture, is required to convert the large volumes of sludge to a saleable product once removed from the pond. The simple management approach of removing sludge could have the added benefit of decreasing the incidence of competition between DNRA and denitrification thereby optimizing the denitrification and anammox processes for N_2 production. If N_2 production could be enhanced to the mean rate reported by Erler et al. [20] from a constructed wetland of 965 $\mu mol\ N\ m^{-2}\ d^{-1}$, then 100% of total daily N inputs would be removed from settlement ponds every day. However, the estimates in the present study are based on a very simplistic understanding of the settlement pond functioning and the model requires better definition of the parameters. For example, accurate rates of NH_4^+ and DON production from the sediments are required to estimate N inputs accurately. Additionally, N_2 production was measured in the dry season in the present study when rates are likely lower than in the wet season. Wet season precipitation lowers the salinity in the ponds to 5‰ in some cases, which favors higher denitrification, lower DNRA and lower NH_4^+ fluxes [56]. Denitrification is further stimulated during periods of heavy precipitation due to increased NO_3^- concentrations from land run-off [43]. An increased understanding of the temporal and spatial variability in N_2 production rates measured using intact core assays, instead of slurry assays, would also allow accurate predictions of N_2 production rates. Slurry assays only generate potential rates of N_2 production and we acknowledge that homogenizing sediments disrupts the sediment profile and can result in different nutrient availability than that which occurs *in situ* [57]. Additionally, an understanding of the rates of competing biogeochemical pathways such as DNRA and assimilation would enhance the accuracy of the model by including N retention rates into the model.

Acknowledgments

We thank all proprietors, managers and staff at Pacific Reef Fisheries (Farm 1), Coral Sea Farm (Farm 2) and Farm 3 for access to the settlement ponds, water quality data and interest in our study. In particular, thank you to Pacific Reef Fisheries for access to records including EPA load calculations. Thank you to Tim Simmonds (AIMS) for assistance with the graphics and to David McKinnon (AIMS) for advice, support and guidance.

Author Contributions

Conceived and designed the experiments: SC DE LT NP. Performed the experiments: SC DE. Analyzed the data: SC DE NP. Contributed reagents/materials/analysis tools: SC DE BE. Wrote the paper: SC DE LT RdN.

References

1. Galloway JN, Townsend AR, Erisman JW, Bekunda M, Cai Z, et al. (2008) Transformation of the nitrogen cycle: Recent trends, questions, and potential solutions. Science 320:

2. Thomas Y, Courties C, El Helwe Y, Herbland A, Lemonnier H (2010) Spatial and temporal extension of eutrophication associated with shrimp farm wastewater discharges in the New Caledonia lagoon. Marine Pollution Bulletin 61: 387–398.

3. Jackson CJ, Preston N, Burford MA, Thompson PJ (2003) Managing the development of sustainable shrimp farming in Australia: the role of sedimentation ponds in treatment of farm discharge water. Aquaculture 226: 23–34.

4. Bartoli M, Nizzoli D, Naldi M, Vezzulli L, Porrello S, et al. (2005) Inorganic nitrogen control in wastewater treatment ponds from a fish farm (Orbetello, Italy): Denitrification versus *Ulva* uptake. Marine Pollution Bulletin 50: 1386–1397.

5. Archer HE, Mara DD (2003) Waste stabilisation pond developments in New Zealand. Water Sci Technol 48: 9–15.

6. Porrello S, Ferrari G, Lenzi M, Persia E (2003) Ammonia variations in phytotreatment ponds of land-based fish farm wastewater. Aquaculture 219: 485–494.

7. Craggs RJ, Sukias JP, Tanner CT, Davies-Colley RJ (2004) Advanced pond system for dairy-farm effluent treatment. New Zealand Journal of Agricultural Research 47: 449–460.

8. Bolan NS, Laurenson S, Luo J, Sukias J (2009) Integrated treatment of farm effluents in New Zealand's dairy operations. Bioresource Technology 100: 5490–5497.

9. Tanner CC, Sukias J (2003) Linking pond and wetland treatment: performance of domestic and farm systems in New Zealand. Water Sci Technol 48: 331–339.

10. de Paula Silva PH, McBride S, de Nys R, Paul NA (2008) Integrating filamentous 'green tide' algae into tropical pond-based aquaculture. Aquaculture 284: 74–80.

11. Erler D, Songsangjinda P, Keawtawee T, Chiayakam K (2007) Nitrogen dynamics in the settlement ponds of a small-scale recirculating shrimp farm (*Penaeus monodon*) in rural Thailand. Aquaculture International 15: 55–66.

12. Jetten MSM, Wagner M, Fuerst JA, van Loosdrecht M, Kuenen JG, et al. (2001) Microbiology and application of the anaerobic ammonium oxidation ('anammox') process. Current opinion in biotechnology 12: 283–288.

13. Strous M, Fuerst JA, Kramer EHM, Logemann S, Muyzer G, et al. (1999) Missing lithotroph identified as new planctomycete. Nature 400:

14. Nicholls JC, Trimmer M (2009) Widespread occurrence of the anammox reaction in estuarine sediments. Aquatic microbial ecology 55: 105–113.

15. Koop-Jakobsen K, Giblin AE (2009) Anammox in tidal marsh sediments: The role of salinity, nitrogen loading and marsh vegetation. Estuaries and Coasts 32: 238–245.

16. Dalsgaard T, Canfield DE, Petersen J, Thamdrup B, Acuña-González J (2003) N_2 production by the anammox reaction in the anoxic water column of Golfo Dulce, Costa Rica. Nature 422: 606–608.

17. Trimmer M, Nicholls JC (2009) Production of nitrogen gas via anammox and denitrification in intact sediment cores along a continental shelf to slope transect in the North Atlantic. Limmol Oceanogr 54: 577–589.

18. Avimelech Y (1999) Carbon/nitrogen ratio as a control element in aquaculture systems. Aquaculture 176: 227–235.

19. Loring DH, Rantala RTT (1992) Manual for the geochemical analyses of marine sediment and suspended particulate matter. Earth Sci Rev 32: 235–283.

20. Erler DV, Eyre BD, Davison L (2008) The contribution of anammox and denitrification to sediment N_2 production in a surface flow constructed wetland. Environmental Science & Technology 42: 9144–9150.

21. Canfield DE, Thamdrup B, Hansen JW (1993) The anaerobic degradation of organic-matter in Danish coastal sediments - iron reduction, manganese reduction, and sulfate reduction. Geochimica Et Cosmochimica Acta 57: 3867–3883.

22. Thamdrup B, Dalsgaard T (2002) Production of N_2 through anaerobic ammonium oxidation coupled to nitrate reduction in marine sediments. Applied and environmental microbiology 68: 1312–1318.

23. Trimmer M, Nicholls JC, Deflandre B (2003) Anaerobic ammonium oxidation measured in sediments along the Thames Estuary, United Kingdom. Applied and environmental microbiology 69: 6447–6454.

24. Dalsgaard T, Thamdrup B (2002) Factors controlling anaerobic ammonium oxidation with nitrite in marine sediments Applied and Environmental Microbiology 68: 3802–3808.

25. Roy D, Hassan K, Boopathy R (2010) Effect of carbon to nitrogen (C:N) ratio on nitrogen removal from shrimp production waste water using sequencing batch reactor. J Ind Microbiol Biotechnol 37: 1105–1110.

26. Avnimelech Y (1999) Carbon/nitrogen ratio as a control element in aquaculture systems. Aquaculture 176: 227–235.

27. Adav SS, Lee DJ, Lai JY (2010) Enhanced biological denitrification of high concentration of nitrite with supplementary carbon source. Applied Microbiology and Biotechnology 85: 773–778.

28. Güven D, Dapena A, Kartal B, Schmid MC, Mass B, et al. (2005) Propionate oxidation by and methanol inhibition of anaerobic ammonium-oxidizing bacteria. Applied and environmental microbiology 71: 1066–1071.

29. Jensen MM, Thamdrup B, Dalsgaard T (2007) Effects of specific inhibitors on anammox and denitrification in marine sediments. Applied and environmental microbiology 73: 3151–3158.

30. Burford MA, Lorenzen K (2004) Modelling nitrogen dynamics in intensive shrimp ponds: the role of sediment remineralization. Aquaculture 229: 129–145.

31. Dong LF, Thornton DCO, Nedwell DB, Underwood GJC (2000) Denitrification in sediments of the River Colne estuary, England. Marine ecology progress series 203: 109–122.

32. Burford MA, Longmore AR (2001) High ammonium production from sediments in hypereutrophic shrimp ponds. Marine ecology progress series 224: 187–195.

33. Anderson MJ, Gorley RN, Clarke KR (2008) PERMANOVA+ for PRIMER: Guide to software and statistical methods. Plymouth: PRIMER-E Ltd.

34. Clarke KR, Warwick RM (2005) Primer-6 computer program. Natural Environment Research Council: Plymouth.

35. Amano T, Yoshinaga I, Yamagishi T, Chu VT, Pham TT, et al. (2011) Contribution of Anammox Bacteria to Benthic Nitrogen Cycling in a Mangrove Forest and Shrimp Ponds, Haiphong, Vietnam. Microbes Environ 26: 1–6.

36. Bulow SE, Rich JJ, Naik HS, Pratihary AK, Ward BB (2010) Denitrification exceeds anammox as a nitrogen loss pathway in the Arabian Sea oxygen minimum zone. Deep-Sea Research I 57: 384–393.

37. Trimmer M, Nicholls JC, Morley N, Davies CA, Aldridge J (2005) Biphasic behavior of anammox regulated by nitrite and nitrate in an estuarine sediment. Applied and Environmental Microbiology 71: 1923–1930.

38. Yin SX, Chen D, Chen LM, Edis R (2002) Dissimilatory nitrate reduction to ammonium and responsible microorganisms in two Chinese and Australian paddy soils. Soil Biol Biochem 34: 1131–1137.

39. Burgin AJ, Hamilton SK (2007) Have we overemphasized the role of denitrification in aquatic ecosystems? A review of nitrate removal pathways. Front Ecol Environ 5: 89–96.

40. King D, Nedwell DB (1984) Changes in the nitrate-reducing community of an anaerobic saltmarsh sediment in response to seasonal selection by temperature. J Gen Microbiol 130: 2935–2941.

41. Brunet RC, Garcia-Gil LJ (1996) Sulfide-induced dissimilatory nitrate reduction to ammonia in anaerobic freshwater sediments. Fems Microbiology Ecology 21: 131–138.

42. Dong LF, Sobey MN, Smith CJ, Rusmana I, Phillips W, et al. (2011) Dissimilatory reduction of nitrate to ammonium, not denitrification or anammox, dominates benthic nitrate reduction in tropical estuaries. Limmol Oceanogr 56: 279–291.

43. Christensen PB, Rysgaard S, Sloth NP, Dalsgaard T, Schwaerter S (2000) Sediment mineralization, nutrient fluxes, denitrification and dissimilatory nitrate reduction to ammonium in an estuarine fjord with sea cage trout farms. Aquatic microbial ecology 21: 73–84.

44. Madrid VM, Aller RC, Aller JY, Chistoserdov AY (2006) Evidence ofthe activity of dissimilatory sulfate-reducing prokaryotes in nonsuledogenic tropical mobile muds. FEMS Microbiology Ecology 57: 169–181.

45. Alongi DM, Tirendi F, Trott LA, Xuan TT (2000) Benthic decomposition rates and pathways in plantations of the mangrove Rhizophora apiculata in the Mekong delta, Vietnam. Mar Ecol-Prog Ser 194: 87–101.

46. Meyer-Reil L-A, Köster M (2000) Eutrophication of marine waters: effects on benthic microbial communities. Marine Pollution Bulletin 41: 255–263.

47. Joye SB, Hollibaugh JT (1995) Influence of sulphide inhibition of nitrification on nitrogen regeneration in sediments. Science 270: 623–625.

48. Jørgensen BB (2010) Big sulfur bacteria. ISME Journal 4: 1083–1084.

49. Dang H, Chen R, Wang L, Guo L, Chen P, et al. (2010) Environmental factors shape sediment anammox bacterial communities in hypernutrified Jiaozhou Bay, China. Applied and environmental microbiology 76: 7036–7047.

50. Risgaard-Petersen N, Meyer RL, Revsbech NP (2005) Denitrification and anaerobic ammonium oxidation in sediments: effects of microphytobenthos and NO3. Aquatic Microbial Ecology 40: 67–76.

51. Li M, Cao HL, Hong YG, Gu JD (2011) Seasonal Dynamics of Anammox Bacteria in Estuarial Sediment of the Mai Po Nature Reserve Revealed by Analyzing the 16S rRNA and Hydrazine Oxidoreductase (hzo) Genes. Microbes Environ 26: 15–22.

52. Hietanen S, Kuparinen J (2008) Seasonal and short-term variation in denitrification and anammox at a coastal station on the Gulf of Finland, Baltic Sea. Hydrobiologia 596: 67–77.

53. Engstrom P, Dalsgaard T, Hulth S, Aller RC (2005) Anaerobic ammonium oxidation by nitirte (anammox): Implications for N2 production in coastal marine sediments. Geochimica et Cosmochimica Acta 69: 2057–2065.

54. Hulth S, Aller RC, Gilbert F (1999) Coupled anoxic nitrification manganese reduction in marine sediments. Geochimica Et Cosmochimica Acta 63: 49–66.

55. IPCC (2001) Climate change 2001: The scientific basis. Cambridge University Press: Cambridge, UK.

56. Giblin AE, Weston NB, Banta GT, Tucker J, Hopkinson CS (2010) The Effects of Salinity on Nitrogen Losses from an Oligohaline Estuarine Sediment. Estuaries and Coasts 33: 1054–1068.

57. Minjeaud L, Michotey VD, Garcia N, Bonin PC (2009) Seasonal variation in di-nitrogen fluxes and associated processes (denitrification, anammox and nitrogen fixation) in sediment subject to shellfish farming influences. Aquatic Sciences 71: 425–435.

3

The Impact of Escaped Farmed Atlantic Salmon (*Salmo salar* L.) on Catch Statistics in Scotland

Darren M. Green*, David J. Penman, Herve Migaud, James E. Bron, John B. Taggart, Brendan J. McAndrew

Institute of Aquaculture, University of Stirling, Stirling, Stirlingshire, United Kingdom

Abstract

In Scotland and elsewhere, there are concerns that escaped farmed Atlantic salmon (*Salmo salar* L.) may impact on wild salmon stocks. Potential detrimental effects could arise through disease spread, competition, or inter-breeding. We investigated whether there is evidence of a direct effect of recorded salmon escape events on wild stocks in Scotland using anglers' counts of caught salmon (classified as wild or farmed) and sea trout (*Salmo trutta* L.). This tests specifically whether documented escape events can be associated with reduced or elevated escapes detected in the catch over a five-year time window, after accounting for overall variation between areas and years. Alternate model frameworks were somewhat inconsistent, however no robust association was found between documented escape events and higher proportion of farm-origin salmon in anglers' catch, nor with overall catch size. A weak positive correlation was found between local escapes and subsequent sea trout catch. This is in the opposite direction to what would be expected if salmon escapes negatively affected wild fish numbers. Our approach specifically investigated documented escape events, contrasting with earlier studies examining potentially wider effects of salmon farming on wild catch size. This approach is more conservative, but alleviates some potential sources of confounding, which are always of concern in observational studies. Successful analysis of anglers' reports of escaped farmed salmon requires high data quality, particularly since reports of farmed salmon are a relatively rare event in the Scottish data. Therefore, as part of our analysis, we reviewed studies of potential sensitivity and specificity of determination of farmed origin. Specificity estimates are generally high in the literature, making an analysis of the form we have performed feasible.

Editor: Howard Browman, Institute of Marine Research, Norway

Funding: This work was supported by funding from the Scottish Salmon Producers' Organisation (SSPO). The funders had no role in study design, data collection and analysis, decision to publish, or preparation of the manuscript.

Competing Interests: The authors have declared that no competing interests exist.

* E-mail: darren.green@stir.ac.uk

Introduction

Since the industry began in the 1960s, production of farmed Atlantic salmon (*Salmo salar* L.) in the North Atlantic gradually increased to reach 1.1×10^6 tonnes in 2009, while annual catch of Atlantic wild salmon has decreased from c. 10000 to 2000 tonnes over the same period [1]. There is concern regarding the large size of the farmed stocks relative to wild fish, particularly over potential adverse impacts of escaped farmed salmon through potential interbreeding with wild fish. In Scotland alone, 1.9 million farmed salmon escaped into the natural environment between 2002–9 [2]. Potential detrimental effects could include increased infestation by sea lice [3], competition for food or other resources, and inter-breeding enabling the spread of farmed genes into the wild population [4], thereby potentially lowering fitness [5,6]. Counteracting these processes, the breeding success of escaped farmed salmon appears low [5]. Escapes can occur at any point in the production cycle from the rearing of juveniles to the smolt stage in fresh water, through ongrowing to marketable size in the sea. Conceivably, with niche overlap between brown trout (*Salmo trutta* L.) and Atlantic salmon, especially in juvenile stages, these competitive effects could extend inter-species. However, there is evidence that brown trout are the more dominant fish [7], potentially reducing this impact. Potential escape routes include storm damage, or holes in nets and cages both in freshwater and

seawater. Routes for escapes and the resulting consequences have been recently reviewed by [8] and [9].

Scottish wild salmon catch has dropped in recent years in farmed areas, coinciding with the rise in salmon farming, located primarily on the west coast, however this is mirrored by a parallel decline in eastern regions without salmon farming, and the rod count alone has remained similar on both coasts (see results section below for examination of the publically available recent data). Rod count for sea trout (the anadromous form of brown trout, *Salmo trutta* morpha *trutta*) has suffered greater decline on the west coast, but this decline predates the establishment of salmon farming there. There may well be confounding factors not taken into account when comparing the East with the West of Scotland through such summary statistics; however, suspicion remains that salmon farming may be a partial cause of the decline. Though our study concerns escapes of farmed salmon, there are several potential mechanisms by which salmon farming could impact wild salmon without this being mediated by escapes, for example, by a rise in the density of sea lice in sea lochs [10]. A significantly higher percent of rod catch reported as farmed salmon in rivers with salmon farms in their sea lochs has been noted [11], alongside reduced freshwater salmon populations in rivers with salmon farms in their mouths [11]. However, such correlation data are insufficient to demonstrate cause and effect.

The River Ewe (Scotland) has been the focus of detailed study, with both salmon farming and a high level of reported escapes in the catch statistics [11,12]. Reported local escapes occurred in 1989 (marine growers), 1990 (smolts), 1992 (a large number of parr and smolts), 1993 (growers) and not again until 1999 (growers, parr, smolts). This matches poorly with the reported rod catch of farmed salmon, which peaked in 1995 and 1997, with lower counts in 1993, 1994 and 1999 [12]. Total rod catch in the Ewe catchment in recent years (including reported farmed salmon) is also within the range experienced prior to the establishment of salmon farming in that catchment. A wider study of monitoring and reporting of escaped farmed salmon in the British Isles found no association between reported escapes and the prevalence of escapes in coastal and freshwater fisheries, and also a weak association between farm production and the prevalence of escapes [13]. Nevertheless, these authors aggregated their data at a regional level, and suggested that a finer geographical scale of study is warranted, as we respond to in the current study.

Some of these previous studies suggested but did not prove links between catch statistics and salmon escapes. Analysis of these data sources is complicated by the potential for confounding factors, and most would not allow effects of salmon escapes *per se* to be distinguished from general effects of salmon farming. Therefore, in this paper, we address a very specific question: whether or not documented escape events can be linked statistically to later changes in catch statistics, either in terms of overall catch, or in terms of the proportion of the catch that are reported as being escaped farmed salmon. Sea trout remain in coastal waters, more directly exposed to potential environmental effects of marine aquaculture, therefore we also analysed the sea trout catch statistics. As a counterpart to this analysis, we considered the likely data quality of catch statistics in terms of accuracy of reporting of the farmed *versus* wild origin of salmon. Our analysis is up to date, using the recently available data on catch and escapee numbers. As a result, our perception of the current trends in salmon and sea trout catch differs somewhat from what would have been concluded even a few years ago.

Materials and Methods

Data sources

Historic (Fig. 1) and recent (2001 to 2009 [14]; Fig. 2) catch data were tabulated against 62 salmon fishery statistical districts in 11 salmon fishery statistical regions (pooled as east and west coasts, Fig. 3). This includes all salmon (including grilse, i.e. salmon returning to freshwater after one winter) caught by rod and line (both retained and released), net and coble (sweep netting using small boats), or fixed engine (e.g. various types of nets, often specific to a local area); and for both wild and farmed caught salmon. The definition of these four catch methods are documented by the Scottish Government [15] and focus on different parts of the water course: fixed-engine fisheries are coastal, outside estuary limits, whereas net-and-coble fisheries may operate in estuaries and lower river reaches. The largest fraction of catch is accounted for by rod-and-line angling, predominantly above tidal limits. Rod-and-line angling is divided into 'catch and retain' and 'catch and release', with the latter becoming an increasingly large proportion of the take for both trout and salmon, as catch size has reduced.

With catch and release (widely implemented in Scotland for salmon since the 1990s, as a conservation tool), there is potential for double counting, but with time-trends in the balance between caught and retained and caught and released, it was assumed pooling the counts was more robust. Catch data for Orkney and

Shetland were sparse and these regions were excluded from further analysis. Catch data for sea trout were treated similarly, with sufficient data for Shetland also included. With trout, effects of identifiable farmed salmon escape events can be studied without potential misidentification of wild (trout) with farmed (salmon).

Reporting of escapes for farmed salmon is mandatory (since 2001), and the available data consisted of count, date, size, and location of escapes by farm name. Escape counts were summed across each statistical district over each calendar year from 2002 to 2009. Additional variables consist of the escapes data lagged by between one to five years, to test for a delayed effect of salmon escapes. Data for both (lagged) escapes *and* catch were available for 2007 to 2009.

Analysis

Two types of models were constructed. In the *proportion escapes* models (1), the proportion of catch n for each district–year consisting of farmed salmon y was regressed against year a, region r, and district d, plus the the incidence of recent escaped farmed salmon in the same district ω, including lag terms. In the *catch statistics* models (2), the total catch per district–year (for both salmon and trout) was related to the same factors and covariates. Models were built using the R software environment, using binomial errors for the proportion escapes model, and Poisson errors for the catch statistics model. Likelihood ratio tests were used to compare nested models; each model was ordered with escape terms later in the list of terms, so as to specifically test for a significant effect of escapee salmon over and above any other local effects. For ease of interpretation, McFadden's pseudo r-square statistics are presented below.

$$\text{logit}\left(\mathbb{E}\left[\frac{y_{a,r,d}}{n_{a,r,d}}\right]\right) = \beta_0 + \beta_{1,a} + \beta_{2,r} + \beta_{3,d} + \sum_{i=0}^{5}\beta_{4,i}\,\omega_{a-i,r,d} + \beta_5 \log(n_{a,r,d} - y_{a,r,d} + 1) \tag{1}$$

$$\log(\mathbb{E}[n_{a,r,d}]) = \beta_0 + \beta_{1,a} + \beta_{2,r} + \beta_{3,d} + \sum_{i=0}^{5}\beta_{4,i}\,\omega_{a-i,r,d} \tag{2}$$

For the *proportion escapes* model alone (1), $\log(n-y+1)$ was fitted as a covariate as a proxy for (otherwise unknown) fishing effort.

As a test of model robustness, a related ANOVA model was fitted in both cases, with appropriate transformation of data. For the proportion escapes model, a weighted least-squares fit was performed on the empirical logit, with response variable

$$\eta = \ln\left(\frac{y + 1/2}{n - y + 1/2}\right). \tag{3}$$

The weighting variable used was the reciprocal of the variance [16], estimated as

$$w = \left(\frac{1}{y + 1/2} + \frac{1}{n - y + 1/2}\right)^{-1}. \tag{4}$$

Explanatory variables were the same as in the generalised linear models.

Each model was fitted to data at the level of the statistical district. Nevertheless, with little evidence of how far escaped

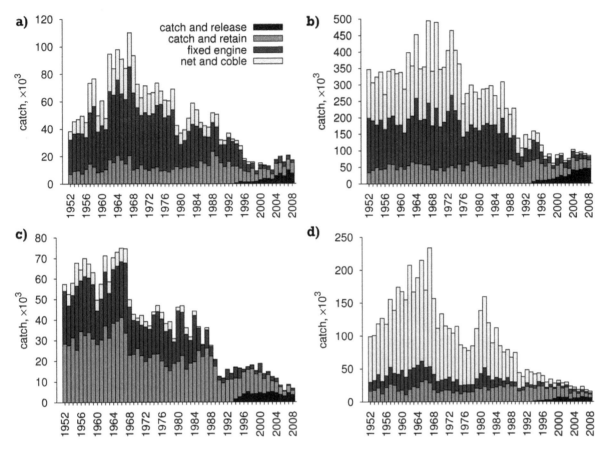

Figure 1. Historical catch data for salmon and sea trout in Scotland. a) west coast salmon; b) east coast salmon; c) west coast sea trout; d) east coast sea trout. East coast: Cape Wrath to Berwick (not including the Northern Isles); west coast: Solway Firth to Cape Wrath plus the Northern Isles. Data with permission from Marine Scotland Science (see Acknowledgements).

salmon disperse, models were also fitted at the statistical region level. As it is unclear which life stages are most likely to impact on wild salmon, models were fitted using either all escapes, or only large marine salmon (over 500 g). Models were also built excluding eastern Scotland, without an active salmon farming industry in the marine stage.

Results

Escape and catch statistics: historic and recent data

Examining the historical catch statistics for the east and west sides of Scotland, where the east side has minimal marine salmon aquaculture, a similar overall long-term downwards trend can be seen in catch statistics (Fig. 1). Some recovery of catch size for salmon can be seen in the last five years on both coasts, though the decrease in sea trout catch on both coasts shows no such halt. An important caveat with these data is the lack of any measure of fishing effort. Nevertheless, there is little sign of an increase in rod catch consequent to declining commercial catch effort, which is a major contributor to the overall decline.

For recent catch data (2001–9), 0.30% of overall catch was identified as of farmed origin, with a higher proportion (2.8%) within the intensely farmed regions (West, North West, Clyde Coast, Outer Hebrides; Fig. 3). The highest catch of farmed-origin salmon was in the West region (5.8%), and the lowest in the East region (0.0045%), where there is no farming activity at all.

From 2002–9, 1.93×10^6 escaped salmon were reported across Scotland in 100 escape events, with considerable geographical (Fig. 2) and annual variation in numbers, from 5.9×10^4 in 2008 to 8.8×10^5 in 2005. Of these, 1.22×10^6 were large salmon (>500 g) at sea, in 77 escape events. Overall, there was no significant correlation between the nationwide proportion of salmon catch reported as farmed, and the numbers of escaped salmon in that or the two preceding years ($p > 0.05$). For older salmon (as opposed to grilse), catch of farmed-origin fish was stratified into two periods: January to April, and May to December. 95% of farmed-origin salmon were reported in the latter period.

Anglers' ability to distinguish farmed-origin salmon

The catch data used in this study are of unknown accuracy, specifically with regards to the specificity and sensitivity of the anglers' ability to identify farmed salmon. To clarify this, reports from the literature [4,12,17–22] were examined to attempt to estimate these parameters. One study [4] sampled salmon from the River Polla (Scotland), known to contain farmed and wild salmon. These were categorised as putative wild or escaped on the basis of morphology. Carotenoid pigment analysis agreed with this categorisation, with 65 of 65 fish with fin deformities containing canthaxanthin, and 14 of 14 fish without such deformities only containing astaxanthin. Bankside assessment of wild/farmed state was of similar success rate, with 18 of 18 wild fish, and 26 of 26

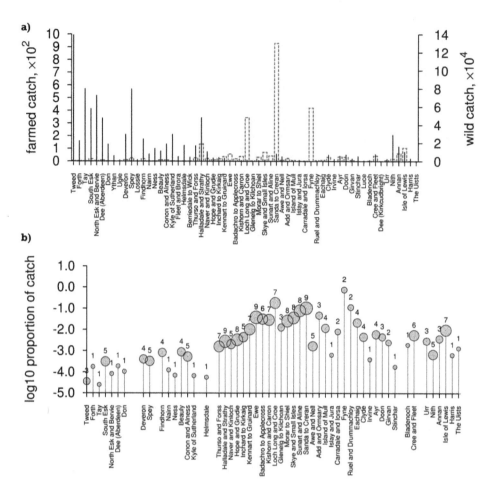

Figure 2. Catch statistics by district (2001 to 2009 data, excluding the Northern Isles). a) Dashed boxes, left axis: farmed catch; lines, right axis: wild catch. b) Proportion of catch of farmed origin, with symbol size indicating number of years (out of 9) excluding districts without catch of salmon of farmed origin. Data with permission from Marine Scotland Science (see Acknowledgements).

farmed fish correctly categorised as farmed or wild. In another study in the River Ewe (Scotland) [12], 95 of 95 wild salmon, and 7 of 10 farmed salmon were correctly identified. And using scale characteristics, a further study [18] confirmed 100 of 101 fish initially classed as of reared origin to have been correctly classified. Several other papers have commented on the difficulty of categorising salmon origin. A study in Greenland produced two datasets [19]: in one, 3 of 272 fish were identified as farmed, but 7 were uncategorisable on the basis of scales; in the second, 6 of 423 fish were identified as farmed, with 6 difficult to categorise. A similar problem was reported in Faroese data [20,21] where 6% of fish were found to be uncategorisable. A Norwegian study [22] compared scale readings with fishers' initial assessment of farmed/wild origin of salmon. They found that 4 of 7 fish initially assessed as wild were correctly reported, as were 373 of 378 fish initially reported as of farmed origin. For a review of papers exploring various morphological and biochemical methods to detect salmon of farmed origin, see [17].

Sensitivity and specificity estimates are presented in Table 1 with binomial confidence intervals obtained from the *binom.profile* function in R. These studies indicate high specificity for detecting farmed-origin fish, though with wide confidence intervals where sample size is restricted. Sensitivity estimates are also high. High specificity supports the 2001–8 Scottish catch statistics, where some regions report vanishingly low proportions of farmed

salmon, however it is unsafe to assume that attribution of fish origin is consistent across districts.

Models for proportion of escapes

In all models of the proportion of escapes, district and region were highly significant. For the district level model, all regions, McFadden's pseudo $r^2 = 0.84$. When included in this model, year of study was significant in a likelihood ratio test ($r^2 = 0.87$). Inclusion of the term for log catch size did not cause a significant reduction in deviance. Including counts of large escapes (0–5-year lags) caused a significant reduction in deviance ($-2 \times$ log-likelihood:(LL)) from 459 to 324 (6 d.f.). This was a better fit than including all escaped salmon ($-2LL = 349$) though with a small difference in deviance.

All these effects are relatively small compared with the null-model deviance of 3626, but significant given the large size of the dataset. In this model, of the individual lag terms, only two were significant in a Wald test ($P < 0.05$), and with contrasting signs. The zero-year term had a coefficient of 1.1×10^{-4}, suggesting a relative odds of a caught salmon being identified as of farmed origin of 3.0 for each 10,000 escaped salmon; The four-year lag term had a coefficient of -1.79×10^{-4}, suggesting a relative odds of 0.16 for each 10,000 escaped salmon. The other lag terms were both insignificant and of inconsistent signs. The equivalent ANOVA model indicated the lag terms to be significant overall

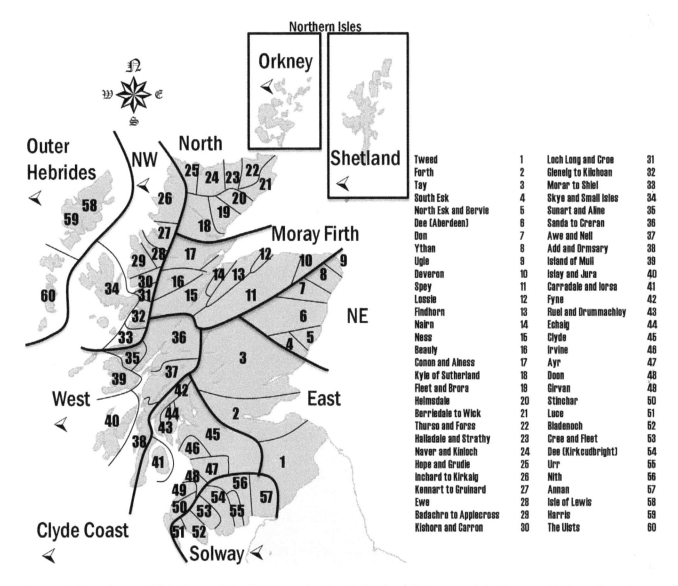

Figure 3. Schematic map of fisheries statistics districts and regions in Scotland. Regions regarded as 'west coast' in the results section are indicated by a left-pointing arrowhead. 'Districts' and 'regions' are not coterminous with other political units of similar name in UK geography.

$(F_{6,111} = 39.4; P < 0.001)$ but a small contributor to overall variance and without any individual lag terms significantly different from zero.

In the regional-level models, region and year remained highly significant factors ($r^2 = 0.87$), as was the covariate term for catch size, if included ($r^2 = 0.89$). The best fit model included the counts of all escaped salmon and their lags (as opposed to large salmon alone), causing a significant reduction in deviance from 271 to 34. All lag terms except the 5-year lag were significant, and all were positive. That with the largest coefficient was for a one-year lag (4.45×10^{-4}). Only the zero-year lag term was significant in the equivalent ANOVA, though all lag terms were of similar magnitude and the same sign as in the logistic regression.

For large escapes, district-level models were repeated for active farming areas only (Outer Hebrides, West, North West, Clyde Coast). Model results were similar to the all-Scotland model in that the zero-year lag term was significantly positive (coef. 9.6×10^{-5}), and the four-year lag term significantly negative (coef. 2.1×10^{-4}). In addition, the one-year lag term was also significantly negative

(coef. -1.4×10^{-4}). As with the all-Scotland model, the equivalent ANOVA model did not identify any lag terms as significantly different from zero.

The prevalence of farmed-origin salmon across the four different catching methods varied, with higher prevalence in fixed engine and net and coble take (24.8% in west-coast salmon farmed regions) compared with rod and line (1.1%), potentially confounding the analyses given geographical and temporal variation in catch methods. Therefore, we fitted models separately to rod-and-line and 'other' fishing methods (which had a 5.8× higher prevalence of farmed-origin salmon overall) at the regional level for actively farmed regions and large escapes. Catch-method data at the district level were not available. Repeating the earlier analysis for regional-level data and large escaped fish for rod-and-line catch only, escapes remained significant in a likelihood ratio test (deviance reduction from 112 to 29, compared with a null deviance of 1560). All coefficients for escape terms were significantly negative (largest coefficient, four-year lag, -5.6×10^{-4}), except that for the zero-year lag which was

Table 1. Sensitivity and specificity estimates for anglers' ability to discern origin of caught salmon.

Reference	true+	false+	false−	true−	sensitivity	specificity
[4]	65	0	0	14	1 (0.95–1)	1 (0.80–1)
...	26	0	0	18	1 (0.88–1)	1 (0.84–1)
[12]	7	0	3	95	0.70 (0.39–0.91)	1 (0.97–1)
[22]	373	5	3	4	0.99 (0.97–0.99)	0.44 (0.17–0.75)
[18]	100	1	n/a	n/a	n/a	n/a
[19]	3	<7		262	>0.7	>0.96
...	6	<6		411	>0.5	>0.99
[21]	n/a	<6%		n/a		>0.94

Positive = farmed fish. 95% confidence intervals are provided.

significantly positive (6.1×10^{-5}). For catch other than rod-and-line, escapes were significant under a likelihood ratio test but no coefficients were significant according to the Wald test.

Deviance residuals were examined to investigate goodness of fit. For the district-level model (all districts, large escaped salmon only), the deviance residuals showed a peaked distribution that deviated from normality (A–D test, $p < 0.001$). This distribution resulted from districts in the dataset having no reports of caught farmed fish over the whole period, resulting in districts with zero residual. Removing these districts led to a complicated result, though residuals showed a more normal distribution. The model using escapes of all salmon was of similar likelihood to that of large escapes only (deviance of 127.0 versus 129.9). The deviance attributed to escapes in both models was significant and similar (103 versus 100 compared with a null deviance of 2167). However, only in the all-escapes model were escape terms significant in a Wald test (all except one-year-lagged escapes), with all coefficients positive, the largest of which was associated with a four-year lag (1.49×10^{-4}).

Models for catch size

In the Poisson regression of salmon catch size at the district level, region and district were highly significant (pseudo $r^2 = 0.97$), explaining as might be expected the majority of the model deviance, because catches differ greatly between districts; including year caused a significant improvement in model fit ($r^2 = 0.98$). Including numbers of large escapes (plus lag terms) caused a significant reduction in deviance from 7926 to 7618 (6 d.f.), a larger reduction compared to including escapes of all sizes (residual deviance 7759). Only the one-year lag term was significant, with a coefficient of (-8.82×10^{-6}) corresponding to a decrease in catch in a district–year of 8.4% per 10,000 escaped salmon. The lag terms were not significant in the equivalent ANOVA model.

As with the proportion of escapes models, the regional-level catch size model gave conflicting results. Region was highly significant ($r^2 = 0.97$), and year significantly improved the model fit on inclusion ($r^2 = 0.98$). All escape terms were negative and significant, with the highest coefficient for the two-year lag term (-5.1×10^{-5}). As with the district-level model, these terms were insignificant in the related ANOVA model.

The same models were fitted for trout catch data, and in both region-level and district-level models, terms accounting for escapes of farmed salmon of all sizes produced a model with a higher likelihood than large escapes alone. Again, region, district, and

year were highly significant, reflecting variability in trout catch (pseudo $r^2 = 0.96$). All terms for escaped salmon were significant when included, reducing model deviance from 6782 to 6081; however, they were not all of like sign: All except the zero-year-lag term were positive, the largest being that for the four-year lag (2.71×10^{-5}), and that for the zero-year lag being -2.63×10^{-6}. The regional-level model gave similar results with coefficients of like sign. As with the other model types, the equivalent ANOVA model indicated fewer significant lag-escape terms, but where significant these were of like sign and similar magnitude to the Poisson regression model.

A significant number of deviance residuals from the Poisson regression in excess of two were found. As a result, an alternate model was fitted with negative binomial errors. For all districts and large escaped salmon, this model proved a better fit (pseudo $r^2 = 0.987$) with the majority of residuals in the range $(-2,2)$. Inclusion of terms for escapes were not significant in a likelihood ratio test ($P = 0.079$).

Discussion

Recaptures reported above account for less than two per thousand of reported escapes, with the fate of the vast majority of escapes unknown. This suggests that escaped salmon either have very low survival in the wild, disperse without returning, or are less readily caught by anglers. Few studies have examined this in Scotland. However, after a simulated escape by the release in 2006 of 678 tagged adult salmon near Ullapool, only five tags were retrieved: two detached, on beaches in Scotland north of the release site, and three on live fish in Scandinavia [23]. It has been hypothesised that escaped salmon in Scotland move east in this way as a combination of instinctive homing behaviour and prevailing current direction [23].

This contrasts with the situation in Norway, where recapture rate of released cultured salmon has been shown to reach as high as 67% [24]. The difference may be in part due to topographical differences between Scotland and Norway, where enclosed fjords exist at much larger sizes than the west coast of Scotland. However, much of this recapture of escaped farmed salmon occurred in Norway in coastal waters, not rivers [22], and recent data show these fish to perform relatively poorly with low survival to maturity due to impaired feeding [25], and loss of migratory performance [26]. Nevertheless after simulated escape of farmed smolts and post-smolts in Norway, tagged fish were recovered after up to three winters at sea [27], though these were small in number compared with those recaptured more quickly, and across wide area of both river (26%) and sea, albeit with the majority close to the site of release.

Our data source does not indicate the distance from river mouth where farmed- and wild-origin fish are caught, that is whether farmed-origin fish are more or less likely to penetrate to the upper reaches. However, there is a strong trend towards a higher prevalence of farmed-origin salmon in fixed-engine and net-and-coble catch (lower down the water course) than in rod-and-line (further up the water course). As a proportion of overall total catch, salmon of farmed origin are comparatively uncommon compared with similar studies in both Norway [28] and eastern North America (Canada and USA) [29], although more comparable if only the non rod-and-line catch (concentrated in coastal areas) is considered. Any comparison between proportions requires care given unknowns of wild population size and catch effort, or even the relative catchability of farmed-origin and wild fish once in the rivers.

The Scottish dataset contains little in the way of stratification by season, however escapes are rare in the catch from the earlier part of the year. (In contrast, the escapes data, aggregated here into years, are described by day of escape.) This is in agreement with findings in Northern Ireland [18] and Norway [28], where escaped farmed salmon tend to enter rivers relatively late in the season. A caveat here is that any seasonal differences in fishing effort by the different methods—in turn concentrated in different sections of the water course—would be confounded with seasonality in appearance of escaped farmed salmon. At shorter timescales beyond the resolution of the Scottish data, a study in Norway [22] reported elevated catch of farmed salmon was detected in fisheries for several weeks after documented escape events; however a considerable 'background' rate of farmed salmon of varied size ranges persisted in the catch, suggesting that in the studied regions of Norway, a 'trickle' of unreported, small escape events may have been an important source of farmed-origin fish in fishery catches [22].

Given such unknowns in salmon biology and behaviour, we have been flexible in our modelling approach. For example, with comparatively little data indicating how salmon may disperse in the open sea (as opposed to enclosed fjords [24]), it is unclear what the appropriate size of geographical area for study should be. One study [11], finding greater depletion of wild stocks in areas with salmon farms, used data at the river level, a finer geographical scale that was available for our study; however, in another [13], with data aggregated at the regional level, no relationship between prevalence of escapes and reported escapes was found. Our analysis asks a subtly different question: we specifically test for an effect of documented escape events, over and above any baseline differences between districts due to other causes. Possible reasons for an increase in catch after escape events could be misidentification of farmed fish as wild, or increased catch effort following known escape events. Thus, any baseline association between escapes and farmed-origin catch are absorbed into the terms for district- and year-level variation.

Our model results, though in places with terms for lagged escapes significantly related to catch size and proportion of escapes, explained a low proportion of the model variation and showed low robustness to changes in model structure, particularly in the case of the more robust ANOVA models where few terms were found significant. In particular, for proportion of catch reported as escapes, under 10% of deviance was explained by escape lag terms even when non-farmed districts were excluded. Effect sizes were relatively small and with contradictory signs when examined at the district and regional level. This partly stems from relatively complicated models with considerable district-to-district

variation, and multiple lag terms for escapes, which were considered necessary due to the long generation time of the species involved. An assumption of both logistic regression and ANOVA is independence of observations. As with many observational studies, there are likely to be uncontrolled grouping variables in our study, such as survey response, individual angler, and sub-district geographical structure.

Despite these caveats of overinterpretation of the model results, some patterns can be ascertained. In district-level models, the proportion of catch reported as of farmed origin was positively associated with local farm escapes in the recent past, but negatively associated at longer time lags. This may be the case if farmed salmon from previous years are more likely to be misidentified as wild fish later. The best-fit model for district levels included escapes of large fish, whereas for regional-level models, all escapes, and with negative coefficients. This is consistent with reported catch (mostly of large fish) being affected more by recent, local fish escapes, with escapes from further back in time being caught over a wider area, and possibly misidentified as wild. The possibility of some form of confounding is also indicated by the difficult-to-explain trout results, where trout catch was found to be positively associated with local escapes of farmed salmon. In addition when only the proportion of farmed fish in the rod-and-line catch was considered, model results were again inconsistent in regional-level models, with negative coefficients.

The historical decline in salmon catch in Scotland fits into the general trend of declining biomass observed in Atlantic salmon across Europe [30]. However, our study relies on secondary data of unknown accuracy, ultimately derived from a large number of questionnaire returns from fisheries (1846 in 2008 alone [14]). Return rate is generally high, though with omissions; for 2008, overall return rate of questionnaires was 93%, with almost all districts with return rates exceeding 80%.

Though we have addressed potential data errors using estimates of potential accuracy from the literature, there are several potential reasons why such parameter estimates may not be appropriate, or even constant between areas. Rod-catch data may poorly estimate occurrence of rod-caught farmed salmon due to both anglers' perceptions and ability to distinguish between salmon types. It can be presumed that given fish of similar possible farm-origin appearance, anglers will be less likely to report these as being of farmed origin when in an area with no history of salmon farming or escapes. Accuracy will also decline over time since escape: particularly for salmon that escape as parr, numbers in the catch statistics may be underestimated [31]. A caveat of this for modelling is that any return of escaping parr caught as adults may be reflected in the model not as escapes, but as a higher total catch. A further concern is that fishing regulations tend to encourage or require catch and release for wild salmon, but retaining escaped farmed salmon is required in some areas, for example the Spey system [32]. This may provide a tasty incentive for characterisation of salmon of unclear origin for a hungry angler. This form of bias would not be so easily identified experimentally by simply testing fishermen for their ability to identify farmed- or wild-origin salmon. Where catch and release occurs for wild salmon, but not for farmed salmon, there is also the potential for the same wild fish to be caught repeatedly, potentially reducing the measured prevalence of farmed salmon, though no data are available on this. Nevertheless, for trout, these sources of bias and misidentification are not present and the analyses should be more robust for this species.

Further possible confounding effects within the data set exist. Catch data primarily pertain to large, adult fish of harvestable size. Therefore, if there are differences in fitness and survivability of farmed and wild-origin fish, these data provide a biased estimate of the prevalence of escapes in smaller, younger fish. Furthermore, except for net and coble, and fixed-engine methods, no record of sampling effort (in terms of time spent fishing) is recorded. Without this, the size of the wild population into which escapes are mingling is difficult to estimate. This is a particular issue when examining data recorded over a longer time series, as changes in catch will reflect not only the biology, but also changes in human habits and industry (for example change in the popularity of angling).

Conclusions

In summary, in this paper we ask a specific question of the large data sets encompassing salmon and trout catch and of recorded salmon escapes from Scottish salmon farms in the last decade—that is whether a statistically significant effect of the recorded salmon escapes can be found in the catch data, over and above the

expected level for the year and district. Our more robust models provide no evidence of depressed catch (either salmon or trout), or firm evidence of elevated prevalence of escapes in the salmon catch in the years immediately following reported escape events.

Acknowledgments

Data documenting historic catch statistics and escape records used in this paper are Crown Copyright, used with the permission of Marine Scotland Science. Marine Scotland is not responsible for interpretation of these data by third parties. With thanks to Jimmy Turnbull and two anonymous reviewers for helpful comments on the manuscript.

Author Contributions

Analyzed the data: DMG DJP. Wrote the paper: DMG DJP HM JEB JBT BJM.

References

1. ICES (2010) Extract of the report of the advisory committee. North Atlantic salmon stocks, as reported to the North Atlantic Salmon Conservation Organization. Technical report. Copenhagen: ICES/CIEM.
2. Marine Scotland (2012) Improved containment – fish farm escapes. Available: http://www.scotland.gov.uk/Topics/marine/Fish-Shellfish/18364/18692. Accessed on 2012 Jul 16.
3. Krkošek M, Lewis M, Volpe J (2005) Transmission dynamics of parasitic sea lice from farm to wild salmon. Proceedings of the Royal Society Series B 272: 689–696.
4. Webb J, Hay D, Cunningham P, Youngson A (1991) The spawning behaviour of escaped farmed and wild adult Atlantic salmo (Salmo salar L.) in a northern Scottish river. Aquaculture 98: 97–110.
5. Fleming I, Hindar K, Mjølnerød I, Jonsson B, Balstad T, et al. (2000) Lifetime success and interactions of farm salmon invading a native population. Proceedings of the Royal Society Series B 267: 1517–1523.
6. Bourret V, O'Reilly P, Carr J, Berg P, Bernatchez L (2011) Temporal change in genetic integrity suggests loss of local adaptation in a wild Atlantic salmon (Salmo salar) population following introgression by farmed escapees. Heredity 106: 500–510.
7. Van Zwol JA, Neff BD, Wilson CC (2012) The effect of competition among three salmonids on dominance and growth during the juvenile life stage. Ecology of Freshwater Fish. doi: 10.1111/j.1600-0633.2012.00573.x.
8. Jensen O, Dempster T, Thorstad E, Uglem I, Fredheim A (2010) Escapes of fishes from Norwegian sea-cage aquaculture: causes, consequences and prevention. Aquaculture Environment Interactions 1: 71–83.
9. Jonsson B, Jonsson N (2006) Cultured Atlantic salmon in nature: a review of their ecology and interaction with wild fish. ICES Journal of Marine Science 63: 1162–1181.
10. Todd C (2007) The copepod parasite (Lepeophtheirus salmonis (Krøyer), Caligus elongatus Nordmann) interactions between wild and farmed Atlantic salmon (Salmo salar L.) and wild sea trout (Salmo trutta L.): a mini review. Journal of Plankton Research 29: 167–171.
11. Butler JRA, Watt J (2003) Assessing and managing the impacts of marine salmon farms on wild Atlantic salmon in western Scotland: identifying priority rivers for conservation. In: Mills D, editor. Salmon at the Edge. Oxford: Blackwell Scientific Communications. pp. 93–118.
12. Butler J, Cunningham P, Starr K (2005) The prevalence of escaped farmed salmon, Salmo salar L., in the River Ewe, western Scotland, with notes on their ages, 21 weights and spawning distribution. Fisheries Management and Ecology 12: 149–159.
13. Walker A, Beveridge M, Crozier W, Ó Maoiléidigh N, Milner M (2006) Monitoring the incidence of escaped farmed Atlantic salmon, Salmo salar L., in rivers and fisheries of the United Kingdom and Ireland: current progress and recommendations for future programmes. ICES Journal 63: 1201–1210.
14. Marine Scotland (2012) Scottish salmon and sea trout fishery statistics. Available: http://www.scotland.gov.uk/Topics/marine/science/Publications/stats/SalmonSeaTroutCatches/. Accessed 2012 Jul 10.
15. The Scottish Government (2011) Scotland's marine atlas: Information for The National Marine Plan. Available: http://www.scotland.gov.uk/Publications/2011/03/16182005/0. Accessed 2020 Jul 10.
16. Gart J (1966) Alternative analyses of contingency tables. Journal of the Royal Statistical Society 28: 164–179.
17. Thorstad E, Fleming I, McGinnity P, Soto D, Wennevik V, et al. (2008). Incidence and impacts of escaped farmed Atlantic salmon Salmo salar in nature. NINA Special Report 36.
18. Crozier W (1998) Incidence of escaped farmed salmon, Salmo salar L., in commercial salmon catches and fresh water in Northern Ireland. Fisheries Management and Ecology 5: 23–29.
19. Hansen L, Reddin D, Lund R (1997) The incidence of reared Atlantic salmon (Salmo salar L.) of fish farm origin at West Greenland. ICES Journal of Marine Science 54: 152–155.
20. Hansen L, Jacobsen J, Lund R (1999) The incidence of escaped farmed Atlantic salmon, Salmo salar L., in the Faroese fishery and estimates of catches of wild salmon. ICES Journal of Marine Science 56: 200–206.
21. Jacobsen J, Lund R, Hansen L, O'Maoileidigh N (2001) Seasonal differences in the origin of Atlantic salmon (Salmo salar L.) in the Norwegian Sea based on estimates from age structures and tag recaptures. Fisheries Research 52: 169–177.
22. Skilbrei O, Wennevik V (2006) The use of catch statistics to monitor the abundance of escaped farmed Atlantic salmon and rainbow trout in the sea. ICES Journal of Marine Science 63: 1190–1200.
23. Hansen L, Youngson A (2010) Dispersal of large farmed Atlantic salmon, Salmo salar, from simulated escapes at fish farms in Norway and Scotland. Fisheries Management and Ecology 17: 28–32.
24. Skilbrei O, Jørgensen T (2010) Recapture of cultured salmon following a large-scale escape experiment. Aquaculture Environment Interactions 1: 107–115.
25. Olsen R, Skilbrei O (2010) Feeding preference of recaptured Atlantic salmon Salmo salar following simulated escape from fish pens during autumn. Aquaculture Environment Interactions 1: 167–174.
26. Skilbrei O (2010) Reduced migratory performance of farmed Atlantic salmon postsmolts from a simulated escape during autumn. Aquaculture Environment Interactions 1: 117–125.
27. Skilbrei O (2010) Adult recaptures of farmed Atlantic salmon post-smolts allowed to escape during summer. Aquaculture Environment Interactions 1: 147–153.
28. Fiske P, Lund R, Hansen L (2006) Relationships between the frequency of farmed Atlantic salmon, Salmo salar L., in wild salmon populations and fish farming activity in Norway, 1989–2004. ICES Journal of Marine Science 63: 1182–1189.
29. Morris M, Fraser D, Heggelin A, Whoriskey F, Carr J, et al. (2008) Prevalence and recurrence of escaped farmed Atlantic salmon (Salmo salar) in eastern North American rivers. Canadian Journal of Fisheries and Aquatic Sciences 65: 2807–2826.
30. Friedland K, MacLean J, Hansen L, Peyronnet A, Karlsson L, et al. (2009) The recruitment of Atlantic salmon in Europe. ICES Journal of Marine Science 66: 289–304.
31. Carr J, Whoriskey F (2006) The escape of juvenile farmed Atlantic salmon from hatcheries into freshwater streams in New Brunswick, Canada. ICES Journal of Marine Science 63: 1263–1268.
32. Strathspey Angling Improvement Association (2012) Membership, permits, rules and conditions of the club. Available: http://www.speyfishing-grantown.co.uk/permitsrules. htm. Accessed 2012 Jul 10.

4

An Automated Microfluidic Chip System for Detection of Piscine Nodavirus and Characterization of Its Potential Carrier in Grouper Farms

Hsiao-Che Kuo[1,2,3,4,5ⁿ], Ting-Yu Wang[1,2ⁿ], Hao-Hsuan Hsu[1,2], Szu-Hsien Lee[6,7], Young-Mao Chen[1,2,3,4,5], Tieh-Jung Tsai[1], Ming-Chang Ou[1], Hsiao-Tung Ku[8,9], Gwo-Bin Lee[6,7,10*], Tzong-Yueh Chen[1,2,3,4,5*]

1 Laboratory of Molecular Genetics, Institute of Biotechnology, National Cheng Kung University, Tainan, Taiwan, 2 Translational Center for Marine Biotechnology, National Cheng Kung University, Tainan, Taiwan, 3 Agriculture Biotechnology Research Center, National Cheng Kung University, Tainan, Taiwan, 4 University Center for Bioscience and Biotechnology, National Cheng Kung University, Tainan, Taiwan, 5 Research Center of Ocean Environment and Technology, National Cheng Kung University, Tainan, Taiwan, 6 Institute of Nanotechnology and Microsystems Engineering, National Cheng Kung University, Tainan, Taiwan, 7 Department of Engineering Science, National Cheng Kung University, Tainan, Taiwan, 8 Research Division I, Taiwan Institute of Economic Research, Taipei, Taiwan, 9 Office for Energy Strategy Development, National Science Council, Taipei, Taiwan, 10 Department of Power Mechanical Engineering, National Tsing Hua University, Hsinchu, Taiwan

Abstract

Groupers of the *Epinephelus* spp. are an important aquaculture species of high economic value in the Asia Pacific region. They are susceptible to piscine nodavirus infection, which results in viral nervous necrosis disease. In this study, a rapid and sensitive automated microfluidic chip system was implemented for the detection of piscine nodavirus; this technology has the advantage of requiring small amounts of sample and has been developed and applied for managing grouper fish farms. Epidemiological investigations revealed an extremely high detection rate of piscine nodavirus (89% of fish samples) from 5 different locations in southern Taiwan. In addition, positive samples from the feces of fish-feeding birds indicated that the birds could be carrying the virus between fish farms. In the present study, we successfully introduced this advanced technology that combines engineering and biological approaches to aquaculture. In the future, we believe that this approach will improve fish farm management and aid in reducing the economic loss experienced by fish farmers due to widespread disease outbreaks.

Editor: Richard C. Willson, University of Houston, United States of America

Funding: This work was supported by grants from Fish Breeding Association Taiwan and Council of Agriculture, Taiwan (94AS-14.2.1-FA-F1, 96AS-14.2.1-BQ-B2[28], and 97AS-14.2.1-BQ-B1[8]). The funders had no role in study design, data collection and analysis, decision to publish, or preparation of the manuscript.

Competing Interests: The authors have declared that no competing interests exist.

* E-mail: ibcty@mail.ncku.edu.tw (TYC); gwobin@pme.nthu.edu.tw (GBL)

ⁿ These authors contributed equally to this work.

Introduction

Nervous necrosis virus (NNV), a piscine nodavirus, is neuropathogenic virus that results in viral nervous necrosis (VNN) and damage throughout the central nervous system [1–3]. NNV affects over 30 different fish species, including economically important marine fish such as groupers, cods, flounders, bass, puffers, and breams [4]. Within these fish species, grouper is a major farming fish species that has suffered from VNN in Taiwan since 1994 [5], and the resulting disease manifestiations are associated with high mortality rates (80–100%) in hatchery-reared larvae and juveniles [6–8]. Large-scale and concentrated fish farming industries currently experience major economic losses due to the spread of NNV-related diseases between individual fish farms.

Understanding the epidemiology of NNV is critical for controlling the spread of disease; however, the transmission pathway of NNV between different fish farms remains a mystery. Piscine nodavirus can be transmitted either vertically from the broodfish via the egg or sperm cells [9–11] or horizontally between individual fish [10]. Furthermore, the virus can persist for long periods of time in subclinically infected fish and remain infectious

[12]. Although ozone has been applied to eliminate any remaining NNV on the surface of the eggs of groupers in seed-producing fish farms to produce virus-free eggs and juveniles, the NNV-related diseases remain an important burden to most fish farms.

The present study sought to apply the microfluidic chip system [13] to investigate NNV infection in grouper fish farms in a large number of samples and to characterize the potential carrier of the virus. In this study, the reverse transcription polymerase chain reaction (RT-PCR) was integrated into the microfluidic chip technology, which not only lowered the cost of virus detection but also shortened the analysis time. The chip is composed of acrylic and glass materials and can only be used for analysis of a single sample. The investigation began in 2002 with the isolation of NNV from different fish farms (Figure S1 and Table S1). An identical NNV strain was found in distant (30–40 km) grouper fish farms, even when the fish eggs were obtained from different suppliers. After these findings, we then started collecting fish samples from 5 grouper fish farms, and 1 of these farms was monitored intensively and sampled continuously for 7 months (October 2008–May 2009). In addition, a number of possible virus carriers were investigated in and around the fish farms, including brine shrimps, birds, *Daphnia* spp., *Ligia* spp., rotifers, *Palaemon* spp., and inlet seawater. A systematic study utilizing the microfluidic chip system was conducted to identify a major carrier of the virus outside of the fish farm and provided important information regarding the mode of virus infection inside of the fish farm.

Results

Performance of RT-PCR on the microfluidic chip

For comparison of the microfluidic chip RT-PCR with conventional RT-PCR, naturally NNV-infected groupers (true positives, VNN syndrome) were collected from 3 grouper fish farms in Cigu, Jiading, and Kunshen, Taiwan. NNV was detected by capillary electrophoresis (CE) and visualized by ethidium bromide staining with microfluidic chip RT-PCR and a slab gel, respectively. Of note, virus was detected in all of the fish with both methods (Table S5). Both of the methods that we used did not reveal VNN-positive signals in healthy groupers (true negatives). Serially diluted RNA templates (1×10^4–0.25×10^1 copies·μL^{-1}) associated with NNV and specific markers were amplified and analyzed. The detection limit of the microfluidic chip was 3 copies·μL^{-1} (starting template) of direct lysis treatment (Figure S4A), which was markedly more sensitive than the detection by conventional PCR and gel visualization. There was a linear relationship between the magnitude of fluorescence signals and the viral copy number (Figure S3A). Measurement of the fluorescence intensity of NNV samples with concentrations ranging from 0–10^5 plaque forming units (PFU)·mL^{-1} revealed that a viral concentration of 5×10^4 PFU·mL^{-1} could be successfully detected by the microfluidic chip (Figure S3B). The detection limit of the end-point detection by fluorescent CE was estimated to be at least 10-fold greater that that of the slab-gel electrophoresis method (Figures S4A).

Viral cDNA fragment separated by capillary electrophoresis (CE)

PCR products amplified from the samples could only be visualized on an agarose gel when the copy number exceeded 50 copies·μL^{-1} (starting template; Figure S4A). All amplicons were cloned and sequenced to confirm that they originated from NNV. The directly lysed RNA extraction method was not as effective as the use of pure RNA for the microfluidic chip template (Figure S4B), but still produced enough RNA for NNV detection. Figure S5 displays an electropherogram of the DNA markers (*Hae*III-digested ΦX174 DNA markers) and resulting RT-PCR products. Eleven DNA marker peaks and a single peak of RT-PCR nodavirus product (203 bps) were successfully separated within 2 min (Figure S5B).

Prevalence of NNV contamination among grouper fish farms

The microfluidic chip method (Tables S5 and S6 and Figures S3, S4, and S5) was field-tested in an epidemiological investigation of 120 fish samples from 5 grouper aquaculture farms in southern Taiwan (see Tables S2, S3, and S4 and Figure S2 for the locations). NNV infection was detected in 89% (Table S6) of the samples surveyed. Four grouper fish samples from Linyuan (Tables S3, S4, Batch I) that were confirmed to be infected with NNV did not display any signs of VNN, which indicated a false positive status or presence of latent NNV infection. However, 3 grouper fish farms in Linyuan that were identified as being NNV infection-free did show signs of VNN (Table S4, Batch II and Batch III), which indicated a false negative status. The 3 false negatives and 10 true negatives produced a negative predictive value (NPV) of

77%. One hundred and three true positives and 4 false negatives resulted in a positive predictive value (PPV) of 96% (Table S6). All of the positive results obtained from microfluidic chip detection were also confirmed by plaque assay.

NNV remained in the fish tank and contaminated newly arrived fish

We continuously monitored the same fish tank in the Linyuan indoor grouper farm (Figure S6) for 7 months (Table S4). There were 3 VNN outbreaks in December 2008 (Batch I), February 2009 (Batch II), and April 2009 (Batch III); the time between batches was approximately 3–4 weeks. The results of Batch II (NNV was detected on the second day of culturing) and Batch III (NNV was detected on the first day of culturing) demonstrated that NNV may still remain in the fish tank even after a 1-month fallow.

Possible carrier of NNV

Due to the high detection rate of NNV measured in this study, it was of utmost importance to determine its carrier between fish farms. Therefore, we tested the possible vectors that might carry NNV between farms. Grouper fish samples (Table 1) were collected from 3 different locations (Cigu, Jiading, and Kunshen) in southern Taiwan and the possible virus carrier around/within grouper fish farms, including brine shrimps, birds (feces), *Daphnia* spp., *Ligia* spp., rotifers, *Palaemon* spp., and inlet seawater (Table 2) were tested with the microfluidic chip system. Table 1 shows that all 3 fish farms were contaminated with NNV. These 3 grouper fish farms utilized either outdoor or semi-outdoor culturing protocols (Figure S6). The results (Table 2) revealed that in 1 out of 5 and 3 out of 5 fish farms in Kunshen and Cigu, respectively, the bird feces were positive for the presence of NNV.

Discussion

It is likely that due to the abnormality of fish farming, i.e., fish reared in greater numbers and closer proximity than those that live in the wild, viruses are more easily transferred between fish in farm environments. An example of this is found in the Japanese flounder, *Paralichthys olivaceus*, in which horizontal transmission of viral hemorrhagic septicemia virus was shown to occur between wild and farmed fish, as well as between infected and uninfected

Table 1. Nervous necrosis virus (NNV) detection results from infected grouper fish from 3 different fish farms.

Location[a]	RT-PCR[b]		Symptoms[c]		After 2 weeks[d]	
	+	−	+	−	+	−
Cigu	3	0	3	0	3	0
Jiading	3	0	3	0	3	0
Kunshen	3	0	3	0	3	0

[a]Locations of the grouper fish farms are shown in Figure S2 and the fish samples were collected in 2008 and 2009.
[b]+, NNV-positive samples by microfluidic chip analysis; −, NNV-negative samples by microfluidic chip analysis.
[c]The first observation of viral nervous necrosis (VNN) clinical signs following sampling; +, groupers displayed VNN clinical signs; −, groupers did not display any clinical signs of disease; Clinical signs of VNN in larval-stage groupers were abnormal schooling and swimming behavior (whirling, spiraling) and loss of appetite.
[d]The tracking observation (disease outbreak) 2 weeks after the first observation from the same grouper fish farms; +, groupers displayed VNN clinical signs; −, groupers did not display any clinical signs of disease.

Table 2. Detection of potential nervous necrosis virus (NNV) carriers.

Location	Samples							
	Bird feces[a]	Rotifers	Brine shrimps[b]	*Daphnia* spp.[c]	*Palaemon* spp.	Infected grouper fish	Inlet sea water	*Ligia* spp.
	(+/−)[d]	(+/−)	(+/−)	(+/−)	(+/−)	(+/−)	(+/−)	(+/−)
Cigu	(3/2)	(0/5)	(0/5)	(0/5)	(0/5)	(5/0)	(0/5)	(0/5)
Jiading	(0/3)	(0/3)	(0/3)	(0/3)	(0/3)	(3/0)	(0/3)	(0/3)
Kunshen	(1/4)	(0/5)	(0/5)	(0/5)	(0/5)	(5/0)	(0/5)	(0/5)

[a]Birds commonly seen in grouper fish farms are *Egretta garzetta* (little egret), *Nycticorax nycticorax* (black-crowned night heron), and *Passer montanus* (tree sparrow).
[b]Brine shrimps used as a feed for larvae.
[c]Common water fleas.
[d](number of positive/number of negative); +, NNV positive; −, NNV negative.

fish [14]. We have shown that the targets of the NNV preferentially replicate not only in the nervous system but also in eye and fin tissues [11]. The results from our previous work raised the possibility that the NNV infection pathway may pass through other organisms exposed to the outside environment. Importantly, before infected fish spread NNV to others, no direct evidence can typically be found regarding how the virus was transmitted to the fish farm initially. The limitation of the sensitivity of previously used detection methods may be the reason for this challenge.

In this study, we applied a recently developed microfluidic chip detection system [15], which exhibited a high analytical sensitivity (3 copies·μL^{-1} of directly lysed sample; Figure S3A) and allowed us to detect virus in small samples. Our epidemiological investigation of 120 samples from 5 grouper aquaculture farms in southern Taiwan (Table 3 and Table S2, S3, and S4) revealed the high prevalence of NNV infection (89% of the samples and 100% of the farms). Grouper aquaculture is prevalent in Southern Taiwan due to its suitable culture environment, and the cities included in our sample (Anping, Cigu, Jiading, Kunshen, and Linyuan) are the main regions of grouper farming in this region. High stocking density and high seawater temperature can also accelerate NNV disease outbreaks and result in high mortality during the rearing period [16]. In our study, hatchery-reared larvae and juvenile groupers were typically reared indoors with a stable, controlled temperature of approximately 28–29°C. Indoor rearing is a common method for intensive rearing during larval and juvenile stages in Taiwan, in which large fiberglass or concrete tanks as large as 100 m^3 are used. Compared to outdoor rearing methods, this approach can yield a higher survival rate and permit easier handling of stock during the early stages of growth. However, the indoor approach also provides an ideal environment (i.e., stable temperature control) for NNV infection. Moreover, there is no standard fish farming procedure for treating dead fish in Taiwan; the dead fish are often dumped into drains or into the sea in the vicinity of the aquaculture farms. This practice could result in the transmission of virus in the seawater by pumping to the aquaculture farms; thus, neighboring facilities could also be at risk.

VNN outbreaks tend to occur approximately 1 week after the first appearance of clinical signs in individual fish. However, there were examples of false positives, such as the 4 samples from Linyuan (collected on October 23, 2008, December 24, 2008, and December 25, 2008) that were identified as being NNV-infected but were actually not (Tables S3, S4). Mortality in NNV-infected grouper aquaculture farms often approaches 100% [6–8], and was approximately 99.7% in the present study. The few groupers that

Table 3. Summary of the Nervous necrosis virus NNV detection results from the Linyuan[a] grouper fish farm by integrated microfluidic chip analysis.

Experiment	RT-PCR[b]		Symptoms[c]		Number of samples
	+	−	+	−	
Batch I	30	8	28	10	38
Batch II	26	1	27	0	27
Batch III	25	2	27	0	27

[a]Locations of the grouper fish farms are shown in Figure S2 and the fish samples were collected in 2008 and 2009.
[b]+, NNV-positive samples by microfluidic chip analysis; −, NNV-negative samples by microfluidic chip analysis.
[c]The observation of viral nervous necrosis (VNN) clinical signs following sampling; +, groupers displayed VNN clinical signs; −, groupers did not display any clinical signs of disease; Clinical signs of VNN in larval-stage groupers were abnormal schooling and swimming behavior (whirling, spiraling) and loss of appetite.

survive these outbreaks may harbor NNV and could be the source of the false positives. However, it is conceivable that the 4 aforementioned grouper fish farms might have been sampled during the NNV incubation period. In addition, 3 grouper fish samples from Linyuan were identified as NNV infection-free, but ultimately displayed symptoms of VNN syndrome (Table S4). This may have been due to sampling errors or artificial errors that occurred during transportation or experiment preparation. In the initial stages of NNV infection, some weakened fish might have already displayed symptoms of VNN while other fish remained uninfected.

The origin and mechanism of initial NNV transmission remained unclear after our preliminary analysis. Therefore, we investigated a number of possible virus carriers in and around the fish farm, including brine shrimps (feed for juveniles), birds (feces), *Daphnia* spp., *Ligia* spp., rotifers (feed for larvae), *Palaemon* spp., and inlet seawater. In our study, we could not detect the virus in inlet seawater, thus the likelihood of acquiring infection through the inlet water was negligible. This finding is in agreement with another study that could not detect any sign of NNV in inlet water [17]. Rotifers and brine shrimp do not appear to be susceptible to nodavirus (Table 3), thus decreasing the likelihood of the feed being a transmission channel. *Daphnia* spp., *Ligia* spp., and *Palaemon* spp. were also tested and showed no indication of carrying NNV (Table 2).

With regard to the other suspects were the wild birds that inhabit areas around fish farms, which have been ignored thus far as possible NNV carriers. Our data suggests that birds are potential carriers of NNV and transfer the virus between fish farms (Table 2). *Egretta garzetta* (little egret) and *Nycticorax nycticorax* (black-crowned night heron) are the most common residents around fish farms, and their food sources consist of fish, batrachians, and insects. The NNV-infected groupers are caught easily for these birds and NNV could survive in the bird's digestive system. It has been shown that nodavirus was very stable under extreme environmental conditions [18]. Once the virus has entered a fish farm or rearing unit, it may be very difficult to exterminate. Therefore, a possible transmission route may be via fish-feeding birds.

Interestingly, observations from the Linyuan grouper farm, which operated under a high security protocol including indoor farming with controlled temperatures and seawater pH and fish eggs treated with ozone, there should not be any NNV contamination in theory. However, this farm was kept virus-free for only 1 month, and was examined for NNV infection for the remaining 6 months. Consequently, we have begun further experiments to identify an alternative transmission pathway for NNV.

In conclusion, the recently developed microfluidic chip system can provide a convenient platform for large-scale NNV detection in grouper fish farms. NNV infection represents a major challenge in grouper farming. Although we only focused on 5 grouper farms in this study, the 100% NNV detection rate in 5 major grouper culturing areas revealed the serious nature of this problem. The high sensitivity of virus (3 copies) detection by this method makes the identification of the virus carrier possible, and the results point to fish-feeding birds as the main suspect of NNV transmission. However, the results from the fish farm that operated under a high security protocol indicated that there may be alternative carriers or transmission pathways involved.

Materials and Methods

Fish cell line and virus

The grouper cell line, GF-1 (Bioresources Collection and Research Center, Taiwan; BCRC 960094) was used for culturing and maintaining NNV. GF-1 was derived from the fin tissue of orange-spotted grouper (*E. coioides*) [19]. The cells were incubated at 28°C in antibiotic-free L15 medium (Life Technologies, Bethesda, MD, USA) supplemented with 5% v/v heat-inactivated fetal bovine serum (FBS) [20]. NNV was obtained from naturally infected grouper (*E. lanceolatus*) juveniles [21]. The virus isolates from diseased fish [2] were cultured in GF-1 cells for 5 days (until they showed a cytopathic effect [CPE]), and the 5-day cultures were submitted for NNV isolation [2].

Virus isolation and purification

Grouper NNV (gNNV) was isolated from naturally infected groupers (*Epinephelus lanceolatus*) collected from Jiading, Taiwan in 2004 [19]. The virus was isolated from fin tissue; the tissues were frozen in liquid nitrogen and homogenated in 10 volumes of L15 medium, centrifuged at 10,000 rpm for 20 min, and the supernatant fraction was passed through a 0.22-μm filter and stored at −80°C until analysis. For collection of viral particles, the isolated virus was cultured in GF-1 cells, and the cells were collected when 90% of the cells displayed a CPE. L15 medium containing GF-1 cells and NNV was mixed with 2.2% NaCl and 5% w/v polyethylene glycol (PEG) 8,000 and centrifuged at 10,000×g at 4°C for 1 h. The pellet was resuspended in 2 mL *N*-tris(hydroxymethyl) methyl-2-aminoethanesulfonic acid (TES) buffer, mixed with an equal amount of Freon 113 and shaken vigorously for 5 min. Supernatants were combined and mixed with 3 mL, 3 mL, and 2 mL of 40%, 30%, and 20% CsCl, respectively. CsCl density gradients were formed by centrifugation in a Beckman SW40Ti rotor (Beckman Coulter, Fullerton, CA) at 35,000 rpm at 4°C for 16 h. Syringes were used to collect 3 mL of the virus-containing fraction, which was diluted 10-fold with TES buffer.

Plaque assay

The plaque morphological assay performed on GF-1 cell monolayers was modified from Kamei et al. (1987) [22]. Serial 10-fold dilutions (10^{-3}, 10^{-4}, 10^{-5}, 10^{-6}, 10^{-7}, and 10^{-8}) of virus were made using L15 (supplemented with 5% FBS) medium and monolayers of GF-1 cells were inoculated in 6-well plates. Two hundred microliters of the virus dilution was added to each well and incubated at 28°C for 1 h. The virus-containing media was removed sequentially from the wells and replaced with 2 mL of diluted agarose medium (0.5% agarose solution in L15 [with 5% FBS]) and incubated at 28°C for 3 to 4 days. The cells were fixed with 2 mL fixing solution (methanol: acetic acid = 3:1 [v/v]) at room temperature for 30 min then stained with 1% crystal violet at room temperature for 30 min. Cells were washed with phosphate buffered saline (PBS). Plates were monitored daily until the number of plaques that were counted did not change for 2 consecutive days. Plaque forming units (PFU) were calculated as follows:

$$PFU mL^{-1}(\text{of original stock}) =$$

$$1/\text{dilution factor} \times \text{number of plaques} \times 1/(\text{mL of inoculums plate}^{-1})$$

Direct lysis, nucleic acid extraction, and RT-PCR microfluidic chip analysis

Before loading the samples onto the chip, the sample (100 mg of homogenized fish) was added to 1 mL of lysis buffer (62.5 mM Tris pH 8.3, 95 mM KCl, 3.8 mM $MgCl_2$, 12.5 mM dithiothreitol, 0.63% NP-40) and centrifuged (12,000×g at 4°C for 15 min). One microliter of extracted RNA (or the supernatant of tissue lysate) was pre-heated at 70°C for 10 min in chamber A (Figure 1) with 7.5 μL reaction mixture (4.5 μL of DEPC treated water, 2 μL of Moloney Murine Leukemia Virus (M-MLV) RT 5×reaction buffer [Promega, Madison, WI, USA] and 0.5 μL of 20 μM of each primer [203-F and 203-R; Blossom, Taipei, Taiwan]). The chamber was cooled to 42°C, and 1.5 μL of the RT reaction mixture (1 μL of 2.5 mM dNTP and 0.5 μL of M-MLV Reverse Transcriptase; 200 U·μL^{-1}; Promega) was automatically transferred from chamber B (Figure 1). Chamber A was maintained at 42°C for 30 min for cDNA synthesis followed by adjustment to 94°C for 2 min for enzyme deactivation. After the RT procedure, 6.5 μL of the reaction mixture containing synthesized cDNA was left in chamber a for the subsequent PCR reaction. Another micropump automatically transferred 3.5 μL of the PCR reaction mixture (1 μL of 2.5 mM dNTP, 1 μL of 10×PCR buffer with 15 mM Mg^{2+} [Violet, Taipei, Taiwan], 0.5 μL of 20 μM of each primer (203-F and 203-R), and 0.5 μL of Taq DNA polymerase; 1000 U; 5 U·mL^{-1} [Violet]) from chamber C (Figure 1) to chamber A. For PCR, chamber A was heated up to 94°C for 1 min (pre-denaturation), and then to 94°C for 10 s, 60°C for 20 s and 72°C for 20 s for 20 thermal cycles. The final cycle was followed by an additional 72°C for 1 min of post-elongation.

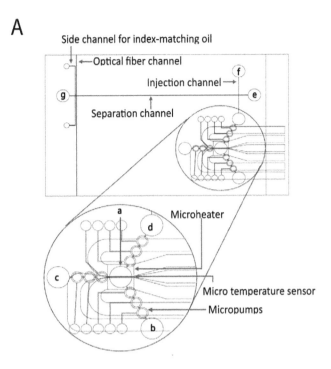

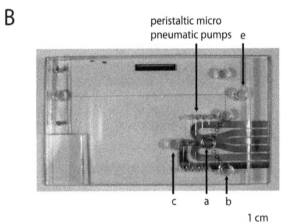

1 cm

Figure 1. Design of integrated microfluidic chip. A. Schematic diagram of the integrated microfluidic chip. The micro RT-PCR module, capillary electrophoresis (CE) module and buried optical fibers were integrated; (a), reaction chamber; (b), RT reagent reservoir; (c), PCR reagent reservoir; (d), sample reservoir; (e), CE sample reservoir; (f), buffer reservoir; (g), waste collection chamber. B. Photograph of the microfluidic chip. The 3 layers of the integrated chip are composed of a polymethylmethacrylate (PMMA) CE chip, a glass micro PCR chip, and Polydimethylsiloxane (PDMS) micropumps. The dimension of chip is $6.5 \times 3.9 \times 0.8$ cm^3. The CE channel is 100 μm in width ×30 μm in depth. The volumes of the PCR and reagent chambers are 11.25 μL.

Finally, the PCR product was automatically transferred to chamber D by the last micropump set. Other samples such as the brine shrimps, bird feces, *Daphnia* spp., *Ligia* spp., rotifers, *Palaemon* spp., and inlet seawater were analyzed by the method described above.

Design of microfluidic chip

The integrated microfluidic chip was designed and fabricated as described in our previous work [13,15,17]. The scheme is shown

in Figure 1. Briefly, the 6.5×4.5 cm^2 chip consists of a micro RT-PCR module, a CE module, and 2 buried optical fibers. The micro RT-PCR module is comprised of microheaters and resistors, a microtemperature sensor, and 3 micropumps and chambers. Therefore, it can perform rapid heating ($20 \pm 0.2°C \cdot s^{-1}$) and cooling ($10 \pm 0.2°C \cdot s^{-1}$) [13,15]. The 100-μm wide and 30-μm deep CE module is comprised of 2 polymethylmethacrylate (PMMA) structures. The lower structure houses the injection, separation, and optical fiber channels. The upper structure contains pre-drilled holes [13,15]. The total volumes of the PCR and reagent chambers are 11.25 μL.

Capillary electrophoresis (CE) on the microfluidic chip

Laser-induced fluorescence technology was used for detection of separated DNA molecules on the CE module of the chip [13,15]. DNA was labeled with florescent dye using laser excitation and emission wavelengths of 491 nm and 509 nm, respectively. The RT-PCR product was kept in the chamber with same volumes of CE buffer as those used for the RT-PCR product (1.75% w/w hydroxypropylmethylcellulose [Sigma-Aldrich, St. Louis, MO, USA] in TBE with 1% v/v YO-PRO-1 fluorescent dye [Molecular Probes, Eugene, OR, USA; 24]) and 7 μL *Hae*III-digested ΦX174 DNA markers (10 ng·μL^{-1}; General Electric (GE) Healthcare, Buckinghamshire, UK) were pumped from chamber F and mixed in chamber E (Figure 1). The mixture in chamber E was driven by electrokinetic forces and injected into the separation channel, where the DNA molecules were simultaneously separated and detected by the buried optical fibers [13,15].

Fish samples and VNN observation

Six to ten juvenile groupers (*Epinephelus coioides* and *E. lanceolatus*, reared for 25–90 days) were randomly collected from each of 5 different grouper aquaculture farms (Anping, Cigu, Jiading, Kunshen, and Linyuan; Figure S2, Table 3, and Table S2, S3, and S4) in southern Taiwan between 2008 and 2009. The samples were pooled together and homogenized in liquid nitrogen for RNA extraction.

The behavior of the grouper fish in the farms (Cigu, Jiading, Kunshen, and Linyuan; Table S3) were recorded by observing the VNN clinical signs for 2 weeks. The fish showing clinical signs were collected and further examined by RT-PCR and microfluidic chip RT-PCR (Figure 1). The clinical signs for identifying VNN were abnormal schooling and swimming behavior (whirling and spiraling), abnormal pigmentation, and loss of appetite.

RNA isolation, cDNA synthesis, and conventional and microfluidic chip RT-PCR

RNA extraction from the homogenated whole fresh fish larvae and pure NNV particles was performed using TRIzol reagent (Molecular Research Center, Cincinnati, OH, USA) according to the manufacturer's instructions. Briefly, for RNA extraction from 100 mg tissue, 1 mL TRIzol regent was added and followed by addition of 200 μL ice-cold chloroform. RNA was precipitated by the addition of 500 μL isopropanol. RNA was subsequently redissolved in 100 μL diethyl pyrocarbonate (DEPC)-treated H$_2$O. RNA that was not used immediately was stored at $-80°C$. Reverse-transcription was performed by M-MLV reverse transcriptase (Promega) according to the manufacturer's protocol. For the RT reaction, 2 μg of the extracted total RNA was used as a template and mixed with random primers (10 μM), dNTPmix (2.5 mM), RT buffer (5×), and reverse transcription transcriptase (200 U·μL^{-1}) in a total volume of 25 μL. One microliter of cDNA from the RT reaction was used as a template and mixed with PCR buffer (10×, 5 μL; Bioman Scientific Co., Ltd., Taipei, Taiwan), dNTPmix (2.5 mM, 4 μL) (Bioman Scientific Co), specific primers pairs (203-F, GACGCGCTTCAAGCAACTC, and 203-R, CGAACACTCCAGCGACACA GCA) [11; (10 μM and 1 μL each)], Bio Taq DNA polymerase (5 U·μL^{-1}, 1 μL; Bioman Scientific Co), and 37 μL of dH$_2$O were used for PCR, which involved 94°C for 5 min and 35 cycles of 94°C for 40 sec, 55°C for 40 sec, 72°C for 40 sec, and 72°C for 5 min. RNA that was isolated from purified virus and cDNA were quantified using an Ultrospec 3300 Pro spectrophotometer (Amersham Biosciences, Piscataway, NJ, USA) and dilutions were made with sheared salmon sperm DNA (5 ng·mL^{-1}) as a diluent.

Direct lysis and nucleic acid extraction for RT-PCR with a microfluidic chip (Figure 1) was used for detecting virus from fish and environmental samples.

Viral copy number calculation

The viral copy number was identified by the molecular weight (1 viral genome = 1.5×10^6 Da[= 4542 (bps) × 330(Da)]) of the virus. For 1 μg·μL^{-1} of viral RNA, there are 6.66×10^{-13} (= $1000 \times 10^{-9}/1.5 \times 10^6$) viral moles, equal to 4.0×10^{11} (= $6.66 \times 10^{-13} \times 6.023 \times 10^{23}$) virus copies.

Supporting Information

Figure S1 Putative coat protein sequences of the 4 isolated NNV strains from 3 fish farms (Table S1). CG051202R2 and CG221002R2 were collected form Cigu, JD170103R2 was collected from Jiading, and KS230102R2 was collected from Kunshen, Taiwan. The amino acid differences are indicated as bold letters. The same virus was presented in 2 distant fish farms. Virus protein sequences for strains CG051202R2 and KS230102R2, isolated from Cigu and Kunshen (30 km apart), are the same. In the other case, virus CG221002R2 and JD170103R2 from Cigu and Jiading (40 km apart) have the identical RNA2 sequences.

Figure S2 The locations of the grouper fish farms in Taiwan that were included in this study. Taiwan is located between the tropical and subtropical regions. The grouper aquacultures are mainly gathered in southern Taiwan due to the preferred warm temperature of grouper fish. Dark circles indicate the 5 major regions of grouper fish farms in which our sampling took place. The white star indicates the location of the National Cheng Kung University. (Modified from a map available at http://mapsof.net under a Creative Commons Attribution-ShareAlike 1.0 License.)

Figure S3 The linear relationship between the magnitude of fluorescence signals and the viral copy number. A. The relationship between the fluorescence signals and the concentration of purified RNA (starting template, 12.5–100 copies·μL^{-1}). B. The detection limit of products analyzed by the capillary electrophoresis (CE) module; fluorescence intensity for different virus concentrations and a threshold line with an amplitude of 1 mV.

Figure S4 The detection of virus on slab-gel electropherograms by using direct lysis method. A. Slab-gel electropherograms of RT-PCR products. Lane M: 100-bp DNA ladders (Yeastern Biotech Corp., Taiwan); Lanes 1–5 contain samples with different viral RNA concentrations of 10^4, 10^3, 10^2, 50, and 25 (copies·μL^{-1}), respectively. B. Comparison of pure RNA (isolated virus) and RNA obtained from the direct lysis method by conventional RT-PCR. Four fish were treated with different RNA extraction methods. The NNV-free control was isolated from a healthy adult grouper's fin tissue for RT-PCR; no NNV was detected by the NNV-specific primer pair (203F and 203R) from this sample. The white arrow indicates the size (203 bp) of the PCR product.

Figure S5 Electropherograms of the RT-PCR products from purified RNA. A. Slab-gel electropherograms for the amplified PCR products from fish tissues using the newly developed micro PCR module. (Lane M: 100-bp DNA ladders; 1: RT-PCR products from brain tissues of *E. lanceolatus*). B. Electropherograms of the RT-PCR products (203 bps) from purified RNA. The minimum concentration detected on the CE module was 12.5·copies μL^{-1}. The mixture of DNA markers and RT-PCR products obtained from the infected grouper resulted in 11 DNA marker peaks and a single peak from the RT-PCR product (203 bps) that were successfully separated within 2 min.

Figure S6 Fish farming protocols. A. Indoor protocol; the temperature and the pH of the seawater are controlled. The fish tank is isolated from outside elements, including fish-feeding birds and sunlight. B. Semi-outdoor protocol; the fish tank is isolated from fish-feeding birds, but sunlight can penetrate inside. C. Outdoor protocol; the culturing tank is in a natural environment.

Table S1 Virus isolates from infected grouper fish in four different fish farms.

Table S2 Examination of the Anping grouper fish farm for nervous necrosis virus (NNV) infection by microfluidic chip analysis.

Table S3 Examination of grouper fish farms in Cigu, Jiading, Kunshen and Linyuan for nervous necrosis virus (NNV) infection by microfluidic chip analysis.

Table S4 Three examinations (Batch I, II, and III) of the Linyuan^a grouper fish farm for nervous necrosis virus (NNV) infection by microfluidic chip analysis.

Table S5 Comparison of microfluidic chip and conventional RT-PCR methods for nervous necrosis virus (NNV) detection.

Table S6 Specificity of the microfluidic chip method.

Acknowledgments

The authors would like to thank all of the fish farmers for kindly providing the fish samples for this study.

Author Contributions

Conceived and designed the experiments: TYC GBL. Performed the experiments: HCK TYW HHH SHL YMC TJT MCO. Analyzed the data: TYC HCK TYW HTK. Contributed reagents/materials/analysis tools: TYC GBL TYW HHH SHL. Wrote the paper: TYC HCK TYW. Assisted in drafting of text and figures of manuscript: TYC HCK TYW.

Revised manuscript critically for important intellectual content: TYC HCK TYW.

References

1. Mori K, Nakai T, Muroga K, Arimoto M, Mushiake K, et al. (1992) Properties of a new virus belonging to nodaviridae found in larval striped jack (Pseudocaranx dentex) with nervous necrosis. Virology 187: 368–371.
2. Kuo HC, Wang TY, Chen PP, Chen YM, Chuang HC, et al. (2011) Real-time quantitative PCR assay for monitoring of nervous necrosis virus infection in grouper aquaculture. J Clin Microbiol 49: 1090–1096.
3. Munday BL, Kwang J, Moody N (2002) Betanodavirus infections of teleost fish: a review. J Fish Dis 25: 127–142.
4. Sano M, NaKai T, Fijan N (2011) Viral nervous necrosis. In: Woo PTK, Bruno DW, editors. Fish Diseases and Disorders. London, UK: CABI 3: 198–207
5. Kai YH, Chi SC (2008) Efficacies of inactivated vaccines against betanodavirus in grouper larvae (Epinephelus coioides) by bath immunization. Vaccine 26: 1450–1457.
6. Munday BL, Nakai T, Nguyen HD (1994) Antigenic relationship of the picornalike virus of larval barramundi, Lates calcarifer Bloch to the nodavirus of larval striped jack, Pseudocaranx dentex (Bloch & Schneider). Aust Vet J 71: 384–385.
7. Munday BL, Nakai T (1997) Nodaviruses as pathogens in larval and juvenile marine finfish. World J Microb Biot 13: 375–381.
8. Skliris GP, Krondiris JV, Sideris DC, Shinn AP, Starkey WG, et al. (2001) Phylogenetic and antigenic characterization of new fish nodavirus isolates from Europe and Asia. Virus Res 75: 59–67.
9. Kai YH, Su HM, Tai KT, Chi SC (2010) Vaccination of grouper broodfish (Epinephelus tukula) reduces the risk of vertical transmission by nervous necrosis virus. Vaccine 28: 996–1001.
10. Ole BS, Audun HN, Trond J, Merete Bjørgan S, Terje S, et al. (2006) Viral and bacterial diseases of Atlantic cod Gadus morhua, their prophylaxis and treatment: a review. Dis Aquat Organ 71: 239–254.
11. Breuil G, Pépin JFP, Boscher S, Thiéry R (2002) Experimental vertical transmission of nodavirus from broodfish to eggs and larvae of the sea bass, Dicentrarchus labrax (L.). J Fish Dis 25: 697–702.
12. Johansen R, Grove S, Svendsen AK, Modahl I, Dannevig B (2004) A sequential study of pathological findings in Atlantic halibut, Hippoglossus hippoglossus (L), throughout one year after an acute outbreak of viral encephalopathy and retinopathy. J Fish Dis 27: 327–341.
13. Lee SH, Ou MC, Chen TY, Lee GB (2007) Integrated Microfluidic Chip for Fast Diagnosis of Piscine Nodavirus; Solid-State Sensors, Actuators and Microsystems Conference, 2007. TRANSDUCERS 2007. International, pp. 943–946.
14. Frerichs GN, Tweedie A, Starkey WG, Richards RH (2000) Temperature, pH and electrolyte sensitivity, and heat, UV and disinfectant inactivation of sea bass (Dicentrarchus labrax) neuropathy nodavirus. Aquaculture 185: 13–24.
15. Huang FC, Liao CS, Lee GB (2006) An integrated microfluidic chip for DNA/RNA amplification, electrophoresis separation and on-line optical detection. Electrophoresis 27: 3297–3305.
16. Tanaka S, Aoki H, Nakai T (1998) Pathogenicity of the Nodavirus Detected from Diseased Sevenband Grouper Epinephelus septemfasciatus. Fish Pathol 33: 31–36.
17. Nerland AH, Skaar C, Eriksen TB, Bleie H (2007) Detection of nodavirus in seawater from rearing facilities for Atlantic halibut Hippoglossus hippoglossus larvae. Dis Aquat Organ 73: 201–205.
18. Lee YF, Lien KY, Lei HY, Lee GB (2009) An integrated microfluidic system for rapid diagnosis of dengue virus infection. Biosens Bioelectron 25: 745–752.
19. Chi SC, Hu WW, Lo BJ (1999) Establishment and characterization of a continuous cell line (GF-1) derived from grouper, Epinephelus coioides (Hamilton): a cell line susceptible to grouper nervous necrosis virus (GNNV). J Fish Dis 22: 173–182.
20. Chen YM, Su YL, Lin JH, Yang HL, Chen TY (2006) Cloning of an orange-spotted grouper (Epinephelus coioides) Mx cDNA and characterisation of its expression in response to nodavirus. Fish Shellfish Immunol 20: 58–71.
21. Ou MC, Chen YM, Jeng MF, Chu CJ, Yang HL, et al. (2007) Identification of critical residues in nervous necrosis virus B2 for dsRNA-binding and RNAi-inhibiting activity through by bioinformatic analysis and mutagenesis. Biochem Biophys Res Commun 361: 634–640.
22. Kamei Y, Yoshimizu M, Kimura T (1987) Plaque assay of Oncorhynchus masou virus (OMV). Fish Pathol 22: 147–152.

5

Modeling Parasite Dynamics on Farmed Salmon for Precautionary Conservation Management of Wild Salmon

Luke A. Rogers[1]*, Stephanie J. Peacock[2], Peter McKenzie[3], Sharon DeDominicis[4], Simon R. M. Jones[5], Peter Chandler[6], Michael G. G. Foreman[6], Crawford W. Revie[7], Martin Krkošek[1,8]

1 Department of Zoology, University of Otago, Dunedin, Otago, New Zealand, 2 Department of Biological Sciences, University of Alberta, Edmonton, Alberta, Canada, 3 Mainstream Canada, Campbell River, British Columbia, Canada, 4 Marine Harvest Canada, Campbell River, British Columbia, Canada, 5 Pacific Biological Station, Fisheries and Oceans Canada, Nanaimo, British Columbia, Canada, 6 Institute of Ocean Sciences, Fisheries and Oceans Canada, Sidney, British Columbia, Canada, 7 Atlantic Veterinary College, University of PEI, Charlottetown, Prince Edward Island, Canada, 8 Department of Ecology and Evolutionary Biology, University of Toronto, Toronto, Ontario, Canada

Abstract

Conservation management of wild fish may include fish health management in sympatric populations of domesticated fish in aquaculture. We developed a mathematical model for the population dynamics of parasitic sea lice (*Lepeophtheirus salmonis*) on domesticated populations of Atlantic salmon (*Salmo salar*) in the Broughton Archipelago region of British Columbia. The model was fit to a seven-year dataset of monthly sea louse counts on farms in the area to estimate population growth rates in relation to abiotic factors (temperature and salinity), local host density (measured as cohort surface area), and the use of a parasiticide, emamectin benzoate, on farms. We then used the model to evaluate management scenarios in relation to policy guidelines that seek to keep motile louse abundance below an average three per farmed salmon during the March–June juvenile wild Pacific salmon (*Oncorhynchus* spp.) migration. Abiotic factors mediated the duration of effectiveness of parasiticide treatments, and results suggest treatment of farmed salmon conducted in January or early February minimized average louse abundance per farmed salmon during the juvenile wild salmon migration. Adapting the management of parasites on farmed salmon according to migrations of wild salmon may therefore provide a precautionary approach to conserving wild salmon populations in salmon farming regions.

Editor: Howard Browman, Institute of Marine Research, Norway

Funding: This research is a product of the Broughton Archipelago Monitoring Program (www.bamp.ca), which is sponsored by Fisheries and Oceans Canada, Coastal Alliance for Aquaculture Reform, Marine Harvest Canada, Mainstream Canada, and Grieg Seafood. The work was also partially supported by funding from a University of Otago Research Grant to MK, and NSERC and Alberta Innovates graduate student scholarships to SJP. The funders had no role in study design, data analysis, decision to publish, or preparation of the manuscript.

Competing Interests: This research is a product of the Broughton Archipelago Monitoring Program (www.bamp.ca), which is partly sponsored by Marine Harvest Canada, Mainstream Canada, and Grieg Seafood. Dr. Peter McKenzie is a Veterinarian and Fish Health Manager employed by the salmon aquaculture company Mainstream Canada. Sharon DeDominicis is an Environmental Sustainability Manager employed by the salmon aquaculture company Marine Harvest Canada. There are no patents, products in development or marketed products to declare.

* E-mail: luke.allister.rogers@gmail.com

Introduction

Global finfish aquaculture production grew by 65% between 2000 and 2010 [1,2], and likely will continue to grow [3,4]. Management of parasites is a challenge for finfish farms and as the density of aquaculture production grows, the risk of infectious disease may increase [5]. In some cases, parasites on finfish farms may pose a greater threat to wild fish than to farmed fish [6,7]. Sea lice (*Lepeophtheirus salmonis*) routinely infect farmed Atlantic salmon (*Salmo salar*) [8], but in British Columbia, Canada, sea lice are seldom a production or health concern for Atlantic salmon on farms [6]. Rather, wild Pacific salmon (*Oncorhynchus gorbuscha, O. keta, O. kisutch*) that migrate near to Atlantic salmon farms may face an elevated risk of sea louse infection [9,10,11,12]. Sea louse infection may be associated with lethal or sub-lethal effects [13,14,15], and infestations have been linked to Pacific salmon population declines [7,9,16,17,18]. Some controlled laboratory experiments, however, show no evidence of increased mortality among juvenile pink salmon infected following artificial exposure [19,20], and there is disagreement about the extent to which productivity in wild stocks is affected [21,22,23,24,25,26]. Sea louse management in British Columbia has sought to limit the exposure of wild Pacific salmon to high densities of sea lice by limiting the average number of motile lice per fish on Atlantic salmon farms [27].

Methods to reduce the exposure of wild salmon to sea lice on farmed salmon include suppressing louse abundance on farmed salmon and separating salmon farms from wild salmon migration routes. Costello [13] reports industry best practices for sea louse control that include the fallowing of salmon farms between stocking cycles, the co-stocking of cleaner-fish (*Labridae* spp.) with farmed salmon to consume lice on infected hosts, the timely administration of paraciticides to prevent sea louse epidemics, and the careful selection of farm sites to reduce transmission among farms or between farmed salmon and wild populations. Fallowing between harvest and stocking, stocking only a single year class, and removal of unhealthy fish are practiced on salmon farms in British Columbia to promote farmed salmon health [28]. Although coordinated fallowing has drawbacks for commercial production, its practice may reduce infection pressure on wild salmon [29,30,31]. No known species of cleaner-fish is available in British Columbia, and importing cleaner-fish may threaten biosecurity [32]. Transmission of sea lice among farms and between farmed and wild salmon may be routine, despite siting precautions [33,34]. Paraciticide treatment administered to farmed salmon remains the most common method to suppress louse abundance on salmon farms. In British Columbia, the only such paraciticide licenced for use on salmon farms is the chemotherapeutant SLICE®, administered in farmed salmon feed [35].

SLICE® is an aquaculture premix of 0.2% emamectin (4″-deoxy-4″epimethy-laminoavermectin B_1) benzoate (EMB) administered orally to farmed salmon. Commercial field trials and use of SLICE® have demonstrated effective suppression of chalimus and motile life stages of sea lice on salmon farms in Norway [36], Scotland [37], Canada [35] and the United States [38]. A typical treatment dosage is 50 µg kg^{-1} fish biomass for seven consecutive days [36,37,39]. Average concentrations of EMB in the blood plasma of farmed salmon appear to vary widely with farm site and season, but show no association with individual fish mass [39]. Although reports of SLICE® efficacy suggest typical reductions in louse abundance of 89–100% compared to pre-treatment levels [36,37,38,40], evidence of sea louse resistance to SLICE® and decreased efficacy of treatment is established or growing in Scotland [40], Atlantic Canada [41,42], Ireland and Norway [35]. In British Columbia, SLICE® has remained effective for louse suppression on salmon farms [35,43], and efforts are being made to tailor the timing of its use to forestall sea louse resistance [44].

The timing of SLICE® treatments on salmon farms is also important to reduce louse abundances along wild salmon migration routes during the annual juvenile wild salmon migration. Juvenile pink (O. gorbuscha) and chum (O. keta) salmon hatch in rivers and migrate immediately from freshwater to open marine waters during March–June each year [11,29]. During the annual migration, these juvenile wild salmon migrate through inshore marine waters in the Broughton Archipelago region of British Columbia and pass near to salmon farms [11,29]. Juvenile pink and chum salmon are at their most vulnerable to negative effects of sea louse infection during the migration due to their small size and recent marine transition [45,46]. Juvenile coho salmon (O. kisutch) are generally larger than migrating juvenile pink and chum salmon because they spend an additional year in freshwater before entering the marine environment, but juvenile coho salmon can face increased sea louse exposure through trophic transmission of sea lice during predation on infected juvenile pink and chum salmon [14]. The timing of treatment to suppress sea louse abundance on salmon farms, therefore, is important to the protection of juvenile Pacific salmon during the annual March–June migration.

Winter treatment may prove effective both to reduce louse abundance on migration routes in advance of the March–June juvenile wild salmon migration [47], and to minimize average annual sea louse abundance on farms [48]. In a study of two salmon farms in the Broughton Archipelago, Krkošek et al. [47] found that maximum reductions in louse abundance on farms lagged SLICE® treatment by 1–3 months, suggesting that treatment to suppress louse abundance prior to the migration ought to take place in January. Sea louse ecology and studies of louse suppression on farms suggest similar timing to utilize

SLICE® most effectively. In his review of sea louse ecology, Costello [13] suggested that treatment during winter is important to reduce louse numbers on farms because female sea lice tend to grow larger and produce more eggs during the winter than during other seasons. Peacock et al. [31] found that an increase over time in the proportion of treatments taking place during October–March was associated with a corresponding decrease in average annual sea louse abundance on farmed salmon and wild juvenile salmon in the Broughton Archipelago. These findings suggest that winter treatment on salmon farms may be important for juvenile Pacific salmon. A framework to understand sea louse dynamics in relation to SLICE® treatment and abiotic factors is therefore desirable to inform the timing of treatment on farms for the conservation management of Pacific salmon.

Mathematical models for sea louse population dynamics on salmon farms can provide valuable insight for sea louse suppression and wild salmon conservation. Although louse population dynamics in the Broughton Archipelago may be sensitive to temperature, salinity, and host density both before and after treatment, previous models for louse population dynamics on farms have not included these effects [47]. Temperature and salinity can influence demographic rates for sea lice on Atlantic salmon [49,50], but effects may be magnified by extremes. For example, effects of temperature on louse abundance are weak or absent in Scotland where winter temperatures rarely fall below 5°C [49,51,52], but are strong and negative in Norway where winter temperatures commonly fall below 2°C [49,53]. Host-density effects, if present, may be local or regional in scale, but modeling louse transmission among farms can be complex. Jansen et al. [53] found a positive association between regional host density and louse abundance on individual farms in Norway, suggesting regional population dynamics, but Revie et al. [51] found no such association in Scotland, suggesting local dynamics only. Our understanding of louse population dynamics may be improved by a model that explicitly accounts for the effects of temperature, salinity, local host density, and treatment on the population growth rates of sea lice on Broughton Archipelago salmon farms.

In this paper we develop a model for sea louse population dynamics across 25 salmon farms in the Broughton Archipelago, British Columbia. We model sea louse population growth rates over time and account for the effects of sea surface temperature and salinity, host density, and SLICE® treatments. The model is applied to seven years of monthly sea louse abundance data spanning 2002–2008. We use the model to assess the optimal timing for SLICE® treatment to reduce motile sea louse abundance in the Broughton Archipelago during the juvenile wild salmon migration, and we use the regulatory limit of three motile sea lice per farmed salmon currently applied in British Columbia [35] as a guideline. The results provide empirical guidelines for the optimal timing of SLICE® treatment for precautionary conservation management of wild salmon within the Broughton Archipelago.

Methods

Data

The data, published previously [25], were collected at 25 salmon farms during 2002–2008 in the Broughton Archipelago (Figure 1). Data were collected monthly at each farm following industry standards similar to those described in Krkošek et al. [47]. Farmed salmon were grouped by cohort, namely all fish at an individual farm during one stock–harvest cycle. At each farm, fish health technicians and aquaculture personnel collected monthly

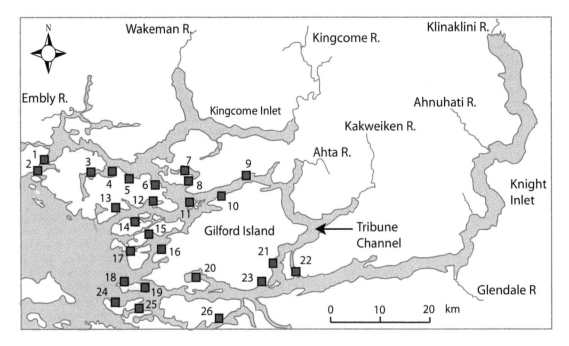

Figure 1. Location of salmon farms in the Broughton Archipelago, BC. Pre-treatment data come from farms 1–4, 6, 8, 9, 11, 13, 15, 17, 18, 20–26. Post-treatment data come from farms 1–4, 6, 9, 11, 13, 17–24. Average temperature and salinity (Table 3) correspond to data from farms 1–9, 11–26.

estimates of the average number of motile sea lice (*L. salmonis*) per farmed salmon, the number of farmed salmon per farm, the age of the cohort, the local sea surface temperature and salinity, and the presence or absence of treatment with SLICE®. From these data, we estimated for each month at each farm the growth rate of the mean abundance of motile sea lice per farmed salmon and the total cohort surface area of farmed salmon.

We chose cohort surface area as a proxy for host density because cohort surface area directly influences sea louse settlement opportunity. The findings of Tucker et al. [54] suggest that for high infection pressure, host surface area is a more accurate predictor of sea louse settlement and survival on Atlantic salmon than is host biomass. Calculations for monthly growth rates and cohort surface area were made as follows.

The monthly growth rates for the average abundance of motile sea lice per farmed salmon (r_t) on a given farm in month t was calculated as $r_t = ln\left[\dfrac{N_{t+1}}{N_t}\right]$ where N_t is the average motile louse count per farmed salmon in month t. The growth rate (r_t) was omitted whenever N_t or N_{t+1} was missing.

Cohort surface area for each month at each farm was calculated using cohort age, estimated individual fish mass, estimated individual surface area by mass, and monthly number of fish per cohort. Individual mass (g) was estimated using age-class growth rates for farmed Atlantic salmon [55] at 8°C average sea temperature. Average individual surface area (cm^2), was estimated from mass, using a surface area formula for hatchery-reared Atlantic salmon, AREA = 14.93(MASS)$^{0.59}$ [56]. The cohort surface area in each month was the product of the estimated surface area per farmed salmon and the number of farmed salmon per cohort.

Cohort surface area, temperature, and salinity were each standardized to a mean of zero and a standard deviation of one over the dataset. This standardization allowed direct comparison among the effect sizes of the covariates. Under the standardiza-

tion, the effect size of a covariate corresponded to the change in the response variable induced by a one-standard-deviation change in that covariate. Without standardization, the effect size of a covariate would have corresponded to the change in the response variable induced by a one-unit change in that covariate, making comparison among covariates measured in different units difficult.

Following standardization, months with missing values for the growth rate (r_t), cohort surface area (A_t), temperature (T_t), or salinity (S_t) were removed. The remaining data were then divided into two sets that were analyzed independently: 250 cohort-months of pre-SLICE®-treatment data representing 51 stock-harvest cycles (19 farms; 2003–2008) and 86 cohort-months of post-SLICE®-treatment data representing 32 stock-harvest cycles (16 farms; 2002–2008). The pre-treatment data covered each cohort on each farm from initial stocking up to but not including the month of first treatment. The post-treatment data covered the 3 months following but not including the month of first treatment for each cohort. The spatial distribution of SLICE® treatments over time specifying the data used in this study may be found as an animation in the supplemental information [Animation S1].

Analysis

We modeled the sea louse population growth rate (r_t) using a linear hierarchical model with random effects for the cohort and farm. The random effects controlled for 1) non-independence among repeated measurements of mean sea louse abundance per fish on a single cohort, and 2) individual farm characteristics, such as flow patterns, that may consistently affect growth rates at a particular site. We then evaluated the influence of cohort surface area, temperature, salinity, and number of months since treatment on sea louse population growth. For the analysis of pre-treatment data, the covariate for the number of months since treatment was excluded. The full linear mixed model was

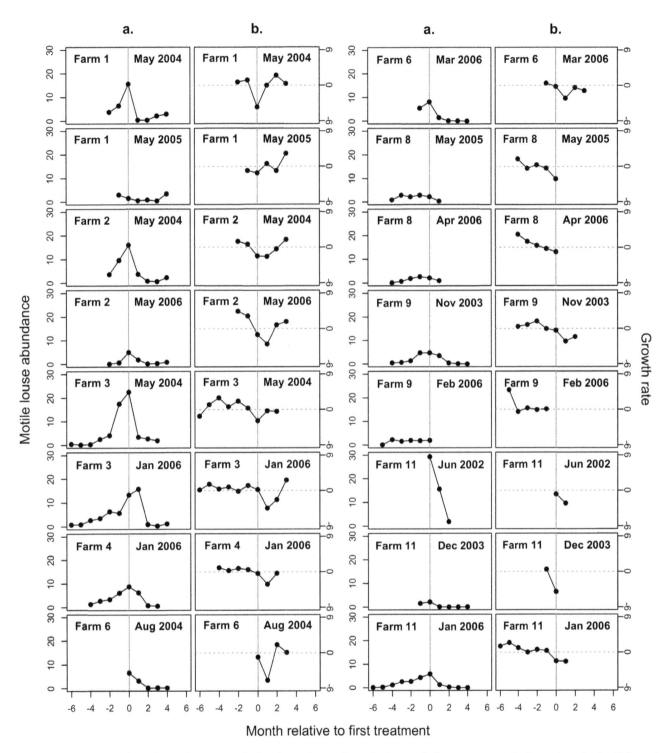

Figures 2. Examples of motile sea louse population dynamics on farms before and after treatment. Sea louse dynamics for 16 farmed salmon cohorts are shown as **a.** average motile louse abundance per farmed salmon, and **b.** growth rate of average motile sea louse abundance per farmed salmon, $r = \ln[N_{t+1}/N_t]$, where N_t is average sea louse abundance per farmed salmon in month t. SLICE® treatment (grey line) was initiated in the month and year specified. Values for the growth rate (r_t) are omitted whenever r_t is undefined (i.e. when $N_t = 0$ or $N_{t+1} = 0$) or missing. Cohorts shown are those with abundance and growth rate data in months immediately preceding or following the first treatment by SLICE® in the pre-treatment or post-treatment data.

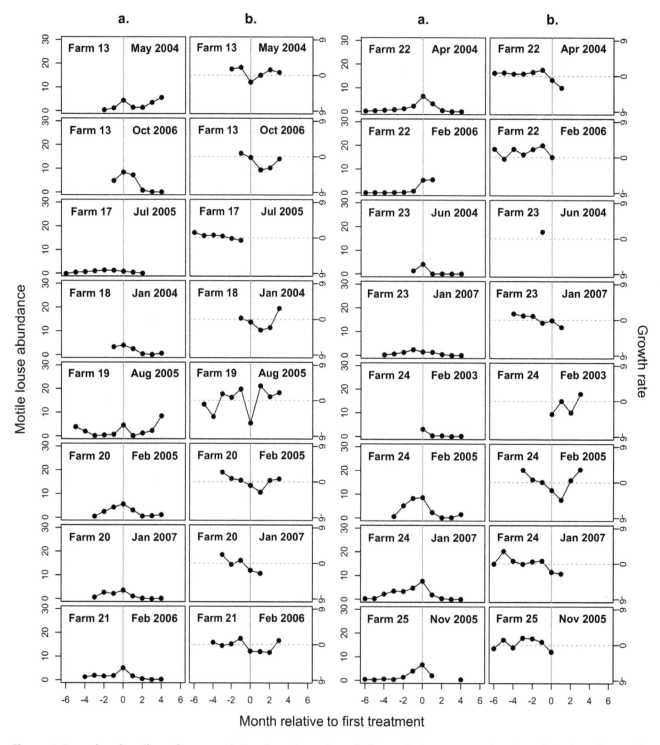

Figures 3. Examples of motile sea louse population dynamics on farms before and after treatment (continued). Sea louse dynamics for 16 farmed salmon cohorts are shown as **a.** average motile louse abundance per farmed salmon, and **b.** growth rate of average motile sea louse abundance per farmed salmon, $r = \ln[N_{t+1}/N_t]$, where N_t is average sea louse abundance per farmed salmon in month t. SLICE® treatment (grey line) was initiated in the month and year specified. Values for the growth rate (r_t) are omitted whenever r_t is undefined (i.e. when $N_t = 0$ or $N_{t+1} = 0$) or missing. Cohorts shown are those with abundance and growth rate data in months immediately preceding or following the first treatment by SLICE® in the pre-treatment or post-treatment data.

Table 1. Pre-treatment and post-treatment model selection statistics.

Data	Model	NLL	AIC	ΔAIC	Weight
Pre-treatment	r+S	321.4	652.8	0.00	0.200
	r	322.5	653.1	0.24	0.178
	r+T	321.8	653.6	0.80	0.135
	r+A	321.9	653.7	0.87	0.129
	r+A+S	320.9	653.9	1.03	0.120
	r+A+T	321.1	654.2	1.37	0.101
	r+S+T	321.3	654.6	1.75	0.084
	r+A+S+T	320.7	655.5	2.66	0.053
Post-treatment	r+S+T+M	132.7	279.3	0.00	0.309
	r+M	135.3	280.7	1.38	0.155
	r+A+S+T+M	132.5	281.1	1.77	0.127
	r+T+M	134.6	281.1	1.80	0.126
	r+S+M	134.8	281.5	2.23	0.102
	r+A+M	135.1	282.1	2.85	0.075
	r+A+T+M	134.3	282.5	3.23	0.062

Models were ranked by Akaike weight, and the collection of best-supported models corresponding to a 95% cumulative weight is shown. Models were generated for the intrinsic sea louse population growth rate (r) and all subsets of the covariates cohort surface area (A), temperature (T), salinity (S), and number of months since treatment (M), with the exception that the covariate for months since treatment (M) is excluded from the pre-treatment models. Columns show the negative log likelihood values (NLL), Akaike Information Criterion (AIC), the AIC differences (ΔAIC), and the Akaike weights.

$$\ln\left[\frac{N_{i,j,t+1}}{N_{i,j,t}}\right] = \left(r+\alpha_j+\beta_{i,j}\right)+aA_{i,j,t}+bT_{i,j,t}+cS_{i,j,t}$$
$$+dM_{i,j,t}+\varepsilon_{i,j,t} \tag{1}$$

where $N_{i,j,t}$ is the average abundance of motile sea lice per farmed salmon in cohort i on farm j during month t; r is the intrinsic rate of growth in sea louse abundance per fish; α_j and $\beta_{i,j}$ are random effects for cohort nested within farm that are normally distributed with means of zero and variances that are estimated from the data; A, T, S, and M are the cohort surface area, sea-surface temperature, salinity, and number of months since treatment respectively; a, b, c, and d are parameters that were estimated from the data to determine the magnitude and direction of the effects of A, T, S, and M on the growth of sea louse abundance per fish, respectively; and $\varepsilon_{i,j}$ is the residual normally-distributed variation.

We fit the full model and models corresponding to all subsets of the covariates to the pre-treatment (8 models) and post-treatment (16 models) data using maximum likelihood [57] in the statistical programming environment R [58]. For each model, we calculated the Akaike Information Criterion (AIC), and ranked the models by their Akaike weights [59]. We used multi-model inference [60] over all models to generate model averaged parameter estimates and standard errors for the intrinsic rate of growth (r) and effect sizes (a, b, c, d) of the covariates [59]. We estimated parameters and standard errors for the pre-treatment and post-treatment data separately.

After estimating parameters and standard errors, we simulated sea louse population dynamics in relation to farm treatment. Simulations were chosen to inform sea louse management for adherence to a maximum average three motile sea lice per farmed salmon guideline [35], and for the minimization of sea louse abundance on salmon farms during the annual juvenile wild salmon migration. For these simulations, we converted equation (1) into a deterministic model for sea louse population dynamics

$$N_{t+1} = N_t \exp[r+aA_t+bT_t+cS_t+dM_t] \tag{2}$$

that incorporates the model averaged estimates for the parameters in equation (1). We then extended equation (2) to a stochastic model for sea louse population dynamics

$$N_{t+1} = N_t \exp[(r+\varepsilon_r)+(a+\varepsilon_a)A+(b+\varepsilon_b)T_t+(c+\varepsilon_c)S_t] \tag{3}$$

by including normal random variables ε_r, ε_a, ε_b, and ε_c with means of zero and standard deviations equal to the estimated standard errors for the intrinsic rate of growth and effect sizes of cohort surface area, temperature, and salinity, respectively.

Using equation (3) and the pre-treatment parameter estimates and standard errors, we simulated scenarios of sea louse population growth in the absence of treatment. Simulations were conducted 10 000 times for different initial sea louse abundances and dates. We used initial conditions of 0.5, 1, and 2 motile sea lice per farmed salmon at dates between December 1st and June 30th. We simulated monthly growth for average motile sea louse abundance until June 30th, the end of the juvenile salmon migration season. Cohort surface area was set to its standardized mean ($A=0$) for each simulation. Values for temperature and salinity were chosen for each month in each simulation by random selection from the standardized observed values for the corresponding month in the combined pre-treatment and post-treatment data. By observing the proportion of simulations for each set of initial conditions that led to an average abundance of three or more motile sea lice per farmed salmon on June 30th, the simulations convey an estimate of the probability that a mean sea louse abundance of 0.5, one, or two motile sea lice per farmed salmon on a particular date will lead to growth that exceeds the three motile sea lice guideline during the juvenile wild salmon migration (March–June). These estimated probabilities may be useful to managers for understanding the probability of exceeding the three motile sea lice per farmed salmon guideline during the juvenile wild salmon migration.

Using equation (2) and the post-treatment parameter estimates, we simulated scenarios of sea louse population growth for four months following treatment. We used the initial condition of three motile sea lice per farmed salmon, corresponding to the regulatory guideline that triggers treatment or harvest [35]. Simulations were conducted with constant temperatures and salinities, and we evaluated the sensitivity of results to a range of 6–12°C and 15–30 PSU. Cohort surface area was set to its standardized mean ($A=0$).

The effect of treatment ended when the post-treatment growth rate became equal to the pre-treatment growth rate, where pre- and post-treatment growth rates are calculated as

$$r_{t,\text{pre}} = r_{\text{pre}}+b_{\text{pre}}T_t+c_{\text{pre}}S_t, \quad \text{and} \tag{4}$$

$$r_{t,\text{post}} = r_{\text{post}}+b_{\text{post}}T_t+c_{\text{post}}S_t+dM_t. \tag{5}$$

When the effect of treatment ended, the months since treatment effect (dM_t) was excluded and the post-treatment parameters (r_{post},

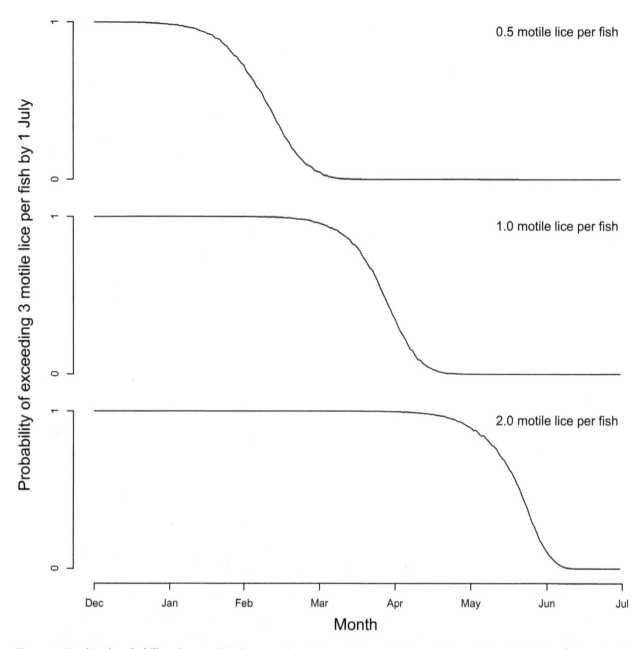

Figure 4. Simulated probability of exceeding three motile sea lice per farmed salmon during the juvenile wild salmon migration in the absence of SLICE® treatment. Probabilities are shown for exceeding a threshold of three motile sea lice per farmed salmon before the end of the juvenile wild salmon migration (June 30th) given average sea louse abundances of 0.5, 1, or 2 motile sea lice per fish in a particular month and in the absence of treatment. Probabilities for each month correspond to the proportion out of 10 000 simulations that met or exceeded three motile sea lice per fish by June 30th. Simulations are based on the stochastic model in equation (3) with random variation around the growth rate (r) and around the parameters for the effects of temperature (b) and salinity (c) Cohort surface area, A, is set to its standardized mean ($A = 0$) and the months since treatment (M) is excluded from the simulation. Values for temperature and salinity are chosen for each month in each simulation by random selection from standardized observed values for that month from the data.

b_{post}, c_{post}) were replaced by the pre-treatment parameters (r_{pre}, b_{pre}, c_{pre}) to ensure that the growth rate in each simulation did not exceed the pre-treatment growth rate.

In a second set of post-treatment simulations using an altered method and equation (2), we simulated the effects of treatment near the March–June juvenile salmon migration. We set the initial condition of three motile sea lice per farmed salmon in a particular month, December, January, February or March, and simulated forwards, varying the temperature and salinity according to the average monthly conditions across farms. As with previous

simulations, when the post-treatment rate reached the pre-treatment growth rate, signaling an end of the treatment effect, we replaced the post-treatment parameters (r_{post}, b_{post}, c_{post}) by the pre-treatment parameters (r_{pre}, b_{pre}, c_{pre}) and excluded the months since treatment effect (dM_t). Insight based on these simulations may be useful to aquaculture managers for timing the treatment of farmed salmon to reduce sea louse populations prior to and during the March–June juvenile wild salmon migration.

Table 2. Pre-treatment and post-treatment model averaged parameter estimates.

Data	Rate (*r*)	Area (*a*)	Temp. (*b*)	Salt (*c*)	Month (*d*)
Pre-treatment	0.35 (0.07)	0.09 (0.08)	−0.06 (0.06)	0.08 (0.06)	–
Post-treatment	−2.91 (0.31)	−0.09 (0.14)	0.22 (0.13)	0.22 (0.14)	1.25 (0.16)

Model-averaged estimates are given for the intrinsic rate of growth of average sea louse abundance per farmed salmon (*r*) and the effect sizes for cohort surface area (*a*), temperature (*b*), salinity (*c*), and number of months since treatment (*d*). Standard errors are given in parentheses. Pre-treatment data span from stocking to the month preceding the first SLICE® treatment; post-treatment data span the three months following the month in which treatment takes place.

Results

On salmon farms in the Broughton Archipelago between 2002 and 2008, the average abundance per farmed salmon of motile sea lice (*Lepeophtheirus salmonis*) typically increased over time until farmed salmon were treated with SLICE® (Figures 2 and 3). Following treatment, the motile sea louse growth rate (r_t) tended to be suppressed below pre-treatment rates for two or three months after which the growth rate tended to return to pre-treatment levels. Following treatment, motile sea louse abundance tended to decline sharply and remained depressed for at least four months.

Comparison of models using AIC revealed model-selection uncertainty, particularly for pre-treatment data (Table 1). Model averaged parameter estimates (and standard errors) for the pre-treatment intrinsic rate of growth (*r*) and effect sizes (*a*, *b*, *c*) of cohort surface area, temperature, and salinity indicated clear positive sea louse population growth prior to treatment. Model-averaged parameter estimates (and standard errors) for the post-treatment intrinsic rate of growth (*r*) and effect sizes (*a*, *b*, *c*, *d*) of cohort surface area, temperature, salinity, and the number of months since treatment indicated clear negative sea louse population growth following treatment, but whose effect depended strongly on the time since treatment (Table 2). The effects of all other covariates on sea louse population growth were small and had large uncertainty, although the effects of temperature and salinity on sea louse population growth were better resolved in post-treatment than pre-treatment data (Table 2).

Stochastic simulations using equation (3) and the pre-treatment parameter estimates and standard errors (Table 2) at standardized mean cohort surface area ($A = 0$) and with temperature and salinity sampled randomly by month indicated that a lack of treatment in the months preceding the March–July juvenile salmon migration could lead to abundances of sea lice on farmed salmon that exceed the guideline of three motile sea lice per farmed salmon during the juvenile salmon migration window (Figure 4). Our simulations suggest that abundances of 0.5 motile sea lice per farmed salmon prior to February 10th, one motile sea louse prior to March 27th, or two motile sea lice prior to May 21st may grow to reach or exceed the three motile sea lice guideline during the juvenile wild salmon migration (March–June) with probability >0.50, assuming no treatment is initiated.

Simulations for post-treatment sea louse population dynamics indicated that treatment effect, measured as growth rate suppression, lasted 2–3 months depending on ambient temperature and salinity (Figure 5). At average salinity (26.9 PSU) and temperatures within 6–12°C, our simulations suggest that the growth rate for sea louse abundance remained below pre-treatment rates for 2.2–3.0 months. At average temperature (8.8°C) and salinity within 15–30 PSU, the simulated growth rate for sea louse abundance remained below pre-treatment levels for 2.6–2.9 months (Figure 5). SLICE® treatment was predicted to depress mean sea louse abundance per farmed salmon below three motile sea lice for at least four months at average temperature (8.8°C) or average salinity (26.9 PSU) (Figure 5).

Seasonal variability in environmental conditions had implications for sea louse management (Table 2 and Table 3). We found that treatment of farmed salmon in January or early February minimized average sea louse abundance per farmed salmon during the entire March–July juvenile wild salmon migration (Figure 6). The effect of treatment before or during December was predicted to wear off and allow pre-treatment growth rates to resume before or during March, early in the juvenile wild salmon migration. Treatment after mid-February was predicted to delay the suppression of sea louse abundance and leave abundances greater than three times the suppression minimum on March 1st, the beginning of the juvenile wild salmon migration.

In all simulations (Figures 3, 4, 5), uncertainty in model predictions propagated as time increased, and long-term forecasting of sea louse population dynamics beyond several months based on our model and parameter estimates was less informative.

Discussion

Precautionary management of parasites on Atlantic salmon farms in British Columbia that suppresses parasite abundance to coincide with the timing of juvenile wild salmon migrations may reduce the risk of infection for wild Pacific salmon. We used a model to understand sea louse population dynamics on salmon farms in the Broughton Archipelago, British Columbia. We found a positive rate of growth for sea louse abundance on farmed salmon in the absence of parasiticide treatment, and a temporarily negative rate of growth following farmed salmon treatment by SLICE®. We found that treatment of farmed salmon by SLICE® tended to depress sea louse abundances for periods commensurate with the duration of the juvenile wild salmon migration. Consequently, the judicious timing of SLICE® treatment on salmon farms may be a viable management option to reduce the transmission of sea lice to migrating juvenile wild salmon. Indeed, we found treatment during January or early February most likely to minimize sea louse abundance on Broughton Archipelago salmon farms during the March–June juvenile wild salmon migration. These findings are consistent with previous studies that identified exponential patterns of sea louse population growth in the absence of treatment [47], and effective sea louse suppression following SLICE® treatment in the Broughton Archipelago [35]. Our findings extend the results of Krkošek et al. [47] to suggest that winter treatment of sea lice, considered among the salmon aquaculture industry best practices [13,48] for annual sea louse suppression on salmon farms [31], may be an effective strategy to reduce sea louse exposure for migrating juvenile wild salmon generally across salmon farms in the Broughton Archipelago.

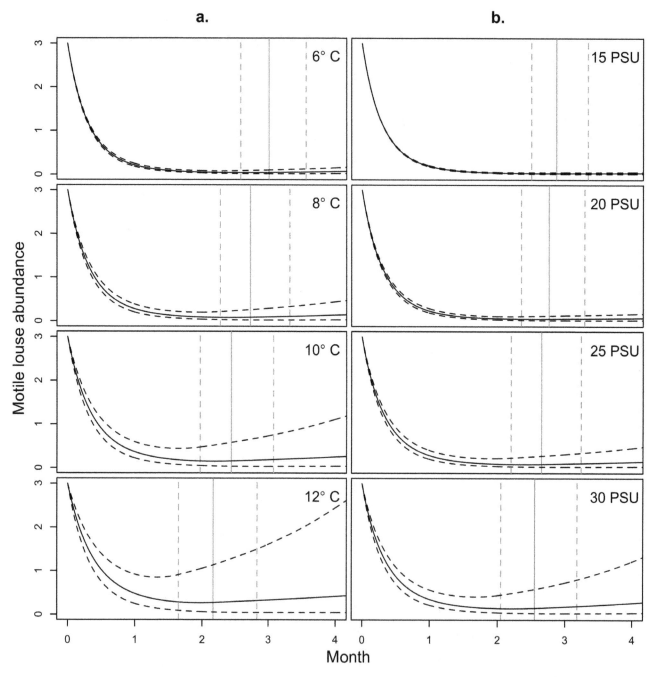

a.

b.

**Figure 5. Simulated decline and recovery of sea louse populations at constant temperature and salinity following SLICE®
treatment.** Simulations are shown for a. salinity held constant at 26.9 PSU and b. temperature held constant at 8.8°C, the average salinity and
temperature recorded for the Broughton Archipelago during the study period (Table 3). To generate the simulations, post-treatment estimates for r,
b, c, and d (Table 2) were used until the effective rate of growth $r_{post}+b_{post}T_t+c_{post}S_t+dM_t$ became equal to the pre-treatment (maximum) rate of
growth $r_{pre}+b_{pre}T_t+c_{pre}S_t$ (Table 2). From this point on the pre-treatment parameters were used, and the effect for the number of months since
treatment was omitted. Cohort surface area was set to its standardized mean ($A=0$) for each simulation. Curved dashed lines correspond to
simulations with parameters set one standard error on either side of the mean parameter estimates. Vertical grey lines correspond to the time of
transition from post-treatment to pre-treatment parameters (i.e. t such that $r_{post}+b_{post}T_t+c_{post}S_t+dM_t = r_{pre}+b_{pre}T_t+c_{pre}S_t$) for the mean (solid) and one-
standard-error-removed (dashed) simulations.

Sea louse population dynamics in the Broughton Archipelago
appear to be mediated by abiotic factors, specifically temperature
and salinity. We detected weak effects for temperature and salinity
on pre-treatment sea louse population growth over the ranges
encountered in our study (Table 3). Sea lice are generally observed
to exhibit higher levels of settlement on hosts, increased survival of

larval stages, and higher rates of development at higher ambient
temperatures and salinities [49,50,61,62]. The data that informed
our parameter estimates were monthly averages that did not
capture seasonal or spatial extremes in temperature and salinity.
These extremes may play an important role in limiting sea louse
population growth. Application of the model to data at a finer

Table 3. Temperature and salinity on Broughton Archipelago salmon farms during 2000–2009.

Month	T (°C)			S (PSU)		
	Mean	SE	Range	Mean	SE	Range
Jan	7.1	(0.6)	6.0–8.0	29.1	(2.8)	21.0–33.0
Feb	7.2	(0.5)	6.5–9.0	28.8	(3.3)	17.0–34.5
Mar	7.4	(0.5)	6.0–9.0	29.5	(3.0)	18.0–34.8
Apr	8.3	(0.6)	7.0–9.8	29.4	(2.9)	18.0–33.5
May	9.3	(0.9)	7.0–11.9	28.0	(4.0)	17.0–34.0
Jun	10.3	(0.9)	8.0–12.5	24.5	(6.0)	10.0–33.0
Jul	10.9	(1.1)	8.7–13.2	22.9	(6.8)	7.3–33.0
Aug	10.9	(1.1)	9.0–14.8	23.3	(6.6)	8.2–33.0
Sep	10.1	(0.9)	8.8–12.3	25.4	(5.8)	7.7–33.0
Oct	9.1	(0.7)	6.7–11.0	25.6	(4.8)	12.3–32.7
Nov	8.1	(0.5)	7.0–10.6	26.8	(4.7)	13.0–33.0
Dec	7.3	(0.6)	5.1–9.8	28.7	(2.7)	21.0–33.0

Monthly averages, standard errors, and range (min–max) are given for recorded temperature and salinity on farms 1–9, 11–26 (Figure 1).

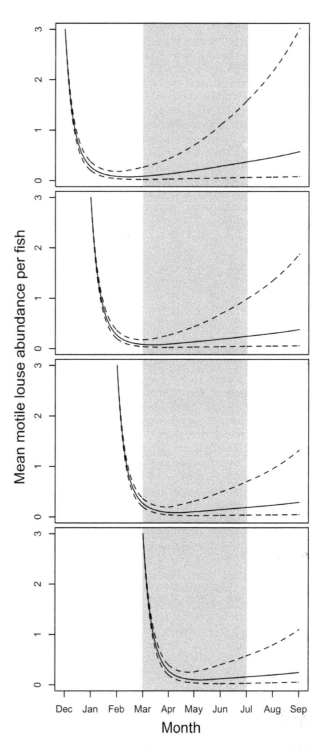

temporal resolution that includes extreme abiotic events and sea louse population responses, which are likely to occur on time scales of days or weeks, could therefore be an informative future extension of this study.

Temperature was found to have a positive effect on sea louse population growth rates following SLICE® treatment. As a result, the temperature effect led to shorter periods of sea louse suppression in warmer waters. At least two mechanisms may be responsible for this effect. First, larval settlement and sea louse development are expected to increase with increased temperature [49,50], leading potentially to shorter generation times and higher rates of population growth relative to populations in cooler waters [13]. Second, the depletion of EMB residue from the body tissue of farmed rainbow trout (*Oncorhynchus mykiss*) appears more rapid at higher temperatures [63]. If EMB depletion from host tissue follows a corresponding pattern in farmed Atlantic salmon, increased temperature could shorten the period during which sea lice populations are suppressed by SLICE®.

We also found that salinity had a positive effect on sea louse population growth rates following SLICE® treatment. As a result, we predicted longer periods of sea louse suppression in less saline waters. This effect may be explained by changes to sea louse demographic rates at low salinity. Hatching, survival, and development rates for sea lice have all been observed to decline at low salinity [61,62]. Surface salinity in the Broughton Archipelago is strongly influenced by an annual runoff of freshwater, leading to reduced surface salinity starting in late spring (Table 3), although this may be less pronounced in the outer regions of the Broughton Archipelago.

Our simulations of sea louse dynamics for average monthly conditions in the Broughton Archipelago point to an optimal timing for SLICE® treatment to suppress sea louse abundance on farms during the juvenile wild salmon migration. Based on our simulations, the optimal time to administer SLICE® treatment is January or early February. Under this regime, sea louse populations on farms are most likely to be suppressed below three motile sea lice per farmed salmon, consistent with regulatory guidelines, for the duration of the March–June juvenile wild

Figure 6. Simulated decline and recovery of sea louse populations at varying temperature and salinity following SLICE® treatment. Simulations correspond to SLICE® treatment initiated in December, January, February, and March in advance of the March–June juvenile wild salmon migration (grey shaded region). Simulations used a combination of post-treatment and pre-treatment parameter estimates as described for Figure 5. Temperature and salinity were varied by month to represent the average monthly conditions across salmon farms during December–September in the Broughton Archipelago. Dashed lines correspond to simulations with parameters set to one standard error on either side of the mean parameter estimates.

salmon migration. A January or early February treatment strategy could likely reduce infection pressure on wild salmon during the juvenile wild salmon migration.

Harm to wild stocks from exposure to sea lice during the juvenile wild salmon migration may change depending on the timing of exposure. Juvenile pink and chum salmon may be more vulnerable to the effects of sea lice early in their migration [64]. Early exposure can also allow time for reproduction of sea lice on juvenile hosts, further amplifying sea louse populations on wild juvenile salmon [33]. The early exposure of pink and chum salmon, two prey sources for juvenile coho salmon, may in turn increase the exposure of juvenile coho salmon to sea lice [14]. Consequently, efforts to reduce impacts to wild salmon should focus on minimizing sea louse exposure for juvenile Pacific salmon during March–April, the first half of the juvenile salmon migration.

Cohort surface area per farm, a proxy for host density, had no consistent effect on sea louse population growth rates. The lack of effect suggests that host surface area did not limit sea louse population growth, likely because sea louse numbers were kept low by SLICE® treatments. The highest average sea louse counts on farms in the Broughton Archipelago typically occurred immediately prior to treatment, and rarely exceeded 10 sea lice per farmed salmon. By contrast, sea louse counts involving Lepeophtheirus salmonis on untreated salmon farms in Scotland have experienced mean sea louse abundances in excess of 30 sea lice per farmed salmon [37]. The observed pattern of mean sea louse population growth truncated by treatment, together with evidence of the potential for much higher sea louse abundances under similar but untreated conditions suggests that sea lice on farmed salmon in British Columbia may be limited primarily by treatment. For this reason, sea louse populations may not approach a carrying capacity, and therefore not be limited by host surface area.

The absence of a strong host-density effect may appear to contradict the recent findings of Jansen et al. [53], who reported that monthly mean counts of sea lice were positively related to the average mass of farmed salmon and local biomass density on salmon farms in Norway. However, we do not expect the relationship that Jansen et al. [53] found for sea louse abundance to hold in our model for growth rates. For example, a positive growth rate that is constant over time will lead to an abundance that increases over time. Therefore, if sea louse population growth were positive on average during a farmed salmon production cycle, when farmed salmon increase in body size, then a positive correlation between fish biomass and sea louse abundance would result even if no effects of fish biomass on the sea louse population growth rate were present. Thus, it is not clear from the present study nor from Jansen et al. [53] that host density has a positive effect on sea louse transmission rates, as would be expected by theory [65].

Dispersal of sea louse larvae due to wind and ocean currents may give rise to sea louse population dynamics that act on a regional scale [34,53,66,67,68], and regional host density may be more relevant to sea louse population dynamics than host density at a single farm [53]. Mechanistic models of sea louse dispersal among farms can account explicitly for regional host-density effects. Wind and ocean currents in the Broughton Archipelago may lead to complex spatial and temporal patterns of larval sea louse dispersal among salmon farms [68] and make the estimation of regional host-density effects difficult. A hydrodynamic model describing the motion of currents in the Broughton Archipelago has been developed [69] and efforts to improve its realism are ongoing [70]. Further work is planned to couple the hydrodynamic and population-dynamic models in order to understand the effects of sea louse dispersal on sea louse dynamics in the Broughton Archipelago.

The present study advances the science of sea louse (L. salmonis) population dynamics in two ways: first, by accommodating explicitly the effects of temperature and salinity on sea louse population growth rates [47], and second, by increasing the spatial and temporal scope of data used to model sea louse population dynamics in the Broughton Archipelago by one order of magnitude over previous studies [47]. Nevertheless, our analysis is subject to several limitations. Data are monthly aggregates and offer only a coarse resolution of the sea louse population dynamics. Although the data represent seven years of sea louse abundance monitoring on 25 salmon farms and correspond to 122 stock-harvest cycles on farms, missing values reduce the usable data to 51 pre-treatment and 32 post-treatment stock-harvest cycles.

While suggestive of timing strategies that could be applied for wild fish conservation management in other regions, the results of the present study are not easily generalizable. British Columbia is atypical among salmon farming regions by virtue of the lack of evidence for sea louse resistance to SLICE® in British Columbia. In jurisdictions where sea louse resistance to SLICE® is established or growing, the timing of SLICE® treatment relative to wild fish migrations may be of reduced value or immaterial to wild fish conservation. Nevertheless, within British Columbia in regions with seasonal temperature and salinity profiles similar to the Broughton Archipelago, there is reason to believe that wild Pacific salmon may benefit generally from winter treatment of salmon farms by SLICE®.

In order to inform adequately the ongoing conservation management of wild Pacific salmon, additional information is required about the long-term efficacy of SLICE® in British Columbia. SLICE® is the only chemotherapeutant that is licenced for use on salmon farms in Pacific Canada [35], and reliance on a single paraciticide can lead to an evolved resistance in parasites and a reduced efficacy [71], as has been observed in Scotland [40], Atlantic Canada [41,42], Ireland, and Norway [35]. Environmental concern over the trace presence of emamectin benzoate in sediment and the tissue of non-target crustaceans [72,73] warrants further study. Nevertheless, our results show that judicious timing for the application of SLICE® on salmon farms in the Broughton Archipelago has the potential to reduce the exposure of juvenile wild pink, chum, and coho salmon to parasitic sea lice. Thus, parasite management on finfish farms to reduce infection pressure on wild fish can help to conserve biodiversity and support wild capture fisheries.

Supporting Information

Animation S1 SLICE® treatments on Atlantic salmon farms in the Broughton Archipelago, British Columbia, Canada from November 1999 to December 2009 (Marty et al. 2010 PNAS). SLICE® treatment is indicated by a red circle at the salmon farm location. Untreated farms are shown by green circles, and farms fade from red to green over the four month efficacy period of SLICE® treatments (see main text). Fallowed farms (i.e., no Atlantic salmon in net pens) are indicated by beige circles with an 'x'. Data that were used in the analysis of sea louse population dynamics before and after treatments (see main text) are circled in thick black. The period of juvenile wild salmon migration is indicated by purple arrows along approximate migration routes.

Acknowledgments

The authors thank Howard Browman and two anonymous referees whose comments improved the manuscript.

Author Contributions

Led collaboration: CWR MK. Revised the manuscript: SJP PM SD SRMJ PC MGGF CWR MK. Conceived and designed the experiments: LAR SJP MK. Performed the experiments: LAR. Analyzed the data: LAR. Wrote the paper: LAR.

References

1. Food and Agricultural Organization of the United Nations (2012) The state of world fisheries and aquaculture. Rome: FAO. 209 p.

2. Food and Agricultural Organization of the United Nations (2002) The state of world fisheries and aquaculture. Rome: FAO.

3. Delgado CL, Wada N, Rosegrant MW, Meijer S, Ahmed M (2003) Fish to 2020: supply and demand in changing global markets. Washington, D.C.: International Food Policy Research Institute.

4. United Nations Department of Economic and Social Affairs Population Division (2011) World population prospects: the 2010 revision, Volume I: comprehensive tables. ST/ESA/SER.A/313.: United Nations, Department of Economic and Social Affairs, Population Division.

5. Murray AG, Peeler EJ (2005) A framework for understanding the potential for emerging diseases in aquaculture. Prev Vet Med 67: 223–235.

6. Saksida S, Constantine J, Karreman GA, Donald A (2007) Evaluation of sea lice abundance levels on farmed Atlantic salmon (Salmo salar L.) located in the Broughton Archipelago of British Columbia from 2003 to 2005. Aquacult Res 38: 219–231.

7. Connors B, Krkosek M, Ford JS, Dill LM (2010) Coho salmon productivity in relation to salmon lice from infected prey and salmon farms. J Appl Ecol 47: 1372–1377.

8. Costello MJ (2009) The global economic cost of sea lice to the salmonid farming industry. J Fish Dis 32: 115–118.

9. Costello MJ (2009) How sea lice from salmon farms may cause wild salmonid declines in Europe and North America and be a threat to fishes elsewhere. Proc Roy Soc B 276: 3385–3394.

10. Morton A, Routledge R, Peet C, Ladwig A (2004) Sea lice (Lepeophtheirus salmonis) infection rates on juvenile pink (Oncorhynchus gorbuscha) and chum (Oncorhynchus keta) salmon in the nearshore marine environment of British Columbia, Canada. Can J Fish Aquat Sci 61: 147–157.

11. Krkosek M (2010) Sea lice and salmon in Pacific Canada: Ecology and Policy. Front Ecol Environ 86: 201–209.

12. Krkosek M, Lewis MA, Morton A, Frazer LN, Volpe JP (2006) Epizootics of wild fish induced by farm fish. Proc Natl Acad Sci USA 103: 15506–15510.

13. Costello MJ (2006) Ecology of sea lice parasitic on farmed and wild fish. Trends Parasitol 22: 475–483.

14. Connors BM, Hargreaves NB, Jones SRM, Dill LM (2010) Predation intensifies parasite exposure in a salmonid food chain. J Appl Ecol 47: 1365–1371.

15. Krkosek M, Connors B, Mages P, Peacock S, Ford H, et al. (2011) Fish farms, parasites, and predators: Implications for salmon population dynamics. Ecol Appl 21: 897–914.

16. Krkosek M, Ford JS, Morton A, Lele S, Myers RA, et al. (2007) Declining wild salmon populations in relation to parasites from farm salmon. Science 318: 1772–1775.

17. Frazer LN (2009) Sea-cage aquaculture, sea lice, and declines of wild fish. Conserv Biol 23: 599–607.

18. Krkosek M, Hilborn R (2011) Sea lice (Lepeophtheirus salmonis) infestations and the productivity of pink salmon (Oncorhynchus gorbuscha) in the Broughton Archipelago, British Columbia, Canada. Can J Fish Aquat Sci 68: 17–29.

19. Jones S, Kim E, Dawe S (2006) Experimental infections with Lepeophtheirus salmonis (Kroyer) on threespine sticklebacks, Gasterosteus aculeatus L., and juvenile Pacific salmon, Oncorhynchus spp. J Fish Dis 29: 489–495.

20. Jones SRM, Fast MD, Johnson SC, Groman DB (2007) Differential rejection of salmon lice by pink and chum salmon: disease consequences and expression of proinflammatory genes. Dis Aquat Organ 75: 229–238.

21. Beamish RJ, Neville CM, Sweeting RM, Ambers N (2005) Sea lice on adult Pacific salmon in the coastal waters of central British Columbia, Canada. Fish Res 76: 198–208.

22. Beamish RJ, Neville CM, Sweeting RM (2007) Response to Dr. Neil Frazer's comment on "Sea lice on adult Pacific salmon in the coastal waters of British Columbia, Canada" by R.J. Beamish et al. (2005). Fish Res 85: 332–333.

23. Frazer LN (2007) Comment on "Sea lice on adult Pacific salmon in the coastal waters of British Columbia, Canada" by R.J. Beamish et al. Fish Res 85: 328–331.

24. Brooks KM, Jones SRM (2008) Perspectives on pink salmon and sea lice: scientific evidence fails to support the extinction hypothesis. Res Fish Sci 16: 403–412.

25. Marty G, Saksida S, Quinn TJ (2010) Relationship of farm salmon, sea lice, and wild salmon populations. Proc Natl Acad Sci USA 107: 22599–22604.

26. Krkosek M, Connors BM, Morton A, Lewis MA, Dill LM, et al. (2011) Effects of parasites from salmon farms on productivity of wild salmon. Proc Natl Acad Sci USA 108: 14700–14704.

27. Saksida SM, Morrison D, Sheppard M, Keith I (2011) Sea Lice Management on Salmon Farms in British Columbia, Canada. In: Jones SRM, Beamish RJ, editors. Salmon lice: an integrated approach to understanding parasite abundance and distribution. Oxford: Wiley-Blackwell. 235–278.

28. Brooks KM (2009) Considerations in developing an integrated pest management programme for control of sea lice on farmed salmon in Pacific Canada. J Fish Dis 32: 59–73.

29. Morton A, Routledge RD, Williams R (2005) Temporal patterns of sea louse infestation on wild Pacific salmon in relation to the fallowing of Atlantic salmon farms. N Am J Fish Manage 25: 811–821.

30. Morton A, Routledge R, McConnell A, Krkosek M (2011) Sea lice dispersion and salmon survival in relation to salmon farm activity in the Broughton Archipelago. ICES J Mar Sci 68: 144–156.

31. Peacock SJ, Krkosek M, Proboscsz S, Orr C, Lewis MA (In press) Cessation of a salmon decline with control of parasites. Ecol Appl.

32. Saksida SM, Marty GD, Jones SRM, Manchester HA, Diamond CL, et al. (2012) Parasites and hepatic lesions among pink salmon, Oncorhynchus gorbuscha (Walbaum), during early seawater residence. J Fish Dis 35: 137–151.

33. Krkosek M, Lewis MA, Volpe JP (2005) Transmission dynamics of parasitic sea lice from farm to wild salmon. Proc Roy Soc B 272: 689–696.

34. Frazer LN, Morton A, Krkosek M (2012) Critical thresholds in sea lice epidemics: Evidence, sensitivity, and subcritical estimation. Proc Roy Soc B 279: 1950–1958.

35. Saksida SM, Morrison D, Revie CW (2010) The efficacy of emamectin benzoate against infestations of sea lice, Lepeophtheirus salmonis, on farmed Atlantic salmon, Salmo salar L., in British Columbia. J Fish Dis 33: 913–917.

36. Ramstad A, Colquhoun DJ, Nordmo R, Sutherland IH, Simmons R (2002) Field trials in Norway with SLICE (R) (0.2% emamectin benzoate) for the oral treatment of sea lice infestation in farmed Atlantic salmon Salmo salar. Dis Aquat Organ 50: 29–33.

37. Stone J, Sutherland IH, Sommerville C, Richards RH, Varma KJ (2000) Commercial trials using emamectin benzoate to control sea lice Lepeophtheirus salmonis infestations in Atlantic salmon Salmo salar. Dis Aquat Organ 41: 141–149.

38. Gustafson L, Ellis S, Robinson T, Marenghi F, Endris R (2006) Efficacy of emamectin benzoate against sea lice infestations of Atlantic salmon, Salmo salar L.: evaluation in the absence of an untreated contemporary control. J Fish Dis 29: 621–627.

39. Berg AGT, Horsberg TE (2009) Plasma concentrations of emamectin benzoate after Slice (TM) treatments of Atlantic salmon (Salmo salar): Differences between fish, cages, sites and seasons. Aquaculture 288: 22–26.

40. Lees F, Baillie M, Gettinby G, Revie CW (2008) The Efficacy of Emamectin Benzoate against Infestations of Lepeophtheirus salmonis on Farmed Atlantic Salmon (Salmo salar L) in Scotland, 2002–2006. PLoS One 3.

41. Westcott JD, Revie CW, Griffin BL, Hammell KL (2010) Evidence of sea lice Lepeophtheirus salmonis tolerance to emamectin benzoate in New Brunswick Canada. Sea Lice 2010–8th International Sea Lice Conference. Victoria, BC.

42. Igboeli OO, Fast MD, Heumann J, Burka JF (2012) Role of P-glycoprotein in emamectin benzoate (SLICE (R)) resistance in sea lice, Lepeophtheirus salmonis. Aquaculture 344: 40–47.

43. Saksida SM, Morrison D, McKenzie P, Milligan B, Downey E, et al. (2012) Use of Atlantic salmon, Salmo salar L., farm treatment data and bioassays to assess for resistance of sea lice, Lepeophtheirus salmonis, to emamectin benzoate (SLICE®) in British Columbia, Canada. J Fish Dis doi: 101111/jfd12018.

44. The SLICE® sustainability project (2010) The SLICE® sustainability project. Intervet Schering-Plough Animal Health. url: aqua.merck-animal-health.com/binaries/SSP_brochure_2012-FINAL_tcm127–209461.pdf. 1–24 p.

45. Jones SRM, Kim E, Bennett W (2008) Early development of resistance to the salmon louse Lepeophtheirus salmonis (Krøyer) in juvenile pink salmon Oncorhynchus gorbuscha (Walbaum). J Fish Dis 31: 591–600.

46. Sutherland BJG, Jantzen SG, Sanderson DS, Koop BF, Jones SRM (2011) Differentiating size-dependent responses of juvenile pink salmon (Oncorhynchus gorbuscha) to sea lice (Lepeophtheirus salmonis) infections. Comp Biochem Physiol D 6: 213–223.

47. Krkosek M, Bateman A, Proboscsz S, Orr C (2010) Dynamics of outbreak and control of salmon lice on two salmon farms in the Broughton Archipelago. Aquacult Environ Interact 1: 137–146.

48. Costello MJ. A checklist of best practices for sealice control on salmon farms; 2003; St Andrews, Canada.

49. Stien A, Bjorn PA, Heuch PA, Elston DA (2005) Population dynamics of salmon lice Lepeophtheirus salmonis on Atlantic salmon and sea trout. Mar Ecol Prog Ser 290: 263–275.

50. Tucker CS, Sommerville C, Wootten R (2000) The effect of temperature and salinity on the settlement and survival of copepodids of Lepeophtheirus salmonis (Kroyer, 1837) on Atlantic salmon, Salmo salar L. J Fish Dis 23: 309–320.

51. Revie CW, Gettinby G, Treasurer JW, Wallace C (2003) Identifying epidemiological factors affecting sea lice Lepeophtheirus salmonis abundance on Scottish salmon farms using general linear models. Dis Aquat Organ 57: 85–95.

52. Revie CW, Gettinby G, Treasurer JW, Rae GH, Clark N (2002) Temporal, environmental and management factors influencing the epidemiological patterns

of sea lice (*Lepeophtheirus salmonis*) infestations on farmed Atlantic salmon (*Salmo salar*) in Scotlandt. Pest Manag Sci 58: 576–584.

53. Jansen PA, Kristoffersen AB, Viljugrein H, Jimenez D, Aldrin M, et al. (2012) Sea lice as a density-dependent constraint to salmonid farming. Proc Roy Soc B 279: 2330–2338.

54. Tucker CS, Sommerville C, Wootten R (2002) Does size really matter? Effects of fish surface area on the settlement and initial survival of *Lepeophtheirus salmonis*, an ectoparasite of Atlantic salmon *Salmo salar*. Dis Aquat Organ 49: 145–152.

55. Austreng E, Storebakken T, Asgard T (1987) Growth-rate estimates for cultured Atlantic salmon and rainbow-trout. Aquaculture 60: 157–160.

56. O'Shea B, Mordue-Luntz AJ, Fryer RJ, Pert CC, Bricknell IR (2006) Determination of the surface area of a fish. J Fish Dis 29: 437–440.

57. Bates D, Maechler M, Bolker B (2011) lme4: Linear mixed-effects models using S4 classes. R package version 0.999375–42. Available: http://CRAN.R-project.org/package=lme4. Accessed 2011 Oct 5.

58. R Development Core Team (2011) R: A language and environment for statistical computing. R Foundation for Statistical Computing, Vienna, Austria. ISBN 3–900051–07–0. Available: http://www.R-project.org/. Accessed 2011 Oct 5.

59. Burnham KP, Anderson DR (2002) Model Selection and Multimodel Inference. New York: Springer.

60. Barton K (2012) MuMIn: Multi-model inference. R package version 1.7.2. Available: http://CRAN.R-project.org/package=MuMIn. Accessed 2012 Mar 7.

61. Brooks KM (2005) The effects of water temperature, salinity, and currents on the survival and distribution of the infective copepodid stage of sea lice (*Lepeophtheirus salmonis*) originating on Atlantic salmon farms in the Broughton Archipelago of British Columbia, Canada. Res Fish Sci 13: 177–204.

62. Johnson SC, Albright LJ (1991) Development, growth, and survival of *Lepeophtheirus salmonis* (Copepoda, Caligidae) under laboratory conditions. J Mar Biol Assoc UK 71: 425–436.

63. Roy WJ, Gillan N, Crouch L, Parker R, Rodger H, et al. (2006) Depletion of emamectin residues following oral administration to rainbow trout, *Oncorhynchus mykiss*. Aquaculture 259: 6–16.

64. Jones SRM, Hargreaves NB (2009) Infection threshold to estimate *Lepeophtheirus salmonis*-associated mortality among juvenile pink salmon. Dis Aquat Organ 84: 131–137.

65. Anderson RM, May RM (1991) Infectious Diseases of Humans. Oxford: Oxford University Press.

66. Murray AG, Gillibrand PA (2006) Modelling salmon lice dispersal in Loch Torridon, Scotland. Mar Pollut Bull 53: 128–135.

67. Amundrud TL, Murray AG (2009) Modelling sea lice dispersion under varying environmental forcing in a Scottish sea loch. J Fish Dis 32: 27–44.

68. Stucchi DJ, Guo M, Foreman MGG, Czajko P, Galbraith M, et al. (2011) Modelling sea lice production and concentrations in the Broughton Archipelago, British Columbia. In: Jones SRM, Beamish RJ, editors. Salmon lice: An integrated approach to understanding parasite abundance and distribution. Oxford, UK: Wiley-Blackwell.

69. Foreman MGG, Czajko P, Stucchi DJ, Guo M (2009) A finite volume model simulation for the Broughton Archipelago, Canada. Ocean Model 30: 29–47.

70. BAMP (2011) Broughton Archipelago Monitoring Program: What's next? Available: http://bamp.ca/pages/whats_next.php. Accessed: 2013 Feb 28.

71. Horsberg TE, Jackson D, Haldorsen R, Burka J, Colleran E, et al. (2006) Sea lice resistance to chemotherapeutants, a handbook in resistance management. EU funded contract QKK2-CT-00809. Available: http://www.rothamsted.ac.uk/pie/search-EU/Handbook.pdf. Accessed 2013 Feb 28.

72. Ikonomou MG, Surridge BD (2012) Ultra-trace determination of aquaculture chemotherapeutants and degradation products in environmental matrices by LC-MS/MS. Int J Environ Anal Chem iFirst: 1–16.

73. Veldhoen N, Ikonomou MG, Buday C, Jordan J, Rehaume V, et al. (2012) Biological effects of the anti-parasitic chemotherapeutant emamectin benzoate on a non-target crustacean, the spot prawn (*Pandalus platyceros* Brandt, 1851) under laboratory conditions. Aquat Toxicol 108: 94–105.

6

Disruption of the Thyroid System by the Thyroid-Disrupting Compound Aroclor 1254 in Juvenile Japanese Flounder (*Paralichthys olivaceus*)

Yifei Dong, Hua Tian, Wei Wang, Xiaona Zhang, Jinxiang Liu, Shaoguo Ru*

Marine Life Science College, Ocean University of China, Qingdao, Shandong Province, The People's Republic of China

Abstract

Polychlorinated biphenyls (PCBs) are a group of persistent organochlorine compounds that have the potential to disrupt the homeostasis of thyroid hormones (THs) in fish, particularly juveniles. In this study, thyroid histology, plasma TH levels, and iodothyronine deiodinase (IDs, including ID_1, ID_2, and ID_3) gene expression patterns were examined in juvenile Japanese flounder (*Paralichthys olivaceus*) following 25- and 50- day waterborne exposure to environmentally relevant concentrations of a commercial PCB mixture, Aroclor 1254 (10, 100, and 1000 ng/L) with two-thirds of the test solutions renewed daily. The results showed that exposure to Aroclor 1254 for 50 d increased follicular cell height, colloid depletion, and hyperplasia. In particular, hypothyroidism, which was induced by the administration of 1000 ng/L Aroclor 1254, significantly decreased plasma TT_4, TT_3, and FT_3 levels. Profiles of the changes in mRNA expression levels of IDs were observed in the liver and kidney after 25 and 50 d PCB exposure, which might be associated with a reduction in plasma THs levels. The expression level of ID_2 mRNA in the liver exhibited a dose-dependent increase, indicating that this ID isotype might serve as sensitive and stable indicator for thyroid-disrupting chemical (TDC) exposure. Overall, our study confirmed that environmentally relevant concentrations of Aroclor 1254 cause significant thyroid disruption, with juvenile Japanese flounder being suitable candidates for use in TDC studies.

Editor: Cheryl S. Rosenfeld, University of Missouri, United States of America

Funding: This work was supported by the National Natural Science Foundation of China (31202001) www.nsfc.gov.cn, Natural Science Foundation of Shandong Province (ZR2012CQ010) www.sdnsf.gov.cn, and Marine Public Scientific Research Funding Project (2012418012) www.soa.gov.cn. The funders had no role in study design, data collection and analysis, decision to publish, or preparation of the manuscript.

Competing Interests: The authors have declared that no competing interests exist.

* Email: rusg@ouc.edu.cn

Introduction

Polychlorinated biphenyls (PCBs) have been listed as one of 21 persistent organic pollutants (POPs) under the Stockholm Convention, due to their recalcitrance to degradation and tendency to biomagnify up the food chain. PCBs are widely studied TDCs that potentially cause various abnormalities in the thyroid system of vertebrates [1–3], especially in amphibians and mammals [4]. Recently, the disturbance of fish thyroid systems by PCBs has received increasing research focus; however, the thyroidal responses of fish to PCBs has shown variable results in different studies [5]. For example, Aroclor 1242 and 1254 (commercial PCBs mixtures) lowered plasma 3,5,3′ -triiodothyronine (T_3) levels without altering plasma thyroxine (T_4) levels when fed to adult coho salmon (*Oncorhynchus kisutch*) [6]. The injection of Aroclor 1254 has been shown to increase plasma T_3 levels and delay the plasma T_4 surge commonly associated with smoltification [7]. These variable effects on thyroid hormone (THs) levels may be related to the physiological stage or age of fish used in different laboratory studies. Most of these studies preferentially used adult fish of sufficient size/age to either obtain adequate blood samples for THs measurement or the assessment of other thyroid indices, while only a few studies have used juvenile fish to assess the thyroid disrupting effects of PCBs [5].

Some researchers recommended that young developing fish should be the focus of future studies on thyroid disruption, because juvenile fish are more dependent on the regulation of THs and more sensitive to TDCs compared to adult fish [5,8]. THs have been linked to a multitude of important functions in early development of fish, such as growth, tissue differentiation, and metamorphosis [8,9]. In Japanese flounder (*Paralichthys olivaceus*), the exogenous administration of THs or elevation of endogenous T_4 levels by thyroid stimulating hormone (TSH) induces advanced metamorphosis, while thiourea (TU, an antithyroid drug) treatment delays the metamorphosis process [10,11]. Exogenous THs also induce the transition of muscle proteins, replacement of erythrocytes, skin pigmentation, and development of the gastric glands in fish [12,13]. These findings indicate that THs are fundamental for the early development and growth of fish, and that TH disruption in juvenile fish may cause growth retardation or abnormal development. Therefore, juvenile fish are assumed to be particularly susceptible to thyroid disruption.

A series of endpoints have been proposed to assess the effects of PCBs on the fish thyroid cascade, and mainly include central controlled effects and peripheral controlled effects [8]. Measurement of the central control of the thyroid cascade may be accomplished *via* thyroid histopathological analysis, in addition to

Table 1. Nucleotide sequences of primers used for real-time PCR and product sizes.

Gene	GenBank Accession No.	Primer sequence (5'–3')	Amplicon size (bp)
ID_1	AB362421	GGTGGTGGACGAAATGAATG	147
		TCCAGTAACGAACGCACCTCT	
ID_2	AB362422	GCACCAGAACTTGGAGGAGAG	142
		GCACACTCGTTCGTTAGACACA	
ID_3	AB362423	TGGCTGGAGCAGTACAGGAG	103
		TGAGGCAGAATGGGCAGA	
5S-rRNA	AB154836	CCATACCACCCTGAACAC	102
		CGGTCTCCCATCCAAGTA	

measurement of plasma total and free THs levels [14–18]. The peripheral control of the conversion of T_4 to T_3 may be assessed *via* a suite of iodothyronine deiodinase (IDs) activities in the liver or other extra-thyroid tissues [19,20]. Three ID isotypes are mainly expressed in teleosts, with these enzymes presenting different catalytic properties [21]. In particular, ID_1 exhibits both outer ring-deiodination (ORD) and inner ring-deiodination (IRD) activities; however, when combined with its preferred substrate, 3,3',5'-triiodo-L-thyronine (rT_3), this enzyme is considered to become even more involved in the degradation of THs, particularly the inactivation of rT_3 to 3,3'-diiodo-L-thyronine (3,3'-T_2). ID_2 activates the ORD pathway, by converting T_4 into T_3. ID_3 catalyses the IRD pathway, which converts T_4 and T_3 into the inactive metabolites rT_3 and 3,3'-T_2, respectively [22–24].

The Japanese flounder is an economically important species that is considered to be an ideal model organism for the study of thyroid disruption. The important roles of THs during the early stage of the development of this flatfish have been extensively demonstrated, particularly during metamorphosis [25–28]. To date, effects of PCBs on the thyroid system of the Japanese flounder remain unclear. This study aimed to obtain an integrated insight into the effects of environmentally relevant concentrations of Aroclor 1254 on the thyroid system of juvenile Japanese flounder. Changes in the development and growth of this fish species were examined, and the tissue levels of PCB congeners were measured. We anticipate that these analyses will indicate the potential suitability of using juvenile Japanese flounder as candidates for use in TDC studies.

Materials and Methods

Ethics statement

The fish were handled according to the National Institute of Health guidelines for the handling and care of experimental animals. The animal utilization protocol was approved by the Institutional Animal Care and Use Committee of the Ocean University of China. All surgery was performed under MS-222 anesthesia, and all efforts were made to minimize suffering.

Animals

Experimental trials were conducted in the marine life science college of Ocean University of China. A total of 360 juvenile Japanese flounder (80 days post hatching) were purchased from a commercial fish farm in China. The fish were raised in 240-L tanks containing 200 L of sand-filtered natural seawater (pH 8.0±0.1; 33 ppt salinity) at an ambient temperature (23±3°C). To minimize the aggressive behavior of juvenile fish, a 24-h dark photoperiod (light/dark cycle, 0/24 h) was maintained, with the tanks only being lit up 10 min before each feeding. Fish were fed a commercial flounder feed (Marubeni Nisshin feed, Chuo-Ku, Japan) 4 times a day (2% total fish weight per tank per day) between 08:00 and 20:00. Fish were allowed to acclimate to experimental conditions for 2 weeks prior to the initiation of experiments. The average wet body weight (W_T) of the fish used in the experiment was 6.21±1.77 g, and the total body length (L_T) was 8.04±1.54 cm.

Table 2. The contents of 7 tracer PCB congeners in juvenile Japanese flounder.

Test item	Control	10 ng/L	100 ng/L	1000 ng/L
PCB28	N/D	N/D	N/D	N/D
PCB52	6.90	10.44	28.53	156.15
PCB101	9.21	20.95	52.33	266.03
PCB118	4.04	12.70	40.98	209.51
PCB153	1.14	7.89	18.48	94.39
PCB138	1.77	6.99	30.18	156.30
PCB180	N/D	0.99	2.078	7.78
Total (ng/kg ww)	21.07	59.99	173.48	890.18

N/D: not detected.

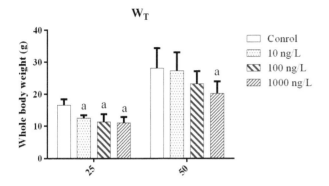

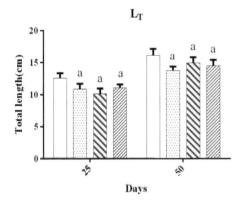

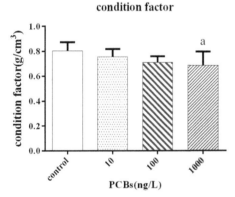

Figure 1. Effect on the total body length, body weight, and condition factor in juvenile Japanese flounder exposed to 0, 10, 100, and 1000 ng/L Aroclor 1254 for 25 and 50 d. The condition factor was calculated at the end of 50 d exposure. [a] $P<0.05$ indicates significant differences between the exposure groups and corresponding control.

Experimental design and fish sampling

Fish were randomly assigned to a control group and 3 treatment groups (size of each group $n = 90$ in each case). Juvenile Japanese flounder were exposed to Aroclor 1254 (AccuStandard Inc, NH, USA, CAS 11097-69-1) at 0 (control), 10, 100, and 1000 ng/L. Aroclor 1254 stock concentrate (1 mg/mL) was made up in ethanol (50 mg Aroclor 1254 was dissolved in 50 mL ethanol). During exposure, two-thirds of the test solutions were changed once per day, and the appropriate amount of seawater and stock solution was added to maintain the specified chemical concentrations.

Fish were deprived of food on the last day of exposure. After 25 and 50 days of exposure, all fish were anesthetised in MS-222 (Sigma, St. Louis, MO, USA), and rinsed with distilled water. The L_T and W_T of the fish in each tank ($n = 9$) were measured to calculate the condition factor ($CF = 100 \times W_T$ (g)$/L_T$ (cm)^3). Blood was collected in heparinised tubes by puncturing the caudal vein within 3 min of netting the fish. After centrifugation, plasma was collected and stored at $-80°C$ until RIA. In particular, at the 25-day sampling point, the plasma of 2–3 fish was pooled ($n = 9$). The liver and kidney tissues ($n = 9$) were isolated, frozen in liquid nitrogen, and stored at $-80°C$ until further processing. For the histology analysis, the thyroid tissues enclosed in the subpharyngeal area that were sampled at the end of 50 d exposure ($n = 9$) were fixed in formalin fixative for 24 h at 4°C, and stained with hematoxylin-eosin.

PCB contaminant analysis

The real concentrations for the 7 tracer PCB congeners in whole fish was measured with GC-MS as described in [29]. At the end of 50 d exposure, 2–3 fish (approximately 50 g in total weight) from each group were randomly sampled, lyophilized, and homogenised in 20 g anhydrous sodium sulphate, and were then placed into a Soxhlet extractor. Samples with 7 types of ^{13}C recovery internal standards (PCB 28, 52, 101, 118, 153, 138, and 180) were extracted by 350 mL hexane/dichloromethane = 1:1 (v/v) for 24 h. After primary purification by gel permeation chromatography, the extracts were then placed in acid silica (44% sulphuric acid, w/w) for further purification and component separation. Hexane (20 mL) was used for the complete elution of PCBs. The final eluate was concentrated to 100 μL under nitrogen and then transferred to a GC vial with ^{13}C-PCB 202 inlet internal standards. The PCB congeners were analysed using an Agilent 6890N/5973i GC-MS system (Agilent Technologies Inc., Palp Alto, USA). The GC-MS analytical parameters have been conducted by referring to Environmental Protection Series: Reference method for the analysis of polychlorinated biphenyls (EPS 1/RM/31, Canada).

Thyroid histological processing

All histopathological endpoints were assayed, as described in [16], with minor modifications. Serial sections were examined under a light microscope until 6–14 follicles/fish were found and photographed. Follicular cell height on the pictures was quantified using Image-Pro plus (version 6.0.0.260). Follicular cell height was determined by obtaining 6 measurements at regular intervals along the follicle perimeter (i.e. 36–84 follicular cell height measurements for each fish and 324–756 measurements for each treatment). A grading system was applied for the hyperplasia evaluation: Grade 1, focal hyperplasia; Grade 2, thyroid follicular cell with less than 50% hyperplasia; and Grade 3, thyroid follicular cell with more than 50% hyperplasia. The average score of 9 fish from each treatment was used, which was based on the sum of the grade of each fish. The sum of the number of colloid deletions (per 10 follicles) was calculated (colloid deletion follicle/10 follicles).

RNA isolation and quantitative RT-PCR

The procedures for RNA extraction and gene expression analysis were performed as previously described by [31]. In brief, total RNA was isolated from the liver and kidney using TRIzol reagent (Invitrogen, Carlsbad, CA, USA) following the manufacturer's instructions. Equal amounts of RNA (1 μg) were reverse-transcribed into cDNA using a PrimeScript RT reagent kit (Takara Bio Inc., Shiga, Japan). Primers were designed for the specific amplification of ID_1, ID_2, ID_3, and 5S-rRNA (an internal control) according to the sequences published in GenBank (Table 1).

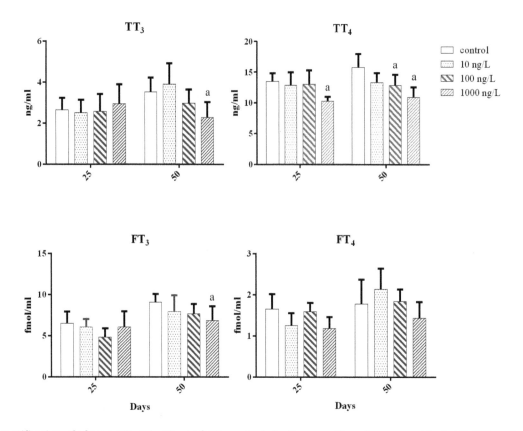

Figure 2. Quantification of plasma TT$_3$, TT$_4$, FT$_3$, and FT$_4$ contents in Japanese flounder exposed to 0, 10, 100, and 1000 ng/L of Aroclor 1254 for 25 and 50 d. [a] $P<0.05$ indicates significant differences between the exposure groups and corresponding control.

All reactions were run on a Eppendorf MasterCycler ep *RealPlex*[4] (Eppendorf, Wesseling-Berz-dorf, Germany). Parallel PCR reactions were conducted to amplify the target gene and 5S-rRNA. Real-time PCR was performed in 20 µL reaction mixtures containing 1× SYBR *Premix Ex Taq* (Takara Bio Inc., Shiga, Japan), 0.4 µM for each primer, 0.4 µL of ROX Reference Dye (Takara Bio Inc., Shiga, Japan), and 4 µL of first-strand cDNA (template). The thermal profile was 95°C for 30 s followed by 40 cycles of 95°C for 5 s and 60°C for 30 s. To ensure that a single product was amplified, melting curve analysis was performed on the PCR products at the end of each PCR run. In addition, 2% agarose gel electrophoresis of the PCR products was performed to confirm the presence of single amplicons of the correct predicted size (not shown). 5S-rRNA transcripts were used as housekeeping genes to standardize the results and to eliminate variations in mRNA and cDNA quantity and quality. 5S-rRNA levels were not affected by any of the experimental conditions in the study. The target gene mRNA abundance in each sample, relative to the abundance of 5S-rRNA, was calculated by the formula $2^{-\Delta\Delta Ct}$ and plotted on a logarithmic scale [31].

Hormone assay

Muscular TT$_3$, TT$_4$, FT$_3$, and FT$_4$ concentrations were measured by radio immunoassay (RIA) (Beijing North Institute of Biological Technology, Beijing, China) according to the manufacturer's instructions. The assay detection limits were 0.05 ng/mL for TT$_3$, 2 ng/mL for TT$_4$, 0.5 fmol/mL for FT$_3$, and 1 fmol/mL for FT$_4$. The inter- and intra-assay coefficients of variation for all the stated hormones were <10% and <15%, respectively.

Statistics

All data are presented as the mean ± standard deviation. Data normality was verified using the Kolmogorov-Smirnov test [32], and homogeneity of variance was checked by Levene's test. If the data failed to pass the test, a logarithmic transformation of the data was performed and retested. Significant differences were assessed between each treatment and the control using one-way analysis of variance (ANOVA), followed by Tukey's multiple comparisons test. $P<0.05$ was considered to be statistically significantly different. All statistical tests were conducted using GraphPad PRISM (Version 6.00) software.

Results

PCB concentrations in Japanese flounder juvenile

The concentrations of 7 tracer PCB congeners in juvenile Japanese flounder are shown in Table 2. A concentration-dependent bioconcentration of Aroclor 1254 was measured in the whole body of all exposure groups. In the 1000 ng/L treatment, the total concentration of measured PCB congeners (including PCB28, PCB52, PCB101, PCB118, PCB153, PCB138, and PCB180) reached 890.18 ng/g ww.

Effects of Aroclor 1254 on the growth of juvenile Japanese flounder

During exposure, mortality rates were below 10% in all groups. As shown in Fig. 1, after 25 days of exposure, Aroclor 1254 significantly reduced W$_T$ and L$_T$ in all treatments. After 50 days of exposure, 10 ng/L and 100 ng/L Aroclor 1254 did not affect W$_T$, but

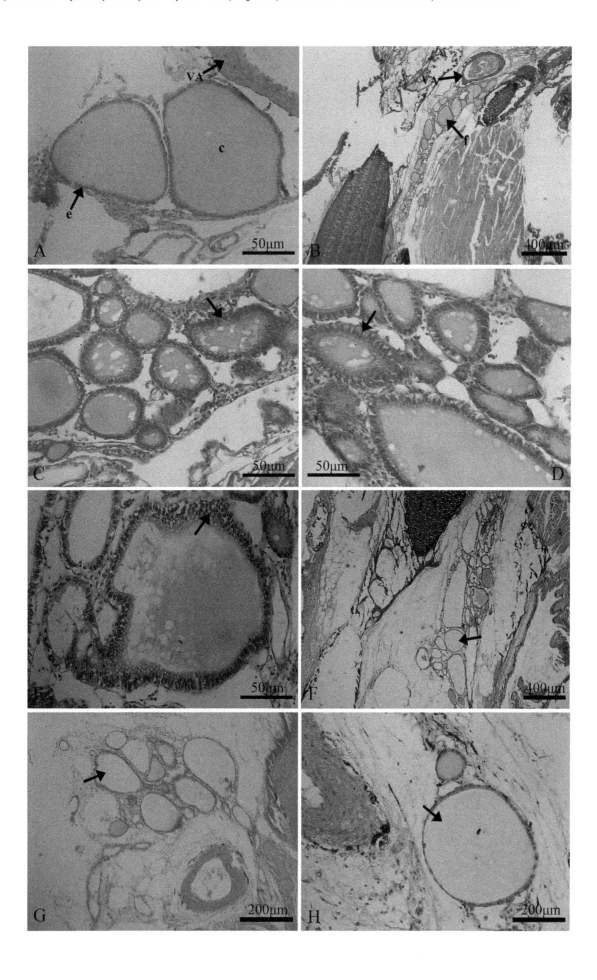

Figure 3. Histological structure of thyroid follicles in juvenile Japanese flounder exposed to 0, 10, 100, and 1000 ng/L Aroclor 1254 for 50 d. (A) and (B) control fish presenting ovoid follicles of variable sizes filled with colloid and lined with squamous follicle cells; (C) and (D) significantly increased epithelial cell height with a little colloid depletion in the lumen after exposure to 100 ng/L Aroclor 1254. (E) Focal hyperplasia in fish exposed to 1000 ng/L. (F) and (G) colloid depletion in fish exposed to 1000 ng/L. (H) Dispersed and reticular colloid in fish exposed to 1000 ng/L. VA = ventral aorta, f = thyroid follicle, c = colloid, and e = thyroid follicle epithelial cell.

significantly inhibited L_T. Exposure to 1000 ng/L Aroclor 1254 for 50 days significantly reduced W_T, L_T, and CF, relative to the control.

Effects of Aroclor 1254 on plasma TT_4, TT_3, FT_4, and FT_3 levels

The effects of Aroclor 1254 on plasma THs levels are shown in Fig. 2. In flounder exposed to different concentrations of Aroclor 1254 for 25 days, the TT_3, FT_3, and FT_4 levels in the plasma were not significantly altered by any of the treatments, whereas plasma TT_4 levels significantly decreased in the 1000 ng/L group. After 50 days of Aroclor 1254 exposure, both plasma TT_3 and FT_3 levels significantly decreased in the 1000 ng/L group, with plasma TT_4 levels showing a dose-dependent decrease, which was significant at concentrations of 100 and 1000 ng/L, while plasma FT_4 levels remained unaltered.

Effects of Aroclor 1254 on thyroid histopathology

The control fish presented oval thyroid follicles of variable sizes that were filled with colloid. In addition, the follicles were line with a single layer of cuboidal to flat follicle epithelial cells (Fig. 3A, B). Representative histopathological abnormalities in Japanese flounder exposed to different concentrations of Aroclor 1254 for 50 days are shown in Fig. 3C–H, including increased epithelial cell height (Fig. 3C, D), hyperplasia (Fig. 3E), and colloid depletion (Fig. 3F, G). Compared to the control group, the colloid observed in the 100 ng/L and 1000 ng/L groups was foamy in appearance, and colloid density decreased (Fig. 3H). For the quantitative analyses, significantly increased levels of follicular epithelial cell height, hyperplasia, and colloid depletion were observed in the 100 and 1000 ng/L Aroclor 1254 treatments (Fig. 4).

Effects of Aroclor 1254 on ID_1, ID_2 and ID_3 mRNA expression in the liver and kidney

As shown in Fig. 5, after 25 days of exposure to Aroclor 1254, ID_1 mRNA levels in the kidney were significantly higher in the 10 and 100 ng/L grouts; however, no significant difference was observed for the liver in any of the treatments. The significant up-regulation of ID_2 and ID_3 mRNA levels was observed in both the kidney and liver of all treatments. In juvenile Japanese flounder exposed to Aroclor 1254 for 50 days, significantly higher ID_1 mRNA levels were obtained in the kidney and liver of the 100 ng/L and 10 ng/L groups, respectively. The transcription of ID_2 mRNA in the kidney was significantly stimulated on exposure to 100 and 1000 ng/L Aroclor 1254, which were significantly upregulated in the liver for all treatments. The transcription levels of ID_3 in the kidney and the liver were not significantly altered by any Aroclor 1254 treatment.

Discussion

Our results showed that exposure to Aroclor 1254 significantly decreased plasma TT_4 and TT_3 levels (Fig. 2). However, interpretation of PCBs on the fish thyroid system is exceedingly complex, and does not appear to elicit consistent, detectable plasma TH responses (Table 3). At least three categories of factors have to be considered: 1) test-species variable, 2) the variable composition of PCB mixtures, and 3) the distinction between exposure and effect due in part to thyroid compensation [5].

In this study, Aroclor 1254 exposure inhibited the L_T, W_T, and CF of juvenile Japanese flounder, which probably led to growth retardation. Crane et al. [15] found that ammonium perchlorate reduces plasma T_4 levels, which inhibited the development of fathead minnow (*Pimephales promelas*) larvae. Schmidt et al. [18] reported that exposure of zebrafish larvae to potassium-perchlorate caused a significant decrease in both plasma T_4 levels and CF. The current study also found that Aroclor 1254 exposure causes plasma T_3 and T_4 levels to decline. Because THs are important in the development and growth of teleosts, particularly during the early life stages, this type of thyroid disruption might inhibit the growth of juvenile Japanese flounder.

However, exposure to PCBs produced different results in adult and juvenile fish. For instance, the study by Schnitzler et al. [29] showed that one PCB mixture induced muscle T_4 levels to decrease in adult sea bass (*Dicentrarchus labrax*), without affecting body length, body weight, or specific growth rates. Iwanowicz et al. [33] reported that the intraperitoneal (*i. p.*) injection of 5 mg/kg Aroclor 1248 caused plasma T_3 levels to decrease in the brown bullhead (*Ameiurus nebulosus*), but had no significant effects on plasma T_4 levels or CF. Following exposure to PCB 126 by *i. p.* injection lower plasma T_4 concentrations was observed in adult lake trout (*Salvelinus namaycush*), whereas it had no effect on fish growth or condition [34]. In adult fish, abundant stores of THs have been found in muscles and other tissues, in addition to the plasma pool, thyroid tissues [35]. These TH stores in extra-thyroidal tissues might be released into the bloodstream or peripheral tissues to compensate thyroid disruption induced by exposure to exogenous compounds. Brown et al. [36] found that muscle T_3 and T_4 contents rapidly reduced in rainbow trout exposed to the PCB 126, with few changes in the histology of thyroid follicles and growth rate. This finding indicates that adult fish have a mechanism to compensate for the thyroid system, enabling them to balance available TH content in peripheral tissues, which does not affect growth. In contrast, the peripheral tissues of juveniles contained relatively low TH levels; therefore, TH deficiency in juveniles might be more likely to trigger a negative feedback regulation compared to adult fish, inducing a series of cascading effects that involve the hypothalamus-pituitary-thyroid (HPT) axis to maintain TH homeostasis. Thus, thyroid tissue might stimulate TH synthesis in juvenile Japanese flounder exposed to Aroclor 1254, based on the observed increase in epithelial cell height, hyperplasia of thyroid follicular epithelial cells, and colloid deletion in the current study. This phenomenon might, to some extent, be attributed to the feedback response to Aroclor 1254 within the thyroid cascade.

The severity of colloid depletion and epithelial cell height are routinely employed markers for identifying thyroid disruption [16]. Crane et al. [15] pointed out that colloid depletion indicates serious injuries, close to the collapse of follicles. In the present study, juvenile Japanese flounder exposed to 100 and 1000 ng/L Aroclor 1254 had significantly greater thyroid follicular epithelial cell height, which reduced colloid area. Many irregularly shaped follicles, some without colloids, were observed, particularly in the highest exposure group. These degenerative changes of the thyroid

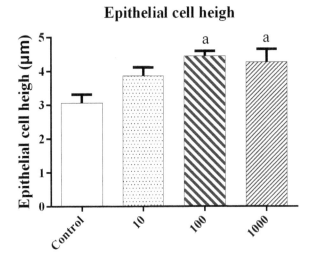

Epithelial cell heigh

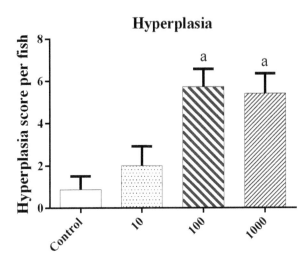

Hyperplasia

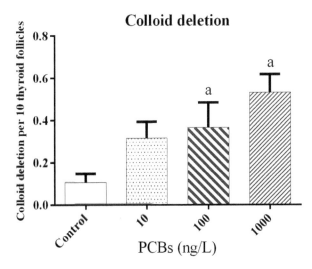

Colloid deletion

Figure 4. Measurement of epithelial cell height, colloid deletion, and hyperplasia of thyroid follicle in juvenile Japanese flounder exposed to Aroclor 1254 for 50 d. [a] $P<0.05$ indicates significant differences between the exposure groups and corresponding control.

tissues might cause hypothyroidism in juvenile Japanese flounder, preventing them from balance the decrease in TT_4 baselines in the 2 Aroclor groups with the highest concentrations (100 and 1000 ng/L) after 50 d exposure.

In particular, changes in thyroid tissue histology caused by Aroclor 1254 exposure were similar to those induced by perchlorate. Perchlorate blocks the iodine uptake of thyroid follicles by competitively inhibiting iodide and sodium/iodine transport proteins from combining; thereby, hindering the synthesis of THs [37]. Consequently, the decline in TH levels might stimulate TSH secretion from the pituitary through the feedback pathway, and eventually cause compensatory hypertrophy, hyperplasia, and colloid reduction of thyroid follicular cells [14,16,18]. In contrast, some inorganic chemicals, like Cd^{2+}, directly damage thyroid follicles by inducing lipid peroxidation; thus, affecting TH synthesis. Therefore, the toxicity mechanism of Aroclor 1254 on thyroid follicles might be similar to that of perchlorate, rather than the direct effect of heavy metals, such as Cd^{2+}. In other words, Aroclor 1254 probably causes plasma TH levels to decrease in juvenile Japanese flounder; thereby, inducing the compensatory hypertrophy and hyperplasia of thyroid follicular cells through negative feedback pathways, to promote TH synthesis.

Previous studies have shown that deiodinase in fish is sensitive to environmental contaminants, such as metals, polychlorinated biphenyls, and pesticides [38–40]. Van der Geyten et al. [41] demonstrated that changes in hepatic ID_1 and ID_2 activities tend to be consistent with that of their mRNA levels, indicating pre-translated regulation, by which deiodinase mRNA levels coincide with deiodinase enzyme activities. In addition, Picard-Aitken et al. [42] suggested that deiodination gene expression could be used as sensitive biomarkers to indicate thyroid disruption in fish on exposure to environmental chemicals. In the present study, the gene expression of IDs in juvenile Japanese flounder was sensitive to exposure to Aroclor 1254. After 25 and 50 d exposure, Aroclor 1254 stimulated the transcription of ID_2 mRNA in the kidney and liver, which would result in more T_4 being converted into T_3. Another study also found that exposure of sea bass to a mixture of Aroclor 1254 and 1248 led to a significant increase in ID_2 activities [29]. ID_2 mRNA expression tended to be the most sensitive and stable indicator for thyroid disruption in the present study, because it showed a dose-dependent increase in all treatment groups after both 25 and 50 days exposure, especially in the liver. However, it is difficult to distinguish whether Aroclor 1254 has a direct or indirect disrupting effect on the thyroid system by triggering compensatory mechanisms within the thyroid system; consequently, it is difficult to explain how the thyroid status of juvenile Japanese flounder exposed to Aroclor 1254 is altered by only a few indicators. Blanton and Specker [8] suggested that the actions of certain xenobiotics at different levels of the fish thyroid cascade could not be independently monitored by any biomarker. However, ID_2 represents one important indicator for interpreting disruption to the thyroid cascade in fish exposed to environmental contaminants.

After 25 d exposure, 10 and 100 ng/L Aroclor 1254 caused ID_1 and ID_3 mRNA expression levels to increase, especially in the kidney. This response would accelerate the metabolism of T_3, which helps maintain plasma THs homeostasis. At the highest dose, the mRNA expression of ID_2 in the kidney and liver was significantly upregulated, while the expression of ID_1 showed no significant change. This result also indicates the presence of a compensatory response to decreased plasma TT_4 levels, to maintain stable plasma TT_3 levels; otherwise, the increased mRNA expression of ID_3 in the kidney and liver might aggravate

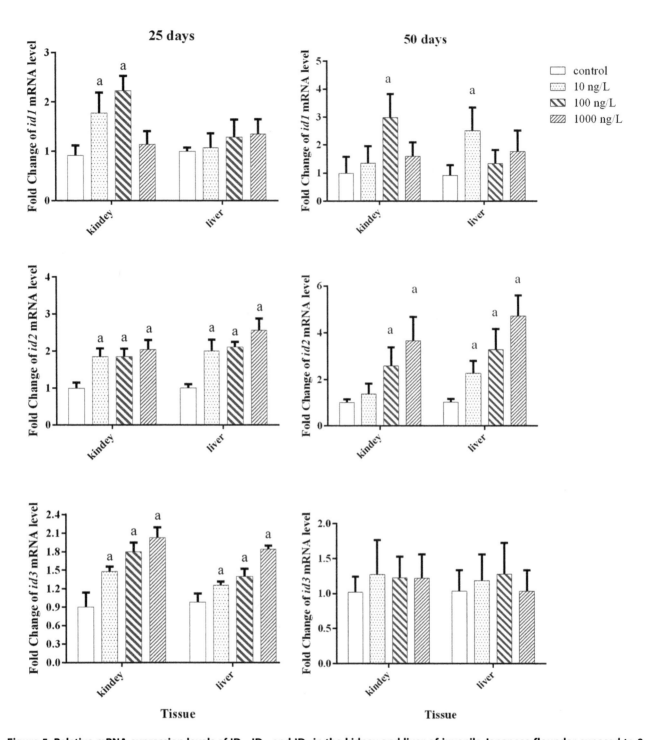

Figure 5. Relative mRNA expression levels of ID₁, ID₂, and ID₃ in the kidney and liver of juvenile Japanese flounder exposed to 0, 10, 100, and 1000 ng/L of Aroclor 1254 for 25 and 50 d. Fold change (y-axis) represents the expression of the target gene mRNA relative to that of the control group (equals 1 by definition). ᵃ P<0.05 indicates significant differences between the exposure groups and corresponding control.

the reduction in plasma TT₄. In studies of tilapia, van der Geyten et al. [43] found that ID₂ activity in the liver and ID₃ activity in the gill decreased with declining T₃ concentrations, which is responsible for balancing the reduction in T₃. Schnitzler et al. [29] suggested that PCB-induced changes in deiodinase activity offset the decline in plasma T₃ levels. Adams et al. [44] suggested that elevated T₄ ORD activity serves as a homeostatic adjustment to offset increased systemic T₃ clearance. After 50 d exposure, the decrease in plasma TT₃ levels at the highest dose was mostly due

to hypothyroidism, which caused a drop in thyroidal T₄ production and secretion; thus, exceeding the regulation ability of IDs, and also resulting in lower FT₃ levels.

Some authors have found that a change in plasma TH levels alters the ID₃ expression. For example, Higgs and Eales [45] found that a decrease in fish T₄ levels leads to a decrease in the metabolic clearance level of T₄; in other words, a decrease in the ID₃ level. A study by Van der Geyten et al. [41] showed that a decrease in the TT₄ and TT₃ levels of tilapia with thyroid

Table 3. The effects of PCBs on fish plasma thyroid hormone homostasis.

Type of PCBs	Dose	Exposure days	Species	Ages of fish	T_4	T_3	Reference
PCB 126	25 µg/kg	210	Salvelinus namaycush	Adult	↑	–	[34]
PCB 126	500 µg/kg	7	Hippoglossoides platessoides	Adult	–	–	[44]
PCB 77	1000 µg/kg	90	Thymallus arcticus	Adult	–	–	[49]
PCB 77	500 µg/kg	7	Hippoglossoides platessoides	Adult	↑	–	[44]
Clophen A50	500 mg/kg	10	Platichthys flesus	Adult	–	–	[50]
Aroclor 1254	150 µg/kg	42	Oncorhynchus kisutch	Adult	→	↑	[6]
Aroclor 1254	1 mg/kg	30	Micropogonias undulatus	Adult	↑	→	[51]
Aroclor 1254	0.5 µg/g	35	Oreochromis niloticus	Adult	–	–	[38]
Aroclor 1248	5 mg/kg	21	Ameiurus nebulosus	Adult	–	→	[33]
1254:1248	50 µg/g	84	Oncorhynchus kisutch	Adult	–	→	[5]
1254:1248	10 µg/g	120	Dicentrarchus labrax	Adult	→	→	[30]

↑, increase; ↓, decrease; –, no effect.

dysfunction caused by methimazole exposure caused ID_1 and ID_2 levels to increase and ID_3 levels to decrease. However, the current study found that a decrease in plasma TH levels did not influence the mRNA expression of ID_3 after 50 days of exposure. Coimbra et al. [38] found that at 21 and 35 days after tilapia were exposed to Aroclor 1254, TT_3 and TT_4 levels showed no significant changes, whereas ID_3 activity levels in the liver significantly increased, while the activity of ID3 increased in the gill after 21 days of exposure. The exact reason for this phenomenon requires further study.

At present, higher exposure concentrations of PCBs are often used to investigate their thyroid-disrupting effects on adult fish (Table 3). Of note, PCB concentrations detected in the environment are far lower than those used in these exposure experiments. For example, the PCBs content of the surface water and sediment in the Minjiang River Estuary, China, are 985 ng/L and 34.39 µg/kg on average, respectively [46]. The total concentration of PCBs ranged from 2.33 µg/kg to 44 µg/kg in the marine sediments in Barcelona, Spain [47], and 10 µg/kg to 899 µg/kg in the surface sediments of Naples Harbour, Italy [48]. Adult sea bass fed with the equivalent of actual environmental concentrations of the mixture of Aroclor 1260 and 1254 only showed reduced muscle T_3 levels, with no significant changes in muscle T_4 levels and thyroid histology; however, exposure to the same contaminants at concentrations 10 times above actual environmental concentrations led to a decrease in both T_3 and T_4 levels in muscles, and caused follicular degeneration [29]. This study found that even environmentally relevant concentrations of Aroclor 1254 caused significant disruption to the thyroid system of flounder juveniles, including changes in thyroid histopathology, altered plasma TH levels, and modulation in the expression levels of IDs mRNA in the liver and kidney. This result supported the hypothesis that juvenile fish are more sensitive to PCBs compared to adult fish, making them suitable candidate animal models for studying TDCs.

Many TDCs, such as sodium perchlorate, have been reported to affect the early growth and development of teleosts [15,16,18]. Mechanisms underlying the effects of PCBs on the early life stages of fish development *via* their thyroid disrupting abilities should be investigated in future studies, not only to delineate the disrupting effects of PCBs at individual and ecological levels, but also to establish some links between the macroscopic effects and the microscopic mechanisms for a more comprehensive ecological risk assessment of these pollutants. In particular, flatfish species, including Japanese flounder, experience a unique and critical process of metamorphosis during development, when the larvae shift from a planktonic to a benthic mode of life, with this process being primarily controlled by the thyroid system. Therefore, the larvae of Japanese flounder may represent an excellent model organism for investigating the effects of PCBs on the thyroid system and fish development in future studies.

Acknowledgments

The authors are grateful to all members in the lab for their help.

Author Contributions

Conceived and designed the experiments: YD SR HT. Performed the experiments: YD JL. Analyzed the data: YD XZ. Contributed reagents/ materials/analysis tools: JL. Contributed to the writing of the manuscript: HT XZ WW SR. Obtained permission for use of fish fertilized eggs: JL.

References

1. Hansen LG (1998) Stepping backward to improve assessment of PCB congener toxicities. Environmental Health Perspectives 106: 171–189.

2. Safe SH (1994) Polychlorinated biphenyls (PCBs): environmental impact, biochemical and toxic responses, and implications for risk assessment. CRC Critical Reviews in Toxicology 24: 87–149.

3. UNEP website. Available: http://www.chem.unep.ch/Legal/ECOSOC/ UNEP%20Consolidated%20List%2010%20May%202010.pdf. Accessed 2014 July 15.

4. Jugan M-L, Levi Y, Blondeau J-P (2010) Endocrine disruptors and thyroid hormone physiology. Biochemical pharmacology 79: 939–947.

5. Brown SB, Adams BA, Cyr DG, Eales JG (2004) Contaminant effects on the teleost fish thyroid. Environmental Toxicology and Chemistry 23: 1680–1701.

6. Leatherland J, Sonstegard R (1978) Lowering of serum thyroxine and triiodothyronine levels in yearling coho salmon, *Oncorhynchus kisutch*, by dietary mirex and PCBs. Journal of the Fisheries Board of Canada 35: 1285–1289.

7. Folmar LC, Dickhoff WW, Zaugg WS, Hodgins HO (1982) The effects of aroclor 1254 and no. 2 fuel oil on smoltification and sea-water adaptation of coho salmon (*Oncorhynchus kisutch*). Aquatic toxicology 2: 291–299.

8. Blanton ML, Specker JL (2007) The hypothalamic-pituitary-thyroid (HPT) axis in fish and its role in fish development and reproduction. Crit Rev Toxicol 37: 97–115.

9. Cyr DG, Eales J (1996) Interrelationships between thyroidal and reproductive endocrine systems in fish. Reviews in Fish Biology and Fisheries 6: 165–200.

10. Inui Y, Tagawa M, Miwa S, Hirano T (1989) Effects of bovine TSH on the tissue thyroxine level and metamorphosis in prometamorphic flounder larvae. General and comparative endocrinology 74: 406–410.

11. Okada N, Morita T, Tanaka M, Tagawa M (2005) Thyroid hormone deficiency in abnormal larvae of the Japanese flounder *Paralichthys olivaceus*. Fisheries Science 71: 107–114.

12. Power D, Llewellyn L, Faustino M, Nowell M, Björnsson BT, et al. (2001) Thyroid hormones in growth and development of fish. Comparative Biochemistry and Physiology Part C: Toxicology & Pharmacology 130: 447–459.

13. Yamano K (2005) The role of thyroid hormone in fish development with reference to aquaculture. Japan Agricultural Research Quarterly 39: 161.

14. Bradford CM, Rinchard J, Carr JA, Theodorakis C (2005) Perchlorate affects thyroid function in eastern mosquitofish (*Gambusia holbrooki*) at environmentally relevant concentrations. Environmental science & technology 39: 5190–5195.

15. Crane HM, Pickford DB, Hutchinson TH, Brown JA (2005) Effects of ammonium perchlorate on thyroid function in developing fathead minnows, *Pimephales promelas*. Environmental health perspectives 113: 396.

16. Liu FJ, Wang JS, Theodorakis CW (2006) Thyrotoxicity of sodium arsenate, sodium perchlorate, and their mixture in zebrafish *Danio rerio*. Environmental science & technology 40: 3429–3436.

17. Mukhi S, Patiño R (2007) Effects of prolonged exposure to perchlorate on thyroid and reproductive function in zebrafish. Toxicological sciences 96: 246–254.

18. Schmidt F, Schnurr S, Wolf R, Braunbeck T (2012) Effects of the anti-thyroidal compound potassium-perchlorate on the thyroid system of the zebrafish. Aquat Toxicol 109: 47–58.

19. Eales J, Brown S (1993) Measurement and regulation of thyroidal status in teleost fish. Reviews in Fish Biology and Fisheries 3: 299–347.

20. Eales JG, Brown SB, Cyr DG, Adams BA, Finnson KR (1999) Deiodination as an index of chemical disruption of thyroid hormone homeostasis and thyroidal status in fish. ASTM SPECIAL TECHNICAL PUBLICATION 1364: 136–164.

21. Orozco A, Valverde-R C (2005) Thyroid hormone deiodination in fish. Thyroid 15: 799–813.

22. Köhrle J (1999) Local activation and inactivation of thyroid hormones: the deiodinase family. Molecular and cellular endocrinology 151: 103–119.

23. Moreno M, Berry MJ, Horst C, Thoma R, Goglia F, et al. (1994) Activation and inactivation of thyroid hormone by type I iodothyronine deiodinase. FEBS letters 344: 143–146.

24. Van der Geyten S, Byamungu N, Reyns G, Kühn E, Darras V (2005) Iodothyronine deiodinases and the control of plasma and tissue thyroid hormone levels in hyperthyroid tilapia (*Oreochromis niloticus*). Journal of endocrinology 184: 467–479.

25. Inui Y, Miwa S (1985) Thyroid hormone induces metamorphosis of flounder larvae. General and comparative endocrinology 60: 450–454.

26. Miwa S, Inui Y (1991) Thyroid hormone stimulates the shift of erythrocyte populations during metamorphosis of the flounder. Journal of Experimental Zoology 259: 222–228.

27. Miwa S, Yamano K, Inui Y (1992) Thyroid hormone stimulates gastric development in flounder larvae during metamorphosis. Journal of Experimental Zoology 261: 424–430.

28. Yamano K, Miwa S, Obinata T, Inui Y (1991) Thyroid hormone regulates developmental changes in muscle during flounder metamorphosis. General and comparative endocrinology 81: 464–472.

29. Schnitzler JG, Celis N, Klaren PH, Blust R, Dirtu AC, et al. (2011) Thyroid dysfunction in sea bass (*Dicentrarchus labrax*): underlying mechanisms and effects of polychlorinated biphenyls on thyroid hormone physiology and metabolism. Aquat Toxicol 105: 438–447.

30. Tian H, Ru S, Bing X, Wang W (2010) Effects of monocrotophos on the reproductive axis in the male goldfish (*Carassius auratus*): Potential mechanisms underlying vitellogenin induction. Aquatic toxicology 98: 67–73.

31. Livak KJ, Schmittgen TD (2001) Analysis of Relative Gene Expression Data Using Real-Time Quantitative PCR and the $2^{-\Delta\Delta Ct}$ Method. methods 25: 402–408.

32. Drezner Z, Turel O, Zerom D (2010) A Modified Kolmogorov-Smirnov Test for Normality. Taylor & Francis 39: 693–704.

33. Iwanowicz LR, Blazer VS, McCormick SD, Vanveld PA, Ottinger CA (2009) Aroclor 1248 exposure leads to immunomodulation, decreased disease resistance and endocrine disruption in the brown bullhead, *Ameiurus nebulosus*. Aquat Toxicol 93: 70–82.

34. Brown SB, Evans RE, Vandenbyllardt L, Finnson KW, Palace VP, et al. (2004) Altered thyroid status in lake trout (*Salvelinus namaycush*) exposed to co-planar 3,3′,4,4′,5-pentachlorobiphenyl. Aquat Toxicol 67: 75–85.

35. Fok P, Eales J, Brown S (1990) Determination of 3, 5, 3′-triiodo-L-thyronine (T₃) levels in tissues of rainbow trout (*Salmo gairdneri*) and the effect of low ambient pH and aluminum. Fish physiology and biochemistry 8: 281–290.

36. Brown SB, Fisk AT, Brown M, Villella M, Muir DC, et al. (2002) Dietary accumulation and biochemical responses of juvenile rainbow trout (*Oncorhynchus mykiss*) to 3, 3′, 4, 4′, 5-pentachlorobiphenyl (PCB 126). Aquatic toxicology 59: 139–152.

37. Leung AM, Pearce EN, Braverman LE (2010) Perchlorate, iodine and the thyroid. Best Practice & Research Clinical Endocrinology & Metabolism 24: 133–141.

38. Coimbra AM, Reis-Henriques MA, Darras VM (2005) Circulating thyroid hormone levels and iodothyronine deiodinase activities in Nile tilapia (*Oreochromis niloticus*) following dietary exposure to Endosulfan and Aroclor 1254. Comp Biochem Physiol C Toxicol Pharmacol 141: 8–14.

39. Li W, Zha J, Li Z, Yang L, Wang Z (2009) Effects of exposure to acetochlor on the expression of thyroid hormone related genes in larval and adult rare minnow (*Gobiocypris rarus*). Aquatic Toxicology 94: 87–93.

40. Zhang X, Tian H, Wang W, Ru S (2013) Exposure to monocrotophos pesticide causes disruption of the hypothalamic-pituitary-thyroid axis in adult male goldfish (*Carassius auratus*). Gen Comp Endocrinol 193: 158–166.

41. Van der Geyten S, Toguyeni A, Baroiller JF, Fauconneau B, Fostier A, et al. (2001) Hypothyroidism induces type I iodothyronine deiodinase expression in tilapia liver. Gen Comp Endocrinol 124: 333–342.

42. Picard-Aitken M, Fournier H, Pariseau R, Marcogliese DJ, Cyr DG (2007) Thyroid disruption in walleye (*Sander vitreus*) exposed to environmental contaminants: Cloning and use of iodothyronine deiodinases as molecular biomarkers. Aquatic toxicology 83: 200–211.

43. Van der Geyten S, Mol K, Pluymers W, Kühn E, Darras V (1998) Changes in plasma T3 during fasting/refeeding in tilapia (*Oreochromis niloticus*) are mainly regulated through changes in hepatic type II iodothyronine deiodinase. Fish Physiology and Biochemistry 19: 135–143.

44. Adams BA, Cyr DG, Eales JG (2000) Thyroid hormone deiodination in tissues of American plaice, *Hippoglossoides platessoides*: characterization and short-term responses to polychlorinated biphenyls (PCBs) 77 and 126. Comparative Biochemistry and Physiology Part C: Pharmacology, Toxicology and Endocrinology 127: 367–378.

45. Higgs DA, Eales J (1977) Influence of food deprivation on radioiodothyronine and radioiodide kinetics in yearling brook trout, *Salvelinus fontinalis* (Mitchill), with a consideration of the extent of l-thyroxine conversion to 3, 5, 3′-triiodo-L-thyronine. General and comparative endocrinology 32: 29–40.

46. Zhang Z, Hong H, Zhou J, Huang J, Yu G (2003) Fate and assessment of persistent organic pollutants in water and sediment from Minjiang River Estuary, Southeast China. Chemosphere 52: 1423–1430.

47. Castells P, Parera J, Santos F, Galceran M (2008) Occurrence of polychlorinated naphthalenes, polychlorinated biphenyls and short-chain chlorinated paraffins in marine sediments from Barcelona (Spain). Chemosphere 70: 1552–1562.

48. Sprovieri M, Feo ML, Prevedello L, Manta DS, Sammartino S, et al. (2007) Heavy metals, polycyclic aromatic hydrocarbons and polychlorinated biphenyls in surface sediments of the Naples harbour (southern Italy). Chemosphere 67: 998–1009.

49. Palace VP, Allen-Gil SM, Brown SB, Evans RE, Metner DA, et al. (2001) Vitamin and thyroid status in arctic grayling (*Thymallus arcticus*) exposed to doses of 3, 3′, 4, 4′-tetrachlorobiphenyl that induce the phase I enzyme system. Chemosphere 45: 185–193.

50. Besselink H, Van Beusekom S, Roex E, Vethaak A, Koeman J, et al. (1996) Low hepatic 7-ethoxyresorufin-O-deethylase (EROD) activity and minor alterations in retinoid and thyroid hormone levels in flounder (*Platichthys flesus*) exposed to the polychlorinated biphenyl (PCB) mixture, Clophen A50. Environmental Pollution 92: 267–274.

51. LeRoy KD, Thomas P, Khan IA (2006) Thyroid hormone status of Atlantic croaker exposed to Aroclor 1254 and selected PCB congeners. Comp Biochem Phyiol C Toxicol Pharmacol 144: 263–271.

7

Screening for Viral Hemorrhagic Septicemia Virus in Marine Fish along the Norwegian Coastal Line

Nina Sandlund[1]*, Britt Gjerset[3], Øivind Bergh[2], Ingebjørg Modahl[3], Niels Jørgen Olesen[4], Renate Johansen[3]

1 Research group Disease and Pathogen transmission, Institute of Marine Research, Bergen, Norway, 2 Research group Oceanography and climate, Institute of Marine Research, Bergen, Norway, 3 Section of Virology, National Veterinary Institute, Oslo, Norway, 4 Section of Virology, Technical University of Denmark, Frederiksberg C, Denmark

Abstract

Viral hemorrhagic septicemia virus (VHSV) infects a wide range of marine fish species. To study the occurrence of VHSV in wild marine fish populations in Norwegian coastal waters and fjord systems a total of 1927 fish from 39 different species were sampled through 5 research cruises conducted in 2009 to 2011. In total, VHSV was detected by rRT-PCR in twelve samples originating from Atlantic herring (*Clupea harengus*), haddock (*Melanogrammus aeglefinus*), whiting (*Merlangius merlangus*) and silvery pout (*Gadiculus argenteus*). All fish tested positive in gills while four herring and one silvery pout also tested positive in internal organs. Successful virus isolation in cell culture was only obtained from one pooled Atlantic herring sample which shows that today's PCR methodology have a much higher sensitivity than cell culture for detection of VHSV. Sequencing revealed that the positive samples belonged to VHSV genotype Ib and phylogenetic analysis shows that the isolate from Atlantic herring and silvery pout are closely related. All positive fish were sampled in the same area in the northern county of Finnmark. This is the first detection of VHSV in Atlantic herring this far north, and to our knowledge the first detection of VHSV in silvery pout. However, low prevalence of VHSV genotype Ib in Atlantic herring and other wild marine fish are well known in other parts of Europe. Earlier there have been a few reports of disease outbreaks in farmed rainbow trout with VHSV of genotype Ib, and our results show that there is a possibility of transfer of VHSV from wild to farmed fish along the Norwegian coast line. The impact of VHSV on wild fish is not well documented.

Editor: Oliver Schildgen, Kliniken der Stadt Köln gGmbH, Germany

Funding: Norwegian Research Council, grants no. 190245 and 224931 (http://www.forskningsradet.no/en/Home_page/1177315753906) RJ. The funders had no role in study design, data collection and analysis, decision to publish, or preparation of the manuscript.

Competing Interests: The authors have declared that no competing interests exist.

* Email: nina.sandlund@imr.no

Introduction

Viral hemorrhagic septicemia (VHS) is a severe virus infection causing great losses in farming of rainbow trout *Oncorhynchus mykiss*. Mortality rates are variable depending on the age of the fish with up to 100% in fry. It is often less in older fish, typically between 30–70% [1,2]. Hence, economical losses could be substantial. VHS causes clinical signs such as haemorrhages in internal organs, pale gills, exophthalmia and darkening of the body [2].

The causative agent is the viral haemorrhagic septicaemia virus (VHSV), an enveloped negative single-stranded RNA virus belonging to the family of Rhabdoviridae and genus *Novirhabdovirus*. VHSV was first thought of as a virus affecting freshwater fish species of continental Europe. The first VHSV isolation from wild fish in the marine environment was in 1979 from Atlantic cod, *Gadhus morhua* [3,4]. Since then the virus has proven to be both widely spread in the northern hemisphere and occurring in more than 80 marine and fresh water fish species [5]. This emphasizes the ability this virus has to adapt to new host species.

The four main genotypes of VHSV (I–IV) and the subtypes (a–e) are geographically distributed; genotype I, II and III are found in Europe while genotype IV occurs in North American and Northern Pacific waters [2,6]. Subtype Ia is highly virulent for rainbow trout and this subtype is found in VHS infected fresh water farms in continental Europe. Subtype Ib has been isolated from several marine species in the Baltic and North Sea (reviewed in [1]) and has caused two outbreaks of VHS in rainbow trout in Sweden [7,8]. Subtype Ic consists of older isolates from Denmark. Subtype Id represent another group of older isolates from the 1960s in Denmark and Norway as well as isolates from the more recent VHS outbreaks in rainbow trout farmed in sea cages in the Åland archipelago in the Baltic Sea [9]. Group Ie is only present in the Black Sea area. The Baltic Sea is also the main reservoir for the genotype II isolates. Genotype III has been isolated from wild marine fish in the North Sea and Skagerrak and has caused disease outbreaks in farmed turbot *Scophthalmus maximus* in Scotland and Ireland, in farmed rainbow trout in Norway [10] and recently also in wild caught farmed wrasse *Labridae* spp [11]. VHSV genotype IVa is widespread in wild marine fishes on both the American and the Asian side of the Northern pacific and has caused disease outbreaks in farmed Atlantic salmon in British Columbia, Canada [12]. It is still debated to what extent VHSV play a role in stock and population variations of Pacific herring

Clupea pallasii ([13], reviewed in [14]). Over the past decade VHSV genotype IVb has become an emerging problem in the Great Lakes region in North America causing high mortality in several wild fish species [13,15–18]. Subtype IVc has been isolated from brackish fishes during mortality events in Atlantic coastal regions of North America including mummichog *Fundulus heteroclitus*, stickleback Gasterosteidae, striped bass *Morone saxatilis* and brown trout *Salmo trutta* [12,19].

Screening surveys of VHSV in wild marine fish have been performed at several separate locations in European waters; the Barents Sea, North Sea, Norwegian Sea, Skagerrak, Kattegat and Baltic Sea [20–24]. VHSV has been detected from the North Sea, coastal areas around Scotland, Skagerrak, Kattegat, the Baltic Sea and Flemish Cap, in the North Atlantic Ocean near Newfoundland and the virus is assumed endemic in these waters (reviewed in [1]). VHSV prevalence of up to 16.7% was found in Atlantic herring *Clupea harrengus* [21,24,25]. But most screening surveys performed on both Atlantic herring and Pacific herring [14,26,27] reports relatively few positive detections, in spite of high number of sampled fish (reviewed in [1]). Recently high prevalence of VHSV was reported in Atlantic herring from the southern part of Norway during the spawning season, however no disease symptoms associated with the findings were reported [28].

Although VHS outbreaks mainly occur in freshwater rainbow trout farms there have been a few reports of outbreaks of VHS in sea farmed rainbow trout in Sweden, Finland, France, Denmark and Norway [1,7,8,10]. The VHS outbreak in Norway occurred in farmed rainbow trout in Storfjorden in 2007 [10]. This was the first detection of VHS in Norway since the eradication of the disease in 1974 and the source of the infection is still unknown. The outbreak in Norway was unexpectedly caused by a marine genotype III VHSV strain and challenge trials confirmed the marine virus-strain as virulent to rainbow trout [10,29]. Before this VHSV genotype III was considered low pathogenic to rainbow trout and other salmonid fish species [30,31]. This makes the outbreak unique and one of the main interests of the current screening was therefore to determine a possible marine reservoir in wild marine fish populations.

The Norwegian fish farming industry, especially the production of Atlantic salmon *Salmo salar*, has grown rapidly over the last decades. The annual production of Atlantic salmon and rainbow trout is now approx 1.2 mill and 70.000 tons respectively according to Statistics, Directorate of Fisheries in Norway (http://www.fiskeridir.no/). The possibility of pathogen transmission between farmed and wild fish is therefore increasing and a major concern. VHSV has so far not been isolated from Atlantic salmon in northern European waters, but VHSV genotype IVa has been isolated from farmed Atlantic salmon on the east coast of Canada (British Columbia) [15]. In general, Atlantic salmon has shown limited susceptibility to VHSV in immersion trails, but using intra peritoneal (i.p.) injection as challenge model has resulted in up to 78% mortality [1,32]. In challenge experiments exposing Atlantic salmon to the genotype III VHSV isolate from Norway, mortality was only experienced in the i.p. injected fish group and not in the immersion trail groups [10]. However, recent work demonstrated that Atlantic salmon were susceptible to VHSV genotype IVa and developed clinical signs after i.p. injection, immersion and when cohabited with VHSV diseased Pacific herring [33]. Transmission of VHSV from infected Atlantic salmon to sympatric Pacific herring was also demonstrated. This shows that VHSV has the ability to adapt to and infect Atlantic salmon. It should be added that Atlantic salmon seems less susceptible to VHSV genotype IVb than IVa [34].

With more than 80 susceptible species in the marine and freshwater environment, VHSV shows high ability of host adaptation and evolution of new strains with increased virulence. VHSV is thereby a potential risk to farming of susceptible species. Evidence support the theory that both wild and farmed fish can function as a reservoir and transmitter of VHSV [11,15,33,35], however more knowledge regarding specific species is needed. The presence of fish species that are asymptomatic carriers of VHSV also needs to be considered. Recently the Shetland Isles experienced an outbreak of VHS in wild caught wrasse held at a marine farm before use as cleaner fish in Atlantic salmon farming. The origin of this VHSV genotype III was likely from the marine environment, as several marine species in and around the locality tested positive for this virus [11]. As use of wrasse as cleaner fish is increasing, this points to another potential route of pathogen transmission to Atlantic salmon.

Screening of wild fish in northern parts of the Atlantic Ocean has earlier mainly been done in the open waters away from the aquaculture facilities. The aim of the present study was to investigate the occurrence of VHSV in marine wild fish in the coastal and fjord areas of Norway by analysing organ samples using both virus isolation and PCR.

Material and Methods

Field collection of fish

Five different research cruises in the fjords and coastal areas in Norway were carried out during 2009–2011. All cruises were part of the Institute of Marine Research annual coastal surveys and the fish samples used for VHSV analysis were randomly selected among the fish from the trawl haul. The Institute of Marine Research is a governmental research institute with given permission to perform research cruises including fish samplings by the Norwegian Government. Hence no additional permits were needed to sample organs from already deceased fish. All sampling and handling of fish were performed by experienced personnel.

Cruise one took place in September/October 2009 targeting the northern coast area. Cruise two was carried out in December 2009 targeting the coast of mid Norway including the Storfjorden fjord system were a VHSV outbreak occurred in farmed rainbow trout in 2007. Cruise three took place on the west coast in May 2010. During cruise four fish was sampled from the coast of north and mid Norway in October/November 2010. Cruise 5 took place in October/November 2011 in mostly the same areas as cruise one. The sampling area and geographical positions of each trawling station are shown in Figure 1.

During cruises one, two, four and five fishing was performed as either bottom or pelagic trawling, with trawling time of approximately 30 minutes. During cruise three only pelagic trawling was performed and it lasted up to 4 hours. Most fish died from suffocation in the trawling process or during handling on deck. All fish were kept in a cooled 4°C room until sampling and maximum time from fishing to sampling was 7–8 hours. During post mortem examination all fish were measured, weighed and any external signs of disease were recorded if present.

Sampling of organs

All tissue samples were processed on board the research vessels. The same sampling procedure was used during all cruises. Organ specimens of spleen, kidney and brain from maximum 5 fish was pooled and diluted 1:10 in transport medium (Eagle's Minimum Essential Medium, pH 7.6, supplemented with 10% newborn bovine serum and 100 µg ml^{-1} gentamicin). The samples were immediately transferred to a −80°C freezer for storage. Gonads

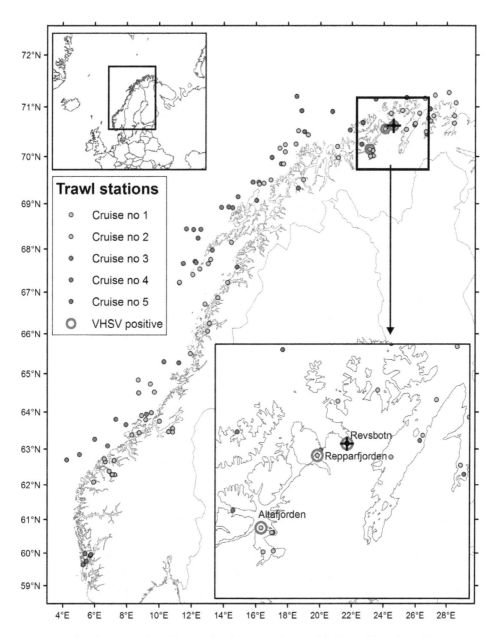

Figure 1. Map showing sampling locations (trawling stations) from cruises 1–5 indicated in colour. The various cruises are presented with different colour codes. Trawling stations with VHSV positive rRT-PCR samples are shown with red circles. The respective latitude/longitude for the locations are; Repparfjorden/Sammelsundet (70.5600/24.0800), Revsbotn (70.62667/24.6150) and Altafjorden (70.14667/23.0950). The black cross shows origin of the one pool sample testing positive by cell culture isolation.

(testes/ovaries) were sampled from sexually mature fish and kept in separate transport medium tubes. In addition individual samples of gills, heart, spleen, kidney, brain and gonads were collected in RNAlater (Sigma, USA), stored at 4°C for 24 hours prior to long term storage at −20°C. In cruise one the individual samples were randomly selected from three of the five pooled fish due to storage and sampling capacity. In cruise two-five individual samples were taken from all fish. Hearts were not individually sampled during cruise two, three and four due to storage capacity. Spleen and brain were diagonally sectioned and one half collected in transport medium and RNAlater for virus isolation and real-time Reverse Transcriptase Polymerase Chain Reaction (rRT-PCR) examination, respectively. The organs were aseptically sampled in this specific order; gills, heart, spleen, kidney and brain. If sexually matured gonads were present, these were sampled after the spleen. Equipment used to collect samples, were cleaned with alcohol, flamed between uses on each organ, and changed between each fish. Disposable gloves and tissue paper were changed between each fish. Equipment was washed and disinfected on daily basis to ensure sterile conditions; using Virkon S (Lilleborg, Norway) for minimum 30 minutes, rinsed in fresh water and boiled in fresh water for 15 minutes.

Following completion of the cruises, all samples were transported to the laboratory on dry ice and stored at either −20°C or −80°C prior to testing.

Virus isolation

Tissue samples pooled in transport medium were homogenized and cleared by low-speed centrifugation, and supernatants inoculated onto subconfluent monolayers of BF-2-cells (ECACC,

Salisbury, UK) in 1:10 and 1:100 dilutions in 24-well tissue culture plates according to the OIE procedure [5]. Inoculated cultures were incubated at 15°C and inspected after 1 week for cytopathic effect (CPE). Culture medium was collected from all wells and passed to new cell cultures. After a further week of incubation, the cultures were again inspected, and supernatant from wells with evident CPE in the second passage was collected, RNA extracted and tested for VHS-virus by rRT-PCR. Pooled organ samples that later were found VHSV positive by rRT-PCR were additionally incubated in BF-2 and EPC-cells three times for 14 days.

RNA extraction and real-time RT-PCR (rRT-PCR)

RNA extraction was performed on homogenized tissue from organ pools (100 μl) and individual organ samples (10–20 mg), and from 150 μl virus supernatant using the automated easyMAG protocol (Biomérieux) or the RNeasy Mini kit (Qiagen). Extracted RNA was measured using NanoDrop ND-1000 (NanoDrop Technologies). The rRT-PCR assay was conducted using 500–1000 ng template RNA with a QIAGEN OneStep RT-PCR kit (QIAGEN Nordic) and nucleoprotein (N) gene primers and probe described by Duesund et al. [29]. The assay was performed with 0.5 μM of each primer and 0.3 μM probe in a 20 μl reaction, with cDNA synthesis at 52°C for 30 min followed by 15 min at 95°C, then 45 cycles of 95°C for 15 s and 60°C for 1 min using the Mx3005p real time PCR system (Stratagene). During the study the laboratory changed the standard VHSV rRT-PCR method to a validated assay with higher analytical and diagnostic sensitivity for all VHSV genotypes [36]. The assay was conducted using the same reaction conditions, but cycling conditions: 30 min a 50°C and 15 min at 95°C, followed by 45 cycles of 94°C for 30 s and 60°C for 1 min. Required positive and negative controls were included in all runs. Samples with specific cycle treshold (Ct) value ≤40 were considered positive.

Sequencing and phylogenetic analysis

VHSV rRT-PCR positive samples were confirmed by sequence analysis of the viral G- and N-protein gene. Partial and full length glycoprotein (G) gene sequence was generated from overlapping sequences using primer sets V2, GB and Gseq [10,37]. Three primers were used to obtain a 1217 bp nucleoprotein (N) gene sequence: N-G1F 5′-GCT CAC AGA CAT GGG CTT CA-3′, N-G2R 5′-TGG ATT GGG CTT CTT CTT-3′, N-G3F 5′-GGC TCA ACG GGA CAG GAA-3′. The RT-PCR was performed using 5 μl extracted RNA, 0.5 μM primer concentration in a 50 μl QIAGEN OneStep reaction, with cDNA synthesis at 50°C for 30 min followed by 15 min at 95°C, then 40 cycles of 95°C for 1 min, 55°C 1 min and 72°C for 90 s. The RT-PCR products were visualised on an agarose gel and purified using the ExoSAP-IT protocol (Usb) prior to sequencing with BigDye Terminator v3.1 Cycle Sequencing kit (Applied Biosystems). Sequences derived were aligned and compared to related VHSV sequences using Vector NTI Advance 11 (Invitrogen). A maximum-likelihood (ML) phylogenetic analysis was conducted using MEGA version 5.0 [38] on the complete G-gene alignment employing the GTR+G model. The obtained G- and N-gene sequences were deposited in GenBank and given accession numbers HM632035–HM632036 and KJ768664–KJ768665 for the isolate from Atlantic herring and silvery pout, respectively.

Results

Fish sampling

During the five cruises, a total of 1927 fish representing 39 different species were caught and sampled at 121 different haul stations. A total overview of the various fish species sampled during the separate cruises, the number of pooled samples and geographical distribution is given in Table 1 and Figure 1. No fish showed any visible signs of clinical disease during the post mortem examinations.

The Atlantic herring caught during the research cruises is part of the Norwegian Spring Spawning herring (NSS) stock. The average length and weight of the herring caught at trawling station Repparfjorden was 15.7 cm (±2.9 cm STDV) and 40 g (±20.1 STDV) and at trawling station Revsbotn 22.7 cm (±3.4 cm STDV) and 124.5 g (±47 STDV). The individual length and weight of the positive fish, included in Table 2, do not differ significantly from the rest of the fish in the same catch (data not shown). Based on analysis of length and age distribution of NSS it can be estimated that the VHSV positive Atlantic herring were 4 years or less [39].

Cell culture isolation

One VHS-virus isolate were obtained from the 453 organ pools tested using BF-2 cell culture inoculation (Tables 1, 2). The positive pool contained organs from five Atlantic herring Clupea harengus collected during cruise one at location Revsbotn in Finnmark county (Figure 1). Full CPE was observed within 2 weeks of incubation, and VHSV was confirmed in the culture supernatant by rRT-PCR (Tables 1, 2). No CPE was observed in the other virus cultures. Organ pools later found VHSV positive by rRT-PCR were additionally sub-cultivated with prolonged incubation time without any detection of CPE.

VHSV rRT-PCR detection

rRT-PCR screening of the 453 pooled samples revealed totally five VHSV positive pools (Table 2), four originating from Atlantic herring and one from silvery pout Gadiculus argenteus, all sampled on three trawling locations in Finnmark county within two days (Figure 1). Each pool consisted of organs from five fish. To follow up these positive findings all available organ samples from the Atlantic herrings and silvery pout represented in the pools (gills, heart, kidney, spleen and brain) were tested individually by rRT-PCR (Table 2). This revealed 1–2 positive fish per pool; respectively five Atlantic herring and one silvery pout.

Generally the Ct values obtained from the individual organs samples in positive fish show that the highest amount of VHSV RNA is present in the heart while gills showed the highest prevalence (Table 2). All individually sampled gills (n = 1369) and hearts (n = 1091) from all cruises were therefore tested by rRT-PCR. Due to low RNA yields on some of the gill samples (n = 183) results are only recorded from 1186 gills to avoid false-negative results (Table 2). Three additional Atlantic herring, two haddock Melanogrammus aeglefinus and one whiting Merlangius merlangus tested positive in gills, and they were all caught in the same trawl as the positive herring. The remaining sampled organs from these individual tested negative.

Sequencing

Unique G and N gene sequences were obtained from the PCR positive whiting, haddock, herring and silvery pout marked in bold in Table 2. This confirmed that they all belonged to genotype Ib and were closely related (99–100% identity, 6 nucleotide difference in the complete G-gene region). The partial sequences from whiting and haddock were too short to be included in the phylogenetic analysis. A ML phylogenetic tree based on complete G-gene sequences group the silvery pout and herring from this study together with other genotype Ib isolates reported from

Table 1. Species and number of fish and pooled organ samples from all five research cruises.

Family / Species	Cruise 1 No. of sampled fish	Cruise 1 No. of pools	Cruise 2 No. of sampled fish	Cruise 2 No. of pools	Cruise 3 No. of sampled fish	Cruise 3 No. of pools	Cruise 4 No. of sampled fish	Cruise 4 No. of pools	Cruise 5 No. of sampled fish	Cruise 5 No. of pools	All cruises Total no. of sampled fish	All cruises Total no. of pools
Ammodytidae												
Small sandeel *Ammodytes tobianus*	10	2									10	2
Lesser sandeel *Ammodytes marinus*	6	2									6	2
Anarhichadidae												
Wolf-fish *Anarhichas lupus*	22	5							4	1	26	6
Argentinidae												
Argentine *Argentina silus*	176	37					9	2	5	1	190	40
Belonidae												
Garfish *Belone belone*					27	7					27	7
Carangidae												
Horse mackerel *Trachurus trachurus*	1	1					3	1			4	2
Clupeidae												
Atlantic herring *Clupea harengus*	170	37*¤					7	2	45	11	222	50*¤
Cyclopteridae												
Lumpsucker *Cyclopterus lumpus*			1	1	17	6					18	7
Gadidae												
Cod *Gadus morhua*	64	14	14	4					32	6	110	24
Blue whiting *Micromesistius poutassou*	104	24					5	1	5	1	114	26
Four-bearded rockling *Rhinonemus cimbrius*	1	1									1	1
Haddock *Melanogrammus aeglefinus*	121	26					5	1	15	3	141	30
Ling *Molva molva*	5	3	1	1			3	1			9	5
Norway pout *Trisopterus esmarkii*	336	68							5	1	341	69

Table 1. Cont.

Family / Species	Cruise 1		Cruise 2		Cruise 3		Cruise 4		Cruise 5		All cruises	
	No. of sampled fish	No. of pools	No. of sampled fish	No. of pools	No. of sampled fish	No. of pools	No. of sampled fish	No. of pools	No. of sampled fish	No. of pools	Total no. of sampled fish	Total no. of pools
Pollack *Pollachius pollachius*			2	1			4	1			6	2
Poor cod *Trisopterus minutus*	12	4	15	3			1	1			28	8
Seith *Pollachius virens*	52	14	29	7			23	5	5	1	109	27
Silvery pout *Gadiculus argenteus*	60	12*									60	12*
Whiting *Merlangius merlangus*	54	12					4	1	13	3	71	16
Lophiidae												
Anglerfish *Lophius piscatorius*	10	6	4	2							14	8
Lotidae												
Tusk *Brosme brosme*	11	9	4	1			9	2			24	12
Blue ling *Molva dypterygia*			3	2							3	2
Merluccidae												
Hake *Merluccius merluccius*	2	1	23	6					4	1	29	8
Osmeridae												
Capline *Mallotus villosus*	79	16							10	2	89	18
Phycidae												
Greater forkbeard *Phycis blennoides*	3	1	3	1			4	1			10	3
Pleuronectidae												
Halibut *Hippoglossus hippoglossus*	1	1									1	1
Lemon sole *Microstomus kitt*	11	3									11	3
American plaice *Hippoglossoides platessoides*	10	2							5	1	15	3
Plaice *Pleuronectes platessa*	17	4	7	2			5	1	5	1	34	8
Witch *Glyptocephalus cynoglossus*	10	4	5	1							15	5

Table 1. Cont.

Family / Species	Cruise 1 No. of sampled fish	Cruise 1 No. of pools	Cruise 2 No. of sampled fish	Cruise 2 No. of pools	Cruise 3 No. of sampled fish	Cruise 3 No. of pools	Cruice 4 No. of sampled fish	Cruice 4 No. of pools	Cruice 5 No. of sampled fish	Cruice 5 No. of pools	All cruises Total no. of sampled fish	All cruises Total no. of pools
Scombridae												
Macerel *Scomber scombrus*					5	1					5	1
Sebastinae												
Small redfish *Sebastes viviparus*	73	16	10	2			5	1			88	19
Golden redfish *Sebastes marinus*	20	5					5	1			25	6
Sebastes sp.	52	11									52	11
Scophthalmidae												
Megrim *Lepidorhombus whiffiagonis*	3	2	1	1							4	3
Squalidae (dogfish sharks)												
Piked dogfish *Squalus acanthias*			5	1							5	1
Sternoptychidae												
Pearlsides *Maurolicus muelleri*			2	1							2	1
Triglidae												
Grey gurnard *Eutrigla gurnardus*	4	2							3	1	7	3
Zoarcidae												
Eelpout *Lycodes vahlii gracilis*	1	1									1	1
Total number	1501	346	129	3	49	14	92	22	156	34	1927	453

Samples were found VHSV positive by virus isolation (¤) and rRT-PCR analysis (*).

Table 2. VHSV rRT-PCR Ct values of pooled and individual organ samples from positive fish.

Trawling location	Species	Length cm	Weight gr	Ct values individual organ samples						Ct values pooled organ samples	
				Gills	Heart	Kidney	Spleen	Brain	Gonads	Pool no.	
Repparfjorden	Herring	16.0	44	28.48	24.75	33.61	31.23	33.03	N.A	63	37.17
Repparfjorden	Herring	17	42	37.44	32.27	-	-	-	♀	63	37.17
Revsbotn	Herring	21.0	87	37.75	-	-	-	-	N.A	66	-
Revsbotn	Herring	25.5	164	38.32	-	-	-	-	N.A	67	-
Revsbotn	Herring	26.0	171	38.76	-	-	-	-	♂	69	-
Revsbotn	Herring	27.0	200	37.68	-	-	-	-	N.A	70	33.17
Revsbotn	Herring	19.5	69	28.06	25.49	29.54	29.83	34.29	N.A	71	36.08
Revsbotn	Herring	17.0	44	28.21	28.96	34.06	31.8	36.53	N.A	**72***	**26.74**
Revsbotn	Haddock	56.0	1152	**36.31**	-	-	-	-	N.A	73	-
Revsbotn	Haddock	44.5	849	**33.90**	-	-	-	-	N.A	74	-
Revsbotn	Whiting	40.0	586	**32.09**	-	-	-	-	♀	75	-
Altafjorden	Silvery pout	17.0	47	37.62	**23.41**	37.9	39.99	38.17	N.A	102	37.65

VHSV rRT-PCR Ct values of separate organ samples from positive individuals, in addition to the corresponding pooled organ sample in which the individual samples were included. Fish length, weight and trawling location, from which each individual were caught, are included. Pool no. = identity pool number Herring = Atlantic herring *Clupea harengus*, silvery pout *Gaduculus argenteus*, haddock *Melanogrammus aeglefinus*, whiting *Merlangius merlangus*.
- = No Ct. N.A = not available.
* = sample positive for VHSV by both virus isolation and rRT-PCR. VHSV sequences for verification were obtained from the samples indicated in bold.

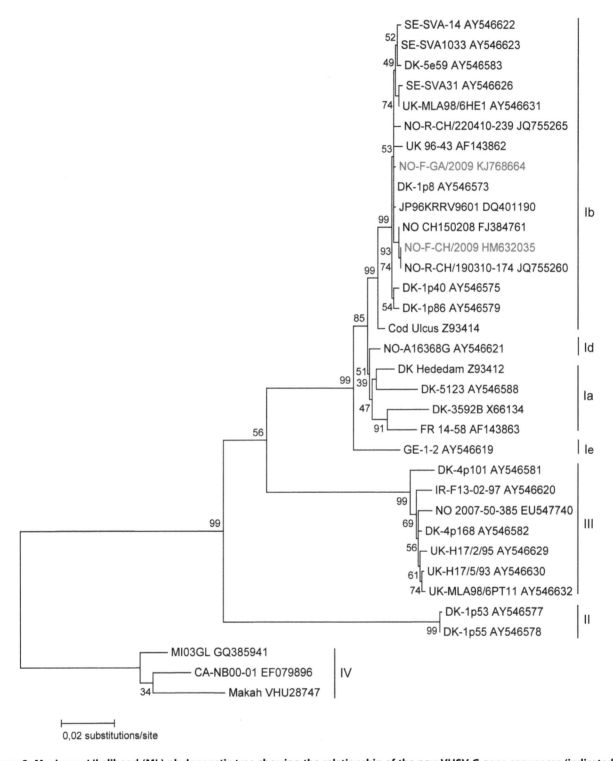

Figure 2. Maximum Likelihood (ML) phylogenetic tree showing the relationship of the new VHSV G-gene sequences (indicated in red) and other VHSV genotype representatives. Sequences are labelled by isolate name and GenBank accession number. The obtained tree was bootstrapped with 500 replicates and genotype IV was used as outgroup.

Atlantic herring and other fish species in the North Sea (Figure 2). The genotype Ib group also includes G-gene sequences detected in Atlantic herring of the Norwegian spring-spawning stock (Acc. no. JQ755260, JQ755265) [28].

Discussion

Sampling during five research cruises was conducted to investigate the presence of VHSV in wild fish along the Norwegian coastline. The present screening survey is the first to look for VHSV in coastal areas and fjord systems in Norway. In total, samples from 12 fish including Atlantic herring *Clupea harengus*, silvery pout *Gadiculus argenteus*, haddock *Melanogrammus aeglefinus* and whiting *Merlangius merlangus* were found positive for VHSV. The VHSV positive fish were sampled on geographically close trawling locations in the northern county of Finnmark. This is the first observation of VHSV this far north, which again points out the large natural marine reservoir for VHSV.

To our knowledge this is the first report of sampling and testing for VHSV in silvery pout. Silvery pout is a small deep water fish of maximum 15 cm length that lives in the Northeast Atlantic and is used for industrial purposes such as fish meal and fish oil. Both Atlantic herring and silvery pout occur in large shoals which can migrate long distances. It is therefore uncertain whether the VHSV-infected fish were infected in Finnmark or elsewhere, and then migrated into the sampling area. It is interesting to note that these two highly different fish species are found in the same area carrying closely related virus isolates (Figures 1, 2).

The haddock and whiting only tested positive for VHSV in gills and it has not been confirmed if these fish where truly infected or only passive carriers of the virus in the gill mucus. None of the fish testing positive for VHSV in this study showed any clinical signs of disease. This is in line with numerous other isolations of VHSV from asymptomatic wild marine fish; including Atlantic herring caught in the English Channel [40], North Sea [21], Baltic Sea [24,25] and Skagerrak and Kattegat [24]. The first marine isolate of VHSV was isolated in Danish coastal waters in 1979 from cod showing "ulcus syndrome", however no evidence of an association between ulcers and VHS have been demonstrated [3,4,21, 24,25,41].

During the five research cruises samples were taken from 39 different species. The number of fish of each species is highly variable due to the fact that both bottom and pelagic trawling was used, and the trawling was performed in various areas at different times of the year. In addition, fish were sampled at random from random hauls and the need for fresh samples had to be prioritised. As high prevalence of VHSV has been found in Atlantic herring on previous screenings, Atlantic herring were prioritised for sampling when available [21,24,25,28].

The 453 pooled organ samples (spleen, kidney and brain) were tested both with cell culturing and rRT-PCR, revealing a higher detection rate with rRT-PCR. Successful virus isolation in cell culture was only obtained from one pool of organs, including samples from five Atlantic herring. This sample was the strongest positive when tested using rRT-PCR (Ct 26), reflecting that overall the viral amount is relatively low. In addition, individual organs (spleen, kidney, brain, heart, gills and gonads) were sampled from most fish. rRT-PCR testing indicates that the highest amount of VHSV RNA was found in the heart and the highest prevalence were detected in gills. Heart samples are suitable for detecting VHSV in various fish species as it is an important target organ for VHSV [18,42–45]. According to OIE (Commissions decision 2001/183) heart and/or brain samples should be included in screenings surveys. Based on our results from this present screening survey and previous findings, it is suggested that heart samples should always be included when sampling marine fish species.

Seven fish, including Atlantic herring, haddock and whiting, tested positive for VHSV by rRT-PCR in gills only. Gills have in various species proven to be a good organ for detecting VHSV carrying fish [28,43,46]. The results obtained by Cornwell et al. [46] further indicate that the sensitivity of detecting VHSV in gill samples might vary between species. The ability of VHSV to infect the gills has also been correlated to virulence [47]. A study testing the viral load of VHSV in various tissues of rainbow trout at various stages during the course of infection showed that fish surviving a VHSV infection had the highest amount of virus in gill and brain tissue [48].

This study detected a low prevalence of VHSV in Atlantic herring (8 positive/222 sampled) and this is in accordance with several other surveys in the waters in the northern part of Europe (reviewed in [1,21,24,25]). High prevalence in Atlantic herring on the west coast of Norway were found during the spawning season in 2010 [28]. Whether this was caused by a generally higher prevalence during spawning seasons or an outbreak in this population is unknown and needs to be further investigated. Isolation of VHSV from Pacific herring *Clupea pallasii* populations is well known [14,26,27]. Studies have come to contradictory conclusions whether VHSV play a role in stock variations of Pacific herring or not ([13], reviwed in [14]).

There are indications that age plays an essential role in the susceptibility of VHSV. Higher prevalence of VHSV has been found in young wild caught Pacific herring compared to older [14,49]. Another explanation can be that herring are infected during the early life stages and that the virus amount decrease over time to a non-detectable level in surviving fish. In captivity asymptomatic Pacific herring larvae and juveniles has developed VHS within a week after confinement [26] indicating a latent virus infection in the fish triggered by and developed in captivity. These findings supports the suggestion that young herring is more susceptible to VHSV infection than older. The age of the Atlantic herring tested in this study was estimated based on length and age distribution of NSSH [39]. This indicates that the VHSV positive Atlantic herring were less than 4 years. It can also be noted that the positive Atlantic herring samples were among the smallest specimen tested from each catch, and these results may be consistent with young herring being more susceptible to VHSV. In the present study the low amount of virus found and lack of clinical signs on the VHSV positive fish indicates that these individuals are asymptomatic carriers of the virus.

Few detections of VHSV from previous screening surveys could be related to methodology. According to Dixon [50] some VHSV isolates prefer BF-2 cells, while others produce highest titers in EPC cells. Isolation of marine VHSV, both of genotype Ib and III, is more successful in BF-2 cells compared to EPC [5], which was therefore the choice of cells used in the present study. Earlier screening surveys in wild marine fish have mostly tested pooled organ samples by cell culture isolation [21,24,25]. One exception is the screening performed by Dixon et al. [23] in which cell culture supernatant or dilutions of homogenized tissue from pooled organ samples were tested by RT-PCR. As in the present study an increased number of positive pooled samples (n = 4) were found using RT-PCR compared to cell culture isolation. Experimental testing of the sensitivity of cell culturing versus PCR-methodology has demonstrated RT-PCR the most sensitive [51,52]. The development of PCR assays with higher sensitivity and a broader detection range to several genotypes of VHSV has made this the preferred method for VHSV detection in most laboratories [36].

Screening for VHSV based on testing of individual organs from marine fish is not common. Pooling of organ samples lower the sensitivity of the detection methods but at the same time allows testing of a larger amount of samples. Testing of all individual organs sampled during this study (n = 6848) was not feasible in our laboratory. The decreased sensitivity of the pooling strategy has been partly compensated by individual testing of all gill and heart samples. All individual organs from all available fish in the VHSV positive pools where tested by rRT-PCR revealing 1–2 positive fish per pool. This illustrates the difficulty of estimating prevalence based on pooled organ samples when virus yield and prevalence is low. The results further indicate that individual organ samples tested by rRT-PCR provide the best estimate for a true prevalence in the population. This difference in methodology may have led to an underestimation of the prevalence of VHSV in the European marine environment as methods with high sensitivity is required to detect carrier fish with low viral titers. Due to sampling and storage capacity during cruise one only three out of the five individuals included in each pool were sampled on RNAlater (n = 943). This could have affected the total number of positive results as additional positive individuals could have been included in the pool.

Sequencing based on the G and N gene revealed that the positive samples belonged to VHSV genotype Ib and were closely related. The ML phylogenetic tree group the strain from silvery pout and herring with isolates occurring in various fish species in the North Sea and Baltic Sea, including strains detected in Atlantic herring of the Norwegian spring-spawning stock [12,28]. No genotype III positive fish were found in the present sampling, but such positive haddock and whiting are known from previous screenings in the North Sea [37]. The positive fish originated from sampling at three close trawling locations. The transmission pattern of VHSV from fish to fish is by direct contact, in water or by ingestions of infected material [53,54]. In theory virus could transmit between fish while in the trawl. This is unlikely since internal organs also tested positive, indicating true carrier status. In addition, with one exception all individuals testing positive for VHSV came from different pool of fish, showing that the possibility of contamination between samples is limited. All positive pools also contain several negative fish showing that contamination was limited.

The risk of inter-species transmission of VHSV is always present in the marine environment. Evidence of VHSV transmission from wild to farmed fish has recently been studied by phylogenetic analysis that verified a closely genetic linkage between VHSV genotype IVa isolates from wild and farmed fish in the British Columbia area [15]. In the fjord systems were the positive marine samples were found in this study both wild and farmed Atlantic salmon are present (data from the Directorate of Fisheries). In addition Altafjorden and Repparfjorden are regulated fjord systems established for the protection of wild salmon. Although Atlantic salmon in general has shown low susceptibility to VHSV, relatively high virus titer (between 1×10^4 and 1×10^6 pfu/g) of VHSV was demonstrated in Atlantic salmon 10 weeks after i.p. injection, immersion and cohabitation challenge with VHSV genotype IV [33]. Three of 12 Atlantic salmon i.p. injected with the genotype III isolate from Storfjorden were still VHSV positive 29 days after infection [10]. This susceptibility and persistency of

VHSV in Atlantic salmon demonstrated by Lovy et al. [33] together with the possibility of transmitting the virus back to Pacific herring indicates that salmon has the ability to serve as a vector and reservoir of VHSV. Although VHSV genotype III isolates does not normally cause disease in anadromous fish species, the outbreak in rainbow trout in Norway in 2007 shows the adaptation capacity of this virus for salmonids [10]. Therefore, it is important to keep farmed fish free of any VHSV by avoiding continuous production of the same fish species for several generations and keep generations at separate locations. The possibility of inter-species transmission of VHSV between fish species in close contact is highly relevant especially with the increased use of cleaner fish as biological control of sea lice in Atlantic salmon farms. Although all Atlantic salmon from a farm site that experienced an outbreak of VHSV on wrasse tested negative for VHSV, the potential of VHSV to adapt and cause disease outbreaks in Atlantic salmon farms, should be taken into account [55].

The fish testing positive for VHSV in the current study appeared asymptomatic carriers. The healthy carriers could represent survivors from an earlier disease outbreak with high mortality rates. Mortality rates in wild fish populations are often not detected due to removal of diseased fish by predators [56]. It is also possible that the virus carrier fish is weakened and thereby more susceptible to other disease problems or predators. Little is known about how VHSV affect the health situation of wild fish populations and further research is needed. Possible transfer of VHSV between wild and farmed fish will always be a potential risk and knowledge about marine reservoirs is therefore essential. It is therefore of major importance to conduct surveillance studies for VHSV to ensure early detection and eradication of the virus from farmed fish. History has taught us that such control is important to avoid adaptation of the virus into more virulent strains.

Acknowledgments

The authors would like to thank Helle Frank Skall and Mike Snow for helpful discussions during the designing of the sampling protocol. The authors are also grateful for the sampling assistance given by Ann Kristin Jøranlid, Rolf Hetlelid Olsen, Trygve Poppe, Arve Kristiansen and to Karen Gjertsen for creating Figure 1.

Author Contributions

Conceived and designed the experiments: NS ØB RJ NJO. Performed the experiments: NS ØB. Analyzed the data: NS RJ BG IM. Contributed reagents/materials/analysis tools: NS BG IM NJO ØB RJ. Contributed to the writing of the manuscript: NS BG IM NJO ØB RJ.

References

1. Skall HF, Olesen NJ, Mellergaard S (2005) Viral haemorrhagic septicaemia virus in marine fish and its implications for fish farming - a review. Journal of Fish Diseases 28: 509–529.
2. Olesen NJ, Skall HF (2013) Viral haemorrhagic septicaemia virus. In: Munir M, editor. Mononegaviruses of veterinary importance Vol I: Pathobiology and molecular diagnosis: CAB International. pp. 323–336.
3. Vestergard Jørgensen PE, Olesen NJ (1987) Cod ulcus syndrome rhabdovirus is indistinguishable from the Egtved (VHS) virus. Bulletin of the European Association of Fish Pathologists 7: 73–74.
4. Jensen NJ, Larsen JL (1979) Ulcus-syndrome in cod (*Gadus morhua*). I. A pathological and histopathological study. Nordisk Veterinaer Medicin 31: 222–228.
5. Anonymous (2014) Manual of Diagnostic Tests for Aquatic Animals 2014. Available: http://www.oie.int/fileadmin/Home/eng/Health_standards/aahm/current/2.3.09_VHS.pdf: World Organisation of Animal Health.
6. Walker PJ, Benmansour A, Calisher CH, Dietzgen R, X FR, et al. (2000) Family Rhabdoviridae. In: van Regenmortel MHV, Fauquet CM, Bishop DHL, Carstens EB, Estes MK et al., editors. Virus taxonomy; Seventh report of the international committee for taxonomy of viruses. San Diego: Academic Press. pp. 563–583.
7. Nordblom B (1998) Report on an outbreak of viral haemorrhagic septicaemia in Sweeden. Sweedish Board of Agriculture, Department of animal Production and Health.
8. Nordblom B, Norell AW (2000) Report on an outbreak of viral haemorrhagic septicaemia in farmed fish in Sweden. Report for the standing veterinary committee. Swedish Board of Agriculture, Department of animal Production and Health.
9. Gadd T, Jakava-Viljanen M, Tapiovaara H, Koski P, Sihvonen L (2011) Epidemiological aspects of viral haemorrhagic septicaemia virus genotype II isolated from Baltic herring, *Clupea harengus* membras L. Journal of Fish Diseases 34: 517–529.
10. Dale OB, Ørpetveit I, Lyngstad TM, Kahns S, Skall HF, et al. (2009) Outbreak of viral haemorrhagic septicaemia (VHS) in seawater-farmed rainbow trout in Norway caused by VHS virus Genotype III. Diseases of Aquatic Organisms 85: 93–103.

11. Hall LM, Smith RJ, Munro ES, Matejusova I, Allan CET, et al. (2013) Epidemiology and control of an outbreak of viral haemorrhagic septicaemia in wrasse around Shetland commencing 2012. The Scottish Government. 1–46 p.

12. Pierce LR, Stepien CA (2012) Evolution and biogeography of an emerging quasispecies: Diversity patterns of the fish Viral Hemorrhagic Septicemia virus (VHSv). Molecular Phylogenetics and Evolution 63: 327–341.

13. Elston RA, Meyers TR (2009) Effect of viral hemorrhagic septicemia virus on Pacific herring in Prince William Sound, Alaska, from 1989 to 2005. Diseases of Aquatic Organisms 83: 223–246.

14. Marty GD, Quinn TJ, Carpenter G, Meyers TR, Willits NH (2003) Role of disease in abundance of a Pacific herring (Clupea pallasi) population. Canadian Journal of Fisheries and Aquatic Sciences 60: 1258–1265.

15. Garver KA, Traxler GS, Hawley LM, Richard J, Ross JP, et al. (2013) Molecular epidemiology of viral haemorrhagic septicaemia virus (VHSV) in British Columbia, Canada, reveals transmission from wild to farmed fish. Dis Aquat Organ 104: 93–104.

16. Thompson TM, Batts WN, Faisal M, Bowser P, Casey JW, et al. (2011) Emergence of Viral hemorrhagic septicemia virus in the North American Great Lakes region is associated with low viral genetic diversity. Diseases of Aquatic Organisms 96: 29–43.

17. Bain MB, Cornwell ER, Hope KM, Eckerlin GE, Casey RN, et al. (2010) Distribution of an invasive aquatic pathogen (Viral Hemorrhagic Septicemia Virus) in the Great Lakes and its relationship to shipping. Plos One 5: 8.

18. Al-Hussinee L, Lord S, Stevenson RMW, Casey RN, Groocock GH, et al. (2011) Immunohistochemistry and pathology of multiple Great Lakes fish from mortality events associated with viral hemorrhagic septicemia virus type IVb. Diseases of Aquatic Organisms 93: 117–127.

19. Gagne N, MacKinnon AM, Boston L, Souter B, Cook-Versloot M, et al. (2007) Isolation of viral haemorrhagic septicaemia virus from mummichog, stickleback, striped bass and brown trout in eastern Canada. Journal of Fish Diseases 30: 213–223.

20. Mortensen HF, Heuer OE, Lorenzen N, Otte L, Olesen NJ (1999) Isolation of viral haemorrhagic septicaemia virus (VHSV) from wild marine fish species in the Baltic Sea, Kattegat, Skagerrak and the North Sea. Virus Research 63: 95–106.

21. King JA, Snow M, Smail DA, Raynard RS (2001) Distribution of viral haemorrhagic septicaemia virus in wild fish species of the North Sea, north east Atlantic Ocean and Irish Sea. Diseases of Aquatic Organisms 47: 81–86.

22. Brudeseth BE, Evensen Ø (2002) Occurrence of viral haemorrhagic septicaemia virus (VHSV) in wild marine fish species in the coastal regions of Norway. Diseases of Aquatic Organisms 52: 21–28.

23. Dixon PF, Avery S, Chambers E, Feist S, Mandhar H, et al. (2003) Four years of monitoring for viral haemorrhagic septicaemia virus in marine waters around the United Kingdom. Diseases of Aquatic Organisms 54: 175–186.

24. Skall HF, Olesen NJ, Mellergaard S (2005) Prevalence of viral haemorrhagic septicaemia virus in Danish marine fishes and its occurrence in new host species. Diseases of Aquatic Organisms 66: 145–151.

25. Mortensen HF, Heuer OE, Lorenzen N, Otte L, Olesen NJ (1999) Isolation of viral haemorrhagic septicaemia virus (VHSV) from wild marine fish species in the Baltic Sea, Kattegat, Skagerrak and the North Sea; pp. 95–106.

26. Kocan RM, Hershberger PK, Elder NE, Winton JR (2001) Epidemiology of viral hemorrhagic septicemia among juvenile pacific herring and Pacific sand lances in Puget Sound, Washington. Journal of Aquatic Animal Health 13: 77–85.

27. Meyers TR, Short S, Lipson K, Batts WN, Winton JR, et al. (1994) Association of viral hemorrhagic septicemia virus with epizootic hemorrhages of the skin in Pacific herring Clupea harengus pallasi from Prince William Sound and Kodiak Island, Alaska, USA. Diseases of Aquatic Organisms 19: 27–37.

28. Johansen R, Bergh Ø, Modahl I, Dahle G, Gjerset B, et al. (2013) High prevalence of viral haemorrhagic septicaemia virus (VHSV) in Norwegian spring-spawning herring. Marine Ecology Progress Series 478: 223–230.

29. Duesund H, Nylund S, Watanabe K, Ottem KF, Nylund A (2010) Characterization of a VHS virus genotype III isolated from rainbow trout (Oncorhychus mykiss) at a marine site on the west coast of Norway. Virology Journal 7: 1–15.

30. King JA, Snow M, Skall HF, Raynard RS (2001) Experimental susceptibility of Atlantic salmon Salmo salar and turbot Scophthalmus maximus to European freshwater and marine isolates of viral haemorrhagic septicaemia virus. Diseases of Aquatic Organisms 47: 25–31.

31. Snow M, Cunningham CO (2000) Virulence and nucleotide sequence analysis of marine viral haemorrhagic septicaemia virus following in vivo passage in rainbow trout Oncorhynchus mykiss. Diseases of Aquatic Organisms 42: 17–26.

32. De Kinkelin P, Castric J (1982) An experimental study of the susceptibility of atlantic salmon fry salmo-salar to viral haemorrhagic septicemia. Journal of Fish Diseases 5: 57–66.

33. Lovy J, Piesik P, Hershberger PK, Garver KA (2013) Experimental infection studies demonstrating Atlantic salmon as a host and reservoir of viral hemorrhagic septicemia virus type IVa with insights into pathology and host immunity. Veterinary Microbiology 166: 91–101.

34. Groocock GH, Frattini SA, Cornwell ER, Coffee LL, Wooster GA, et al. (2012) Experimental infection of four aquacultured species with viral hemorrhagic septicemia virus type IVb. Journal of the World Aquaculture Society 43: 459–476.

35. Schönherz AA, Lorenzen N, Einer-Jensen K (2013) Inter-species transmission of viral hemorrhagic septicemia virus (VHSV) from turbot (Scophthalmus maximus) to rainbow trout (Onchorhynchus mykiss). Journal of General Virology 94: 869–875.

36. Jonstrup SP, Kahns S, Skall HF, Boutrup TS, Olesen NJ (2013) Development and validation of a novel Taqman-based real-time RT-PCR assay suitable for demonstrating freedom from viral haemorrhagic septicaemia virus. Journal of Fish Diseases 36: 9–23.

37. Einer-Jensen K, Ahrens P, Forsberg R, Lorenzen N (2004) Evolution of the fish rhabdovirus viral haemorrhagic septicaemia virus. Journal of General Virology 85: 1167–1179.

38. Tamura K, Peterson D, Peterson N, Stecher G, Nei M, et al. (2011) MEGA5: Molecular evolutionary genetics analysis using maximum likelihood, evolutionary distance, and maximum parsimony methods. Molecular Biology and Evolution 28: 2731–2739.

39. Silva FFG, Slotte A, Johannessen A, Kennedy J, Kjesbu OS (2013) Strategies for partition between body growth and reproductive investment in migratory and stationary populations of spring-spawning Atlantic herring (Clupea harengus L.). Fisheries Research 138: 71–79.

40. Dixon PF, Feist S, Kehoe E, Parry L, Stone DM, et al. (1997) Isolation of viral haemorrhagic septicaemia virus from Atlantic herring Clupea harengus from the English Channel. Diseases of Aquatic Organisms 30: 81–89.

41. Smail DA (2000) Isolation and identification of Viral Haemorrhagic Septicaemia (VHS) viruses from cod Gadus morhua with the ulcus syndrome and from haddock Melanogrammus aeglefinus having skin haemorrhages in the North Sea. Diseases of Aquatic Organisms 41: 231–235.

42. Iida H, Mori K, Nishizawa T, Arimoto M, Muroga K (2003) Fate of viral hemorrhagic septicemia virus in Japanese flounder Paralichthys olivaceus challenged by immersion. Fish Pathology 38: 87–91.

43. Sandlund N, Johansen R, Fiksdal IU, Einen A-CB, Modahl I, et al. (Submitted) Susceptibility and pathology in juvenile Atlantic cod Gadus morhua to a marine viral haemorrhagic septicaemia virus isolated from diseased rainbow trout Oncorhynchus mykiss. Diseases of Aquatic Organisms.

44. Isshiki T, Nishizawa T, Kobayashi T, Nagano T, Miyazaki T (2001) An outbreak of VHSV (viral hemorrhagic septicemia virus) infection in farmed Japanese flounder Paralichthys olivaceus in Japan. Diseases of Aquatic Organisms 47: 87–99.

45. Nishizawa T, Savas H, Isidan H, Ustundag C, Iwamoto H, et al. (2006) Genotyping and pathogenicity of viral hemorrhagic septicemia virus from free-living turbot (Psetta maxima) in a Turkish coastal area of the Black Sea. Applied and Environmental Microbiology 72: 2373–2378.

46. Cornwell ER, Bellmund CA, Groocock GH, Wong PT, Hambury KL, et al. (2013) Fin and gill biopsies are effective nonlethal samples for detection of Viral hemorrhagic septicemia virus genotype IVb. Journal of Veterinary Diagnostic Investigation 25: 203–209.

47. Brudeseth BE, Skall HF, Evensen Ø (2008) Differences in virulence of marine and freshwater isolates of viral hemorrhagic septicemia virus in vivo correlate with in vitro ability to infect gill epithelial cells and macrophages of rainbow trout (Oncorhynchus mykiss). Journal of Virology 82: 10359–10365.

48. Oidtmann B, Joiner C, Stone D, Dodge M, Reese RA, et al. (2011) Viral load of various tissues of rainbow trout challenged with viral haemorrhagic septicaemia virus at various stages of disease. Diseases of Aquatic Organisms 93: 93–104.

49. Hershberger PK, Kocan RM, Elder NE, Meyers TR, Winton JR (1999) Epizootiology of viral hemorrhagic septicemia virus in Pacific herring from the spawn-on-kelp fishery in Prince William Sound, Alaska, USA. Diseases of Aquatic Organisms 37: 23–31.

50. Dixon PF (1999) VHSV came from the marine environment: Clues from the literature, or just red herrings? Bulletin of the European Association of Fish Pathologists 19: 60–65.

51. Knüsel R, Bergmann SM, Einer-Jensen K, Casey J, Segner H, et al. (2007) Virus isolation vs RT-PCR: which method is more successful in detecting VHSV and IHNV in fish tissue sampled under field conditions? Journal of Fish Diseases 30: 559–568.

52. Hope KM, Casey RN, Groocock GH, Getchell RG, Bowser PR, et al. (2010) Comparison of quantitative RT-PCR with cell culture to detect viral hemorrhagic septicemia virus (VHSV) IVb Infections in the Great Lakes. Journal of Aquatic Animal Health 22: 50–61.

53. Schönherz AA, Hansen MHH, Jorgensen HBH, Berg P, Lorenzen N, et al. (2012) Oral transmission as a route of infection for viral haemorrhagic septicaemia virus in rainbow trout, Oncorhynchus mykiss (Walbaum). Journal of Fish Diseases 35: 395–406.

54. Kurath G, Winton J (2011) Complex dynamics at the interface between wild and domestic viruses of finfish. Current Opinion in Virology 1: 73–80.

55. Munro ES, Allan CET, Matejusova I, Murray AG, Raynard RS (2013) An outbreak of viral haemorrhagic septicaemia (vhs) in wrasse cohabiting with Atlantic salmon in the Shetland Isles, Scotland. Report on the 17th Annual Meeting of the National Reference Laboratories for Fish Diseases Copenhagen, Denmark, May 29–30, 2013: 22–23.

56. Bergh Ø (2007) The dual myths of the healthy wild fish and the unhealthy farmed fish. Diseases of Aquatic Organisms 75: 159–164.

8

Chitosan-Nanoconjugated Hormone Nanoparticles for Sustained Surge of Gonadotropins and Enhanced Reproductive Output in Female Fish

Mohd Ashraf Rather, Rupam Sharma, Subodh Gupta, S. Ferosekhan, V. L. Ramya, Sanjay B. Jadhao*

Central Institute of Fisheries Education, Versova, Mumbai, India

Abstract

A controlled release delivery system helps to overcome the problem of short life of the leutinizing hormone releasing hormone (LHRH) in blood and avoids use of multiple injections to enhance reproductive efficacy. Chitosan- and chitosan-gold nanoconjugates of salmon LHRH of desired size, dispersity and zeta potential were synthesized and evaluated at half the dose rate against full dose of bare LHRH for their reproductive efficacy in the female fish, *Cyprinus carpio*. Whereas injections of both the nanoconjugates induced controlled and sustained surge of the hormones with peak (P<0.01) at 24 hrs, surge due to bare LHRH reached its peak at 7 hrs and either remained at plateau or sharply declined thereafter. While the percentage of relative total eggs produced by fish were 130 and 67 per cent higher, that of fertilised eggs were 171 and 88 per cent higher on chitosan- and chitosan-gold nanoconjugates than bare LHRH. Chitosan nanoconjugates had a 13 per cent higher and chitosan gold preparation had a 9 per cent higher fertilization rate than bare LHRH. Histology of the ovaries also attested the pronounced effect of nanoparticles on reproductive output. This is the first report on use of chitosan-conjugated nanodelivery of gonadotropic hormone in fish.

Editor: Valentin Ceña, Universidad de Castilla-La Mancha, Spain

Funding: The authors have no support or funding to report.

Competing Interests: The authors have declared that no competing interests exist.

* E-mail: sbjadhao@hotmail.com

Introduction

Natural spawning grounds are undergoing extreme changes as a result of anthropogenic activities, pollution and climate change, leading to decline in fish populations. Moreover, inadequate knowledge of the reproductive and breeding process of several fish species is one of the reasons for inability to overcome depleted fish stock and biodiversity loss. It is a hindrance to realize the full potential of intensification of aquaculture worldwide. Aquaculture production of the world will be 65–85 million tons by 2020 and 79–110 million tons by 2030 as compared to 55 million tons in 2009 [1]. Thus lion's share of this production will be contributed by fish farming in captive condition. But many cultivable fishes do not reproduce spontaneously in captivity. This leads to inadequate supply of good quality seeds for intensification of aquaculture. Aquaculture production depends on captive broodstock in which full control of biological processes can be exercised through environmental, hormonal and genetic manipulations. In captive condition fishes may exhibit different reproductive responses due to abiotic factors prevailing in the cultured system, which results in stress-induced immune and endocrine disequilibrium. In all fishes reproductive failure such as incomplete vitellogenesis is mostly observed in females, and hence final oocyte maturation and ovulation do not occur [2]. Sometimes, in spite of having oocyte maturation and ovulation, voluntary spawning does not take place. This is due to the complex mating behaviour required for successful spawning and the absence of the natural environmental conditions such as substrate, hydrology and temperature [3].

Thus, there is a need to develop methods for controlling reproductive processes in fishes reared in captivity so that problems related to synchronization; egg collection and seasonal reproduction can be overcome. Nanobiotechnology has enormous potential to revolutionize the field of reproductive biology and overcome barriers to successful reproduction in a number of species, especially via the use of controlled hormonal nanodelivery systems.

The reproductive process can be regulated through controlled release of hypothalmo-pituitary hormones or gonadotropins (GtH) such as follicle stimulating hormone (FSH) and luteinizing hormone (LH). The release of these hormones is controlled by gonadotropin releasing hormone (GnRH). Ovulation and spawning induction therapies have been developed employing pituitary homogenates, purified gonadotropic preparations and more recently, synthetic agonists of the gonadotropin-releasing hormone or other hormones [4–6]. All these therapies are expected to enhance surge in GtH. However, sufficient ovulation and spawning does not occur on account of either short lifetime of the GnRH in blood caused by rapid degradation by both specific endopeptidases and non-specific exopeptidases present in pituitary, kidney and liver [7], or insufficient effect of these therapies (effective only in 20–30 percent of females). One way to conquer this limitation and to attain long lasting surge in GtH in the blood is to use multiple injections of GnRH or analogues, but again, multiple injections can place stress on fish. Another approach which has been successfully employed to enhance long lasting surge of GtH in blood is to use a sustained release delivery system

[2] through the implantation method. The nanoconjugated delivery of peptides enhances half life of the biomolecule leading to controlled and sustained delivery, as nano-carriers make them able to pass through biological barriers. The coating of biomolecules on nanoparticles of inert metals like gold and silver or encapsulation of nanoparticles of biopolymers provide protection from rapid degradation, targeted delivery and control of the release of bioactive agents.

Chitosan, being cheaper, is one among several compounds which has received a great deal of attention for the preparation of micro and nano-particles for parenteral, nasal, ophthalmic, transdermal and implantable delivery of drugs, proteins, peptides, and gene materials [8]. Chitosan [α (1$\rightarrow$4) 2-amino 2-deoxy β-D-glucan] is a cationic polysaccharide obtained from the deacetylation of chitin. It has unique properties such as biocompatibility, biodegradability, low-immunogenicity and non-toxicity [9]. Chitosan nanoparticles are shown to be an attractive alternative to liposomes for the delivery of peptides, proteins, antigen oligonucleotides and genes, since it has the advantages of longer shelf life and generally a higher drug carrying capacity [10] as chitosan exhibits pH dependent solubility [11].

Gold nanoparticles provide non-toxic carriers for drug and gene delivery applications. With these systems, the gold core imparts stability to the assembly, while the monolayer of active compound (such as hormone in this case) allows tuning of surface properties such as charge and hydrophobicity. Nanoparticles of gold are excellent candidates for making nanoconjugates of biomolecules [12]. Many researchers have shown that biologically active substances with amine function can bind strongly with gold nanoparticles [13]. They are biocompatible, bind readily to a large range of biomolecules such as amino acids, proteins, enzymes and DNA, and expose large surface areas for immobilization of such biomolecules [14]. The ability to modulate the surface chemistry of gold nanoparticles by binding suitable ligand has important applications in many areas including drug delivery. Recently, a rapid and mild method for synthesizing magnetite-gold nanoparticles using chitosan was investigated [15].

Gonadotropin-releasing hormone (GnRH), also known as luteinizing-hormone-releasing hormone (LHRH), is a tropic peptide hormone responsible for the release of follicle stimulating hormone (FSH) and LH from the anterior pituitary. GnRH is degraded by proteolysis within a few minutes. The GnRH and its analogues [16] and its controlled release formulations [2] have been extensively used for manipulating reproduction of fishes. The estradiol loaded chitosan nanoparticle delivery has been shown to be more effective in the treatment of various ailments such as estrogen deficient hyperlipidemic condition [17] and long term oestrogen replacement therapy and Alzheimer's disease [18] in rats. However, to the best of our knowledge to date, use of chitosan for fabrication of reproductive hormone nanoparticles has not been reported. The aim of the present study was to prepare nanoconjugates of salmon LHRH with chitosan and chitosan-gold for assessing their efficacy in sustaining serum hormone levels including functional response (ovulation, total eggs and fertile eggs) in test animals. The test species chosen for the purpose was *Cyprinus carpio* (common carp) for its ease of breeding, high fecundity and short breeding season.

Materials and Methods

Chemicals

Chitosan from shrimp shell (degree of deacetylation >85 percent, MW 200kDa), salmon luteinizing hormone releasing hormone (amino acid sequence: Glp-His-Trp-Ser-Tyr-Gly-Trp-Leu-Pro-Gly-NH$_2$), gold powder and pentasodium tripolyphosphate were purchased from Sigma-Aldrich Corporation (St. Louis, MN). Commercial enzyme immunoassay (EIA) kits were purchased from Omega Diagnostics Ltd, Scotland, U.K. All chemicals were of reagent grade.

Preparation of Chitosan Nanoparticles

Chitosan nanoparticles were prepared based on the ionic gelation method [19] of chitosan and tripolyphosphate (TPP) anion with little modification. Briefly, 2 mg of chitosan were dissolved in 100 ml aqueous acidic solution to obtain the cation of chitosan. This aqueous solution was prepared with 80 ml of water, 15 ml of TPP and 5 ml of acetic acid. The solution was subjected to the constant magnetic stirring for 10 min. The pH of the solution was adjusted to 6.5. During the process involving chemical reaction, chitosan undergoes ionic gelation and precipitates to form spherical particles.

Synthesis of Chitosan- Conjugated Gold Nanoparticles

Chitosan-conjugated gold nanoparticles were synthesized according to the Turkevich [20] procedure of reduction of chloroauric acid with slight modifications. Various chitosan concentrations (0.01%, 0.1% and 0.2%) were used to determine the effect of chitosan concentration on the formation of gold nanoparticles. In each of these solutions 50 µl of chloroauric acid (HAuCl$_4$) at a concentration of 1 mg/ml was added. The whole solution was heated at 60–80°C under constant magnetic stirring for 20 min to yield a ruby-red solution. The absorbance of the solution was measured at 520 nm.

Nano-conjugation of the Salmon Luteinizing Hormone Releasing Hormone

For conjugation of the nanoparticles, a high pressure homogenization process was used. A stock solution of salmon's LHRH was prepared with 1 mg/1 ml of water. From this stock solution, three different volumes of 50µl, 100 µl and 200 µl were added to chitosan and chitosan-gold nanoparticle solutions. Then concentration of the protein, i.e., LHRH in the solution was measured at 660 nm [21]. After that the solution was homogenized at 35,000 rpm for 10 min and kept for overnight in refrigerator at 4°C. The next day, the solution was centrifuged again at 2000 rpm. Supernatant was collected and the concentration of protein was again determined [21]. The entrapment efficiency (EE) of the hormone in chitosan-LHRH and chitosan-gold-LHRH was calculated using the formula

$$EE = \frac{TotalLHRH\text{-}FreeLHRHinsupernatant}{TotalLHRH} \times 100$$

Characterization of Chitosan and Chitosan-gold LHRH Nanoparticles

The particle size and zeta potential of the nanoparticles were measured using Beckman Coulter Delsa Nano C- NanoParticle Size Analyzer (Brea, CA). The instrument which uses photon correlation spectroscopy gives the particle size in the range of 0.6 nm to 7 µm. A drop of the nanoconjugate suspension sample (about 5µl) was diluted with 5 ml of deionized water. The cuvette was well-shaken by hand and placed immediately inside the sample holder of the particle size analyzer. Once the required intensity was reached, analysis was performed to obtain the mean particle size and polydispersity index (PDI) of the sample. Measurement of the zeta potential gives an idea about the stability of any colloidal system and it was determined based on an electrophoretic light scattering (ELS) technique. The surface morphology (roundness, smoothness and formation of aggregates) was studied by transmission electron microscope (TEM).

Ethics Statement

The research undertaken complies with the current animal welfare laws in India. The care and treatment of animals used in this study were in accordance with the guidelines of the CPCSEA [(Committee for the Purpose of Control and Supervision of Experiments on Animals), Ministry of Environment & Forests (Animal Welfare Division), Govt of India] on care and use of animals in scientific research. The study was approved by the Board of Studies and authorities of the Central Institute of Fisheries Education (Deemed University), Mumbai-61. As the experimental fish *Cyprinus carpio* is not an endangered fish, the provisions of the Govt of India's Wildlife Protection Act of 1972 are not applicable for experiments on this fish.

Animal Procurement, Rearing and Experimental Conditions

The animals (*Cyprinus carpio*) were procured from Aquaculture Farm, Pen, Raigad Dist, Maharashtra (India) and were stocked in a circular tank (1000 L) after giving a prophylactic dip treatment in $KMnO_4$ solution (50 mg/L) for 2 min. They were maintained for two months in a backyard hatchery of the Central Institute of Fisheries Education, Mumbai, India prior to the experiment and were fed twice a day with a diet containing 35% crude protein. Water in the tanks was aerated round the clock. One-half of the water was exchanged every week. Water quality parameters in the tank were recorded during the study [22].

Animal Work and Experimental Treatments

Sexually mature male (average body weight 550g) and female (average body weight 900g) fishes were used for the study. The animal experiment in this study included four treatments with three replicates. Each replicate contained two matured male and two female brood fish in each tank. The first group was maintained as a control in which fishes were not given any injection of hormone. In second group, female broods were given intramuscular injection of salmon LHRH @ 0.2 ml/kg body weight of fish. It was expected that the efficacy of conjugated LHRH nanoparticles would surpass the bare LHRH, hence a solution of chitosan- and chitosan-gold conjugated LHRH was injected at 0.1 ml/kg of fish (half the dose of the bare hormone) in the third and fourth groups, respectively. Solution of each injectable preparation contained 20 µg LHRH and 10 mg domperidone per ml. The injections were given using a 1.0 ml syringe fitted with 22 gauge needle. Blood was collected at the 0th, 3rd, 7th and 24th hour post-injection in all the groups. However, to avoid stress during blood collection, blood was collected from a single fish on 0th hr and 7th hr while another fish was used for 3rd and 24th hr in each replicate. Post-injection period of 24 hrs was selected for the studies because successful ovulation in this fish [23] and also in majority of tropical fishes can be accomplished within 24 hrs by administration of a single dose of any hormonal preparation including but not limited to LHRH [24]. Various hormones were analysed by enzyme immunoassay (EIA) kit as per manufacturer's instructions. The ELISA plates were read using Biotek Microplate Instrumentation (Winooski, VT). After spawning, eggs were collected and counted as per standard fish hatchery practice [25].

Collection, Preparation of Tissue Samples and Histology

Fish were anaesthetized with clove oil (50 µl/l)) until they remained motionless. When there was cessation of heartbeat/respiration and the fish were unresponsive to external stimuli and had lost all reflexes, fish were removed from water. Ovary was dissected out and fixed in aqueous Bouin's fluid for 24 hours. After fixing, tissues were washed in 70% alcohol in various changes till the yellow colour of picric acid was removed. Further, the ovaries were dehydrated in ascending grades of ethanol, cleaned in xylene and embedded in paraffin wax (58–60°C congealing point). From the paraffin block, section of 6–8 µm thickness were prepared using a rotary microtome and stretched on albumenised slide. The slides were fixed at 60°C overnight. The next morning, sections were deparaffinised in three changes of xylene and dehydrated in descending grades of alcohol to distilled water. The slides were stained in haematoxylin for 20 minutes, differentiated in 1% of alcohol and blued in ammonia water. After washing, sections were stained with eosin (working) for 10 minutes. Dehydrated and cleaned sections were then mounted in DPX [26] and observed under microscope to record gonadal changes through microphotography.

Statistical Analysis

Data on serum hormone levels was subjected to analysis of variance (ANOVA), followed by Duncan's Multiple Range Test with the help of SPSS-16.0 version software. All the data analysis was expressed as mean ± standard error.

Results

Functional Structures and Physicochemical Characterization of Nanoparticles

Due to the reaction involving complex formation between oppositely charged species (negatively charged groups of the pentasodium tri-polyphosphate and the positively charged amino groups of chitosan), chitosan undergoes ionic gelation and precipitates to form spherical particles. The functional structure of chitosan nanoparticles is depicted in Fig. 1. In case of chitosan-gold nanoparticles, according to Turkevich, varying the ratio of gold salt to chitosan concentration in the medium helps in achieving size-controlled synthesis of nanoparticles and thus the formation of intense ruby red colour in beaker containing 0.2 per cent chitosan and 50 µg gold was indication of nano-size particles (Fig. 2), which were used for the animal experimentation. Thus, the solution containing 0.2% chitosan and 50 µg of gold was used for synthesizing LHRH preparation that was used for animal studies. Schematic diagram showing the functional structure of chitosan-gold-LHRH nanoparticles is shown in Fig. 3. It depicts chitosan encapsulation of gold nanoparticles and subsequent conjugation of LHRH on chitosan encapsulated gold nanoparticles. The distribution of particle size for chitosan-LHRH and chitosan-gold-LHRH nanoparticles is given in Fig. 4a and b. Chitosan-gold-LHRH nanoparticles were bigger (P<0.05) in size than chitosan-LHRH nanoparticles (Table 1). In general, both the particles were compact, spherical in structure, well dispersed and pretty stable as evinced by TEM and physico-chemical characteristics like polydispersity index (0.335 to 0.45) and zeta potential (−33.14 to −34.95 mV) (Table 1). Chitosan nanoparticles had 69 per cent, whereas chitosan-gold nanoparticles had 60 percent, entrapment efficiency for LHRH.

Serum Hormone Levels

Significant effects of treatment (P<0.01) using nanoconjuagated hormone were discernible where increased levels of the hormone

Figure 1. Schematic diagram of the functional structure of chitosan nanoparticles formed during ionic gelation process. (A) Sodium tripolyphosphate, (B) chitosan, (C) chitosan nanoparticles.

were recorded. Both the nanoconjugated preparations of LHRH induced sustained and controlled release of most of the hormones with highest (P<0.01) surge at 24 hrs. In case of bare hormone, surge due to LHRH reached its peak at 7 hrs and either remained at plateau for 17α, 20β-dihydroxy-4- pregnen-3-one (17α, 20β- OHP) and testosterone, or sharply declined for LH, FSH and 17β-estradiol (E2) (Fig. 5).

Fertilization Rate

As fishes were reared in water appropriate for carp breeding (dissolved O_2:4 to 6 mg/L, temperature: 23 to 27°C, pH-7.2–8.3,

0.01% Chitosan + 50μg of gold

0.1% Chitosan + 50μg of gold

0.2% Chitosan + 50μg of gold

Figure 2. Formation of chitosan gold nanoparticles. Varying the ratio of gold salt to chitosan in the medium helps in achieving size controlled synthesis of nanoparticles. Nanoparticles formed in the last beaker were used for animal experimentation.

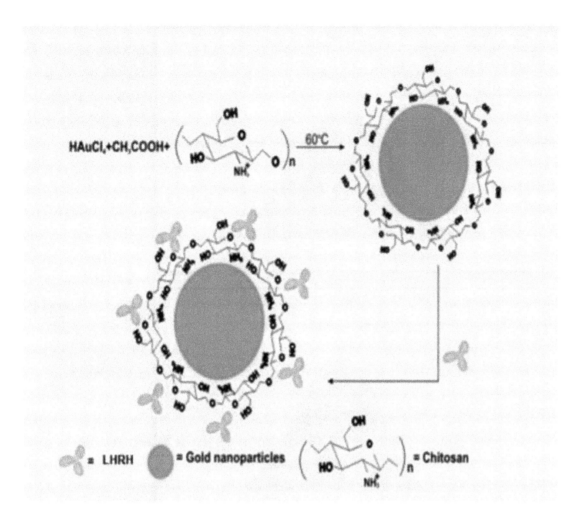

Figure 3. Schematic diagram showing the functional structure of chitosan encapsulated gold nanoparticles and subsequent conjugation of LHRH on chitosan encapsulated gold nanoparticles.

total hardness: 243–254 mg/L, ammonia: 0.11–0.23, nitrite: 0.002–0.003 and nitrate: 0.02–0.04), and the hormonal preparations were effective, the spawning and fertilisation response was good. Out of total 348, 253.3 and 151.2 thousands eggs released following injections of bare LHRH, chitosan-nanoconjugated and chitosan-gold nanoconjugated LHRH 74%, 87% and 83% eggs were fertilised, respectively (Fig. 6). Thus, the pronounced effect of chitosan LHRH nanoparticles was evinced in terms of reproductive output.

Histological Study of Gonads

The histology photographs of ovaries from different groups are given in Fig. 7. Histological studies of gonads in control fish showed no spawning and the matured ova were seen with yolk globules. The fish injected with LHRH showed partial spawning with many fully matured ova. However, complete spawning occurred as evidenced by the presence of spent ova and empty follicles in the ovaries of both the chitosan-LHRH and chitosan-gold LHRH nanoparticle-injected fish.

Discussion

The present work was carried out to develop the nanodelivery of the hormone in fish for obtaining a long lasting surge of GnRH/ LHRH. This nanoconjugation increases the stability of the hormone, which otherwise has short lifetime in the blood [2]. Such nanodelivery has quite a pronounced advantage over multiple injections. Various chitosan-conjugated nanoparticles of LHRH have been developed for treating the diseases such as steroid hormone disorders [27] for terrestrial species. However, this is the first report on development of chitosan-nanoconjugated hormone for its application in fish reproduction. Two types of salmon LHRH nanoparticles were developed using chitosan and chitosan-gold as a nanocarrier. The ionotropic gelation method used in the present study has already been successfully employed to prepare chitosan nanoparticles for the delivery of peptides and proteins [28].

During synthesis, intense ruby red colour formation in case of chitosan-gold nanoparticles was a clear indication of development of nanoparticles. A similar observation was also made by Turkevich [20]. In the present work, chitosan-gold nanoparticles were seen changing colour from whitish to purple and then ruby red colour. In a study by Bhumkar et al. [29] chitosan concentrations of 0.1% and above employed for reduction and synthesis of gold nanoparticles showed no aggregation, while lower concentrations of 0.01% to 0.05% resulted in aggregated particles. Hence, they used 0.2% chitosan reduced gold nanoparticles for loading of insulin because of the best zeta potential, stability (6 months) and viscosity of the various gold nanoparticles prepared

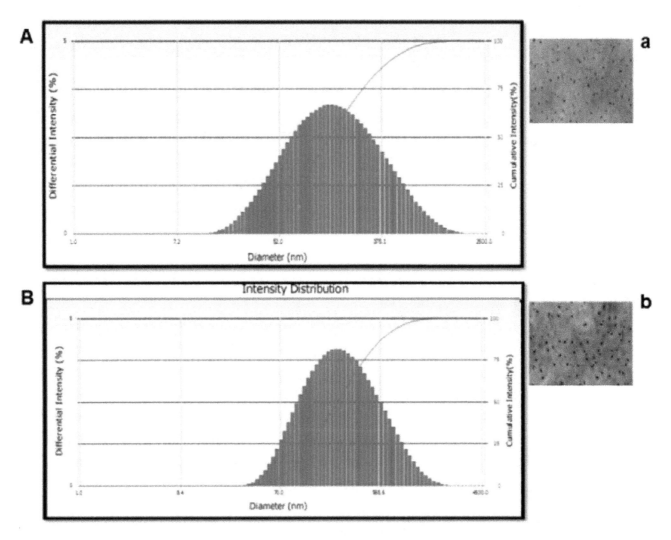

Figure 4. Particle size distributions of chitosan-LHRH (A) and chitosan-gold-LHRH nanoparticles (B) and their respective TEM images (a, b).

from these, which showed effective glucose control in diabetes in rats. In another study [30], out of three different chitosan concentrations (0.1, 0.2 and 0.3 per cent), optimum encapsulation and overall release of protein was found with nanoparticles prepared from 0.2% chitosan. In yet another study, hepatocyte growth factor incorporated (0.2%) chitosan nanoparticles were found to augment the differentiation of stem cells into hepatocytes for the recovery of liver cirrhosis in mice [31]. Thus, our selection

of 0.2% chitosan nanoparticles corroborate with published reports [29–31]. The chitosan and chitosan-gold preparations of LHRH were nanosize as evidenced by the particle size analyzer and TEM image. The larger (P<0.05) size of chitosan-gold nanoparticles with 9% less entrapment efficiency than chitosan nanoparticles was obvious because of the usage of gold in addition to chitosan. The mean size of estradiol-loaded chitosan nanoparticles prepared [18] using the same technique was 269 nm with a zeta potential of

Table 1. Morphological and physico-chemical characteristics of salmon luteinizing-releasing hormone (LHRH) nanoparticles at 25°C in diluent water.

LHRH nanoparticles	Chitosan-conjugated	Chitosan- gold-conjugated
Size (nm)	114±10.3	192.5*±19.1
Zeta potential[†] (mV)	−33.14±6.67	−34.95±7.5
Polydispersity Index[†]	0.335	0.47
Entrapment efficiency (%)	69.00	60.00

[†]Size and zeta potential are expressed as mean± SE of n = 3.
*P<0.05 (by student t test).

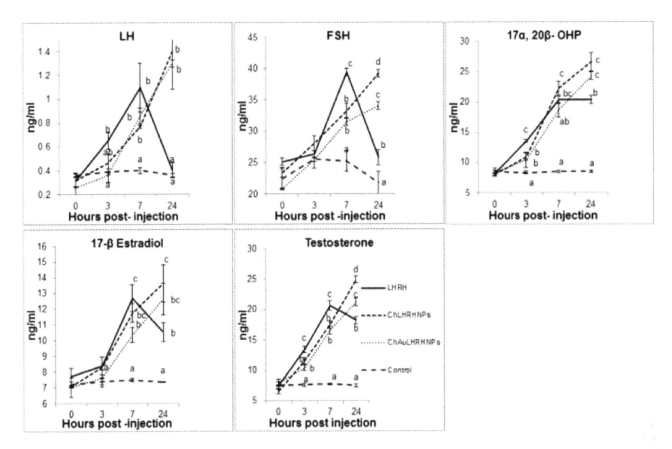

Figure 5. Serum levels of various reproductive hormones in female *Cyprinus carpio* **following injection of different preparations of salmon luteinising hormone-releasing hormone (LHRH) during 24 hr period.** Fishes were injected with either no active compounds (control) or with bare LHRH, chitosan LHRH nanoparticles (ChLHRHNPs) and chitosan-gold LHRH (ChAuLHRHNPs). Means bearing different superscript letters (a, b, c, d) differ significantly (P<0.01) at that hour point (vertical comparison). Each point is mean ±SE of three observations.

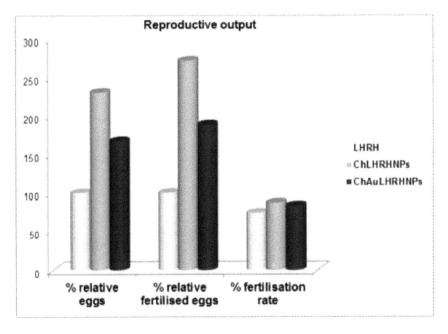

Figure 6. Reproductive output of the *Cyprinus carpio* **injected with different preparations of salmon luteinising hormone-releasing hormone (LHRH).** Percent relative eggs and fertilised eggs are expressed as relative to 100% in bare LHRH preparation. Fishes were injected with bare LHRH, chitosan LHRH nanoparticles (ChLHRHNPS) and chitosan-gold LHRH (ChAuLHRHNPs).

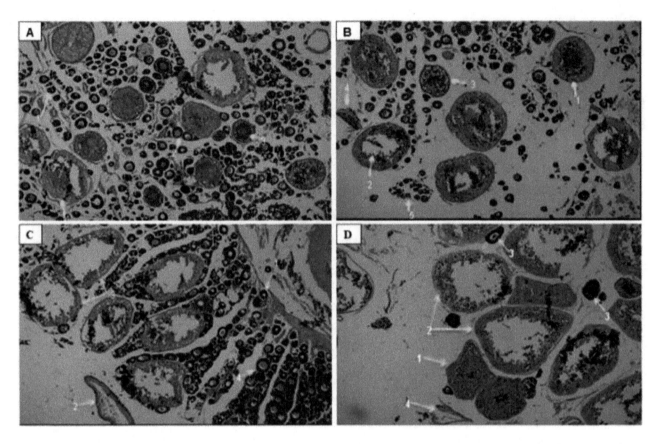

Figure 7. Histology of ovary from *Cyprinus carpio* injected with different preparations of salmon luteinising hormone- releasing hormone (LHRH). A]. from control fishes: 1.Ovigerous lamellae with immature oocyte; 2. mature oocyte; 3. maturing oocyte; 4. atretic oocyte. B] from fish injected with bare LHRH. 1. mature/ripe oocyte; 2. Partially spent oocyte; 3. cortical alveoli stage oocyte; 4. spent follicle; 5. immature oocytes in nest C] from fish injected with chitosan LHRH nanoparticles. 1. Oocytes nested in lamellae 2. spent follicle 3. spent oocytes 4. ovigerous lamellae with immature oocytes D] from fish injected with chitosan gold LHRH nanoparticles 1. mature oocytes 2. spent oocytes 3. immature oocytes 4. discharged follicle.

+25.4 mV and entrapment efficiency of 67%. The nanoparticles developed in our study were much smaller in size and more stable than these reported [18;30]. Our result is comparable with reported [21] particle size of 115 nm with 51% entrapment for estradiol loaded PLGA nanoparticles. While polydispersity index (PDI) of chitosan-gold-LHRH nanoconjugates was slightly higher (0.47) than chitosan-LHRH nanoconjugates (0.33), the PDI and functionality was well in accordance with published reports [30,32]. Out of twenty nanoformulations prepared with different concentrations of plasmid DNA and chitosan for expression of interleukin-12 (IL-12) having antitumor effect, the chitosan-DNA nanoparticles that had minimal cytotoxicity and suggested as suitable candidate for IL-12 gene delivery had mean particle size of 381.83 ± 82.77 nm, PDI of 0.44 ± 0.066 and encapsulation efficiency of 82.17 ± 5.61 [32]. In studies by Dounighi et al. [30], out of three different nanoproteins, the best (0.2%) chitosan-loaded protein nanoparticles (in terms of encapsulation and protein release) were 370 ± 34.7 nm in size and had PDI of 0.429, while PDI of nanoparticles with 0.1% chitosan concentration was not even within the acceptable range, as it formed aggregates with large diameters.

The present hormonal and reproductive output studies suggest that the common carp pituitary is receptive to bare and nanoconjugated LHRH, which is evidenced by hormone levels in treated fish. Spawning experiments under routine hatchery conditions with a single hormonal injection [23] resulted in an initial rise in the level of maturational gonadotropin (GtH) 3 hrs after treatment reaching peak at 14 hrs, when full ovulation took place, as reflected by the presence of expelled eggs on the bottom of the tank. This rise was associated with increased levels of 17β-estradiol (E2) and 17α, 20β-OHP. Physiological increases in serum level of gonadotropins (FSH, LH), estadiol and 17α, 20β-OHP in *C. carpio* in the current study is in agreement with that of Drori et al. [23]. Moreover, the effect of the size of the nanoparticles on endogenous hormone levels (LH, FSH, 17α, 20β-OHP, estradiol and testosterone) and duration of release has already been reported [17]. The observed response to nanoconjugated preparations of LHRH in terms of the levels of these hormones is in agreement with the previous studies. Compared to the bare LHRH-administered group where hormonal peaks was found to decrease after 7 hrs post injection, sustained and controlled release of the endogenous hormone showing enhanced peak at 24 hrs were indicative of resistance of chitosan- and chitosan-gold nanoparticles against enzymatic degradation. High drug loading and prolonged drug release are the advantages of using chitosan for delivery of biomolecules [33]. Chitosan nanoparticles are promising carriers for protein delivery [34] because of the solubility of chitosan in an aquatic medium resulting in better permeability. A similar permeation enhancing effect of chitosan on human insulin has been reported [35–36]. It was expected that the efficacy of nanoform of LHRH would surpass the bare LHRH. Indeed, despite injecting half the dose of bare LHRH in nanoconjugated forms (either with chitosan or chitosan-gold), elevated and sustained surge of steroid hormones, especially gonadotropins (FSH and LH), were observed. This resulted in an increase in follicle maturation and subsequent ovulation, leading to high reproductive output in the form of total and fertilized eggs. The nano-form of LHRH could penetrate deep into tissues through fine capillaries, cross the epithelial lining, and is taken up efficiently by the cells [37]. Elevated testosterone and estradiol levels during vitellogenic growth and ovulation found in this study are consistent with published reports in many species [38–40]. Plasma 17α, 20β- OHP levels in nano LHRH-injected fish followed the same trend as estradiol and testosterone. This is consistent with many studies with teleosts that suggest the involvement of 17α, 20β-OHP in final oocyte maturation [41]. Our results on sustained elevated levels of LH over a period of 24 hrs corroborate well with a published report [42], in which enhanced reproductive output (milt production) in spermiating males of European sea bass (*Dicentrarchus labrax*) for six weeks with controlled-release of GnRH analogue was found. Compared to bare LHRH preparation, the per cent relative eggs (total and fertilised) produced by fish injected with chitosan LHRH nanoconjugates (130 and 171 per cent) and chitosan-gold nanoconjugates were more (67/88 per cent). Sustained delivery systems of gonadotropin-releasing hormone analogue [43] have been successfully used to synchronize females for reproduction with better quality egg production, a 1.75 times increase in fertilization and a 2.4 times greater hatching rate in female yellowtail flounder (*Pleuronectes ferrugineus*).

No adverse effect of gold nanoparticles was discernible through histological examination. In spite of repeated administration of higher dose in mice, no toxicity of gold nanoparticles was found in mice [44]. Bhumkar et al. [29] reported improved pharmacodynamics with transdermal application of chitosan reduced gold nanoparticles loaded with insulin in controlling the postprandial hyperglycemia. And unlike in diabetes, which calls for repeated administration of chitosan-gold insulin nanoparticles, there is greater safety with a single injection of chitosan-gold hormonal nanoparticles. Moreover, these can be used in diseases for which only a few injections are required to treat.

Conclusions

The chitosan- and chitosan-gold nanoconjugates of salmon leutinizing hormone releasing hormone increased the surge of gonadotropin levels in *Cyprinus carpio*. The sustained release of these nanoconjugated hormones resulted in good reproductive output with no toxic effect. This establishes the potential of chitosan- and chitosan-gold conjugated hormone nanoparticles as a new means of manipulating reproduction in fish. The low cost chitosan-conjugated LHRH may be useful for overcoming reproductive problems in fishes. Such nanodelivery has quite a pronounced advantage over multiple injections. This is the first comprehensive report on development of chitosan- and chitosan-gold nanoconjugated hormone for its application in animal reproduction.

Acknowledgments

The authors are grateful to Dr W.S. Lakra, Director, Central Institute of Fisheries Education and former Director, Dr. Dilip Kumar, for providing support and necessary facilities for carrying out this experiment. We are thankful to Mr Umesh C. Taralkar, Application Specialist, Beckman Coulter's Particle Characterization Laboratory, Tritech Instruments (India) Pvt. Ltd, Mumbai, for his help in analyzing nanoparticles. Ms Theresa Sumrall deserves special thanks for a very careful and critical reading of the revised manuscript shortly before its submission.

Author Contributions

Conceived and designed the experiments: MAR RS SG. Performed the experiments: MAR RS SG. Analyzed the data: MAR SBJ. Contributed reagents/materials/analysis tools: RS SG SF VLR SBJ. Wrote the paper: MAR SBJ.

References

1. Hall SJ, Delaporte A, Philips MJ, Beveridge M, Keefe MO (2011) Blue Frontiers: Managing the environmental costs of aquaculture. The WorldFish Center, Penang, Malaysia.
2. Zohar J (Eilat IL) (1994) Manipulation of ovulation and spawning in fish. US Patent 5288705.
3. Marino PG, Longobardi EA, Finoia MG, Zohar Y, Mylonas CC (2003) Induction of ovulation in captive-reared dusky grouper, *Epinephelus marginatus* (Lowe, 1834) with a sustained-release GnRHa implant. Aquaculture 219: 841–858.
4. Crim LW, Bettles S (1997) Use of GnRH analogues in fish culture in: Fingerman M, Nagabhushamam R, Thompson MF. (Eds.) Recent Advances in Marine Biotechnology: 1, 369–382. New Delhi Oxford and IBH Publishing,
5. Zohar Y (1988) Gonadotropin releasing hormone in spawning induction in teleosts: basic and applied considerations, in: Zohar, Y. Breton, B. (Eds.), Reproduction in Fish: Basic and Applied Aspects in Endocrinology and Genetics. 47–62 INRA Press, Paris.
6. Zohar Y, Mylonas CC (2001) Endocrine manipulations of spawning in cultured fish: from hormones to genes. Aquaculture 197: 99–136.
7. Goren A, Zohar Y, Fridkin E, Elhanati E, Koch Y (1987) Degradation of gonadotropin-releasing hormone in the gilthead seabream, *Sparus aurata*: Cleavage of native salmon GnRH, mammalian LHRH and their analogs in the pituitary. Gen Com Endocrinol 79: 291–305.
8. Chun W, Xiong F, Lian SY (2007) Water-soluble chitosan nanoparticles as a novel carrier system for protein delivery. Chin Sci Bull 52: 883–889.
9. Thanou M, Verhoef JC, Junginger HE (2001) Chitosan and its derivatives as intestinal absorption enhancers. Adv Drug Deliv Res 50: 91–101.
10. Janes KA, Calvo P, Alonso M (2001) Polysaccharide colloidal particles as delivery systems for macromolecules. Adv Drug Deliv Res 47: 83–97.
11. Kim TH, Park IK, Nah JW, Choi YJ, Cho CS (2004) Galactosylated chitosan/ DNA nanoparticles prepared using water-soluble chitosan as a gene carrier. Biomaterials 253: 783–3792.
12. Hughes GA (2005) Nanostructure-mediated drug delivery. Nanomed: Nanotechol Biol Med 1: 22–30.

13. Joshi H, Shirude PS, Bansal V, Ganesh KN, Sastry M (2004) Isothermal titration calorimetry studies on the binding amino acid delivery system to gold nanoparticles. J Phys Chem 108: 11535–11540.
14. Joshi HM, Bhumkar DR, Joshi K, Pokharkar V, Sastry M (2006) Gold nanoparticles as carriers for efficient transmucosal insulin delivery. Langmuir 22: 300–305.
15. Salehizadeh H, Hekmatian E, Sadeghi M, Kennedy K (2012) Synthesis and characterization of core-shell Fe_3O_4-gold-chitosan nanostructure. J Nanobiotechol 10: 3.
16. Mikolajczyk T, Roelants I, Epler P, Ollevier F, Chyb J, et al. (2002) Modified absorption of sGnRH-a following rectal and oral delivery to common carp, *Cyprinus carpio*. Aquaculture 203: 375–388.
17. Mittal G, Sahana DK, Bhardwaj V (2007) Estradiol loaded PLGA nanoparticles for oral administration: Effect of polymer molecular weight and copolymer composition on release behavior *in vitro* and *in vivo*. J Control Release 11: 977–85.
18. Wang, X Chi, N Tang X (2008) Preparation of estradiol chitosan nanoparticles for improving nasal absorption and brain targeting. Eur J Pharm Biopharm 70: 735–740.
19. Calvo P, Remunan-Lopez, C Vila-Jato JL, Alonso MJ (1997) Novel hydrophilic chitosan-polyethylene oxide nanoparticles as protein carrier. J Applied Poly Sci 63: 125–132.
20. Turkevich J (1985) Colloidal gold. Part I- Historical and preparative aspects, morphology and structure. Gold Bulletin 18: 86–90.
21. Lowry OH, Rosebrough NJ, Farr AL, Randall RJ (1951) Protein measurement with the Folin phenol reagent. J Biol Chem 193: 265–275.
22. APHA (1998) Standard Methods for the Examination of Water and Wastewater, in: Clesceri LS, Greenberg AE, Eaton AD (Ed), 20th ed. Public Health Association, American Water Works Association, Water Environment Federation, Washington DC.
23. Drori S, Ofir M, Sivan BL, Yaron Z (1994) Spawning induction in common carp (*Cyprinus carpio*) using pituitary extract or GnRH superactive analogue combined

with metoclopramide: analysis of hormone profile, progress of oocyte maturation and dependence on temperature. Aquaculture 119: 393–407.

24. Marte CL (1989) Hormone-induced spawning of cultured tropical finfishes. In *Advances in tropical aquaculture*. Tahiti, Feb.20-March 4. *Aquacop. IFREMER Actes de Colloque*, 9: 519–539.

25. Chondar S (1994) Induced carp breeding, third ed. CBS Publishers and Distributors, 64–65. New Delhi, India,

26. Luna LG (1968) Manual of Histologic staining Methods of the armed forces institute of pathology, third ed. McGraw Hill Book company, New York 38–39.

27. Leuschner C, Kumar CS, Hansel W, Soboyejo W, Zhou J, et al. (2006) LHRH-conjugated magnetic iron oxide nanoparticles for detection of breast cancer metastases. Breast Cancer Res Treat 99: 163–176.

28. Chae SY, Jang MK, Nah JW (2005) Influence of molecular weight on oral absorption of water soluble chitosans. J Control Release 102: 383–394.

29. Bhumkar DR, Joshi HM, Sastry M, Pokharkar VB (2007) Chitosan reduced gold nanoparticles as novel carriers for transmucosal delivery of insulin. Pharm Res 24: 1415–26.

30. Dounighi MN, Eskandari R, Avadi MR, Zolfagharian H, Sadeghi MMA, et al. (2012) Preparation and *in vitro* characterization of chitosan nanoparticles containing *Mesobuthus eupeus* scorpion venom as an antigen delivery system. J Venom Anim Toxins 18: 44–52.

31. Pulavendran S, Rose C, Manda AB (2011) Hepatocyte growth factor incorporated chitosan nanoparticles augment the differentiation of stem cell into hepatocytes for the recovery of liver cirrhosis in mice. J Nanobiotechnol 9: 15.

32. Hallaj-Nezhadi S, Valizadeh H, Dastmalchi S, Baradaran B, Jalali MB, et al. (2011) Preparation of chitosan-plasmid DNA nanoparticles encoding interleukin-12 and their expression in CT-26 colon carcinoma cells. J Pharm Pharm Sci. 14(2): 181–95.

33. Bhattarai N, Ramay HR, Chou SH, Zhang M (2006) Chitosan and lactic acid-grafted chitosan nanoparticles as carriers for prolonged drug delivery. Int J Nanomedicine 1: 181–187.

34. Chen Y, Siddalingappa B, Chan PH, Benson HA (2008) Development of a chitosan-based nanoparticle formulation for delivery of a hydrophilic hexapeptide, dalargin. Biopolym 90: 663–70.

35. Sadeghi AMM, Dorkoosh FA, Avadi MR, Weinhold M, Bayat A, et al. (2008) Permeation enhancer effect of chitosan and chitosan derivatives: Comparison of formulations as soluble polymers and nanoparticulate systems on insulin absorption in Caco-2 cells. Eur J Pharm Biopharm 07: 270–278.

36. Yu JM, Li YJ, Qiu LY, Jin Y (2008) Self-aggregated nanoparticles of cholesterol-modified glycol chitosan conjugate: Preparation, characterization, and preliminary assessment as a new drug delivery carrier. Eur Polym J 44: 555–565.

37. Vinogradov SV, Bronich TK, Kabanov AV (2002) Nanosized cationic hydrogels for drug delivery: preparation, properties and interactions with cells. Adv Drug Deliv Res 54: 223–233.

38. Kobayshi MK, Hinayu AI (1989) Involvement of steroid hormones in pre-ovulatatory gonadotropins surge in female gold fish. Fish Physiol Biochem 7: 141–146.

39. Nagahama Y (1983) The functional morphology of teleost gonads. In: Hoar WS, Randall DJ, Donaldson, EM. (Eds.), Fish Physiology Vol. IX, Part A: 223–275 Reproduction. Academic Press, Orlando.

40. Pankhurst NW, Carragher JF (1991) Seasonal endocrine cycles in marine teleosts. in: Scott AP, Sumpter JP, Kime DE, Rolfe MS. (Eds.), Reproductive Physiology of Fish (1991) Fish Symp 91, Sheffield, 131–135.

41. Goetz FW (1983) Hormonal control of oocyte final maturation and ovulation in fish. in: Hoar, WS Randall, DJ Donaldson, EM (Eds.), Fish Physiology. Vol. IX, Part B: Reproduction Academic Press, Orlando, Florida, 117–170.

42. Mananos E, Carrillo M, Sorbera LS, Mylonas CC, Asturiano JF, et al. (2002) Luteinizing hormone and sexual steroid plasma levels after treatment of European sea bass with sustained-release delivery systems for gonadotropin-releasing hormone analogue. J Fish Biol 60: 328–339.

43. Larsson DGJ, Mylonaz CC, Zohar Y, Crim LW (1997) Gonadotropin-releasing hormone analogue (GnRH-A) induces multiple ovulations of high-quality eggs in a cold-water, batch-spawning teleost, the yellowtail flounder (*Pleuronectes ferrugineus*). Can J Fish Aquat Sci 54: (9) 1957–1964.

44. Lasagna-Reeves C, Gonzalez-Romero D, Barria MA, Olmedo I, Clos A, et al. (2010) Bioaccumulation and toxicity of gold nanoparticles after repeated administration in mice. Biochem Biophys Res Commun 393: 649–655.

9

Flavobacterium plurextorum sp. nov. Isolated from Farmed Rainbow Trout (Oncorhynchus mykiss)

Leydis Zamora[1], José F. Fernández-Garayzábal[1,2]*, Cristina Sánchez-Porro[4], Mari Angel Palacios[3], Edward R. B. Moore[5], Lucas Domínguez[1], Antonio Ventosa[4], Ana I. Vela[1,2]

1 Centro de Vigilancia Sanitaria Veterinaria (VISAVET), Universidad Complutense, Madrid, Spain, 2 Departamento de Sanidad Animal, Facultad de Veterinaria, Universidad Complutense, Madrid, Spain, 3 Piszolla, S.L., Alba de Tormes, Salamanca, Spain, 4 Departamento de Microbiología y Parasitología, Facultad de Farmacia, Universidad de Sevilla, Sevilla, Spain, 5 Culture Collection University of Gothenburg (CCUG) and Department of Infectious Disease, Sahlgrenska Academy of the University of Gothenburg, Göteborg, Sweden

Abstract

Five strains (1126-1H-08[T], 51B-09, 986-08, 1084B-08 and 424-08) were isolated from diseased rainbow trout. Cells were Gram-negative rods, 0.7 µm wide and 3 µm long, non-endospore-forming, catalase and oxidase positive. Colonies were circular, yellow-pigmented, smooth and entire on TGE agar after 72 hours incubation at 25°C. They grew in a temperature range between 15°C to 30°C, but they did not grow at 37°C or 42°C. Based on 16S rRNA gene sequence analysis, the isolates belonged to the genus *Flavobacterium*. Strain 1126-1H-08[T] exhibited the highest levels of similarity with *Flavobacterium oncorhynchi* CECT 7678[T] and *Flavobacterium pectinovorum* DSM 6368[T] (98.5% and 97.9% sequence similarity, respectively). DNA–DNA hybridization values were 87 to 99% among the five isolates and ranged from 21 to 48% between strain 1126-1H-08[T], selected as a representative isolate, and the type strains of *Flavobacterium oncorhynchi* CECT 7678[T] and other phylogenetic related *Flavobacterium* species. The DNA G+C content of strain 1126-1H-08[T] was 33.2 mol%. The predominant respiratory quinone was MK-6 and the major fatty acids were iso-$C_{15:0}$ and $C_{15:0}$. These data were similar to those reported for *Flavobacterium* species. Several physiological and biochemical tests differentiated the novel bacterial strains from related *Flavobacterium* species. Phylogenetic, genetic and phenotypic data indicate that these strains represent a new species of the genus *Flavobacterium*, for which the name *Flavobacterium plurextorum* sp. nov. was proposed. The type strain is 1126-1H-08[T] (= CECT 7844[T] = CCUG 60112[T]).

Editor: Dongsheng Zhou, Beijing Institute of Microbiology and Epidemiology, China

Funding: This work was funded by projects CENIT 2007-2010 (ACUISOST) of the Spanish Office for Science and Technology (CDETI), CGL2010-19303 of the Spanish Ministry of Science and Innovation and P10-CVI-6226 from the Junta de Andalucía. ERBM was supported by funding of Västra Götaland Region projects VGFOUREG-30781, 83080 and 157801. The funders had no role in study design, data collection and analysis, decision to publish, or preparation of the manuscript.

Competing Interests: One of the authors, Dr. M.A.Palacios is affiliated to Piszolla, S.L., 37800 Alba de Tormes (Salamanca), Spain. She is the technical manager of the company that identified infectious problems in one of their fish farms and she has collaborated with us for describing the new *Flavobacterium* species.

* E-mail: garayzab@vet.ucm.es

Introduction

The genus *Flavobacterium* is the type genus of the family *Flavobacteriaceae* accommodating Gram-negative, non-endospore-forming, aerobic, oxidase-positive, non-fermenting, predominantly gliding, yellow-pigmented bacteria [1,2]. The genus, initially described to accommodate seven species, has considerably expanded with the description of many new species. Currently it includes 99 species, many of them described during the last five years. [3]. Members of the genus *Flavobacterium* can been isolated from a number of diverse habitats such as soil, water, sludge, plants, food products such as fish, meat, poultry, milk or lactic acid beverages [2,4]. Most species are non-pathogenic, although a number of species have been associated with different clinical infections, being freshwater fish the animals most prone to flavobacterial infections [5]. Some *Flavobacterium* species, mainly *Flavobacterium columnare*, *Flavobacterium branchiophilum* and *Flavobacterium psychrophilum*, are well-recognized fish pathogens responsible for important economic losses in the fish farming industry [6,7]. However, several other

species such as *Flavobacterium hydatis*, *Flavobacterium jhonsoniae*, *Flavobacterium succinicans*, *Flavobacterium chilense*, *Flavobacterium araucananum* or *Flavobacterium oncorhynchi* have been also associated with infections in fish [1,4,5,8–10]. Additionally, a number of new *Flavobacterium* species also have been described from the water of aquaculture facilities [11–13]. This plethora of *Flavobacterium* species could reproduce the diversity of flavobacteria associated with fish or fish surrounding environments. Some of these species could be considered commensal and opportunistic pathogenic bacteria [4], which point out the necessity for an accurate identifications of those strains of *Flavobacterium* spp. isolated from fish or fish farm environments. However, such identifications are extremely difficult based exclusively on biochemical criteria [4,8,14] and must be complemented with chemotaxomic and genetic methods [4,5].

In this article, we report the phenotypic, genotypic and phylogenetic characterization of five novel *Flavobacterium*-like strains isolated from diseased trout. Based on the presented findings, a new species of the genus *Flavobacterium*, *Flavobacterium plurextorum* sp. nov., is proposed.

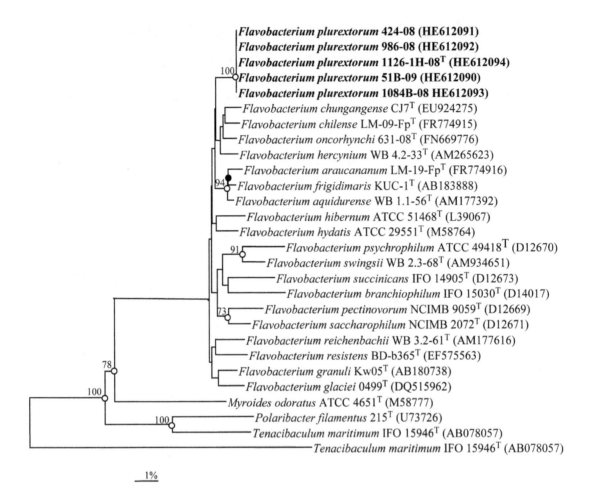

Figure 1. Phylogenetic tree based on 16S rRNA gene sequence comparisons, obtained with the neighbour-joining algorithm, showing the relationships of *Flavobacterium plurextorum* sp. nov. with related species. *Flexibacter flexilis* ATCC 23079^T was used as an outgroup. Bootstrap values (expressed as a percentage of 1,000 replications) greater than 70% are given at the nodes. Solid circles indicate that the corresponding nodes (groupings) are also obtained on the maximum-likelihood tree. Open circles indicate that the corresponding nodes (groupings) are also obtained on the maximum-likelihood and parsimony trees. Sequence accession numbers are indicated in brackets. Bar, 1% sequence divergence.

Materials and Methods

The present work does not include any experimental infections trial with farmed trout, just trout exclusively were used to identify microbiologically the etiological agent of the bacterial septicemia. Therefore, we did not consult with the IACUC and no specific national regulations for these procedures are available. Nevertheless, in order to ensure the welfare and ameliorate suffering of trout during transportation to the laboratory and euthanasia, trout were handled according to guidelines of relevant international organisms such as OIE (http://www.oie.int/doc/ged/D7821.PDF) and AVMA (https://www.avma.org/KB/Policies/Documents/euthanasia.pdf) and they were further necropsied under aseptic conditions. In addition, these procedures were approved by the responsible of animal welfare of the UCM Animal Health Department. The trout were sacrificed for the purpose of the study and the sacrifice was approved by the Technical Manager (Mari Angel Palacios, DVM, PhD) of the fish farm located in the west of Spain.

Trout and Strain Isolation

A clinical episode of septicemia occurred in a rainbow trout (*Oncorhynchus mykiss*) farm located in the central region of Spain.

Affected trout were submitted by the Technical Manager of the fish farm to the Animal Health Surveillance Centre (VISAVET) of the Universidad Complutense (Madrid, Spain) for a confirmatory microbiological diagnosis.

Five Gram-negative, rod-shaped bacteria were isolated from liver (strains 986-08 and 424-08), gills (strains 1084B-08 and 51B-09) and eggs (1126-1H-08^T) of five different trout. The strains were recovered in two different years (2008 and 2009) and they were isolated on tryptone glucose extract agar (TGE; Difco) after incubation at 25°C for 72 hours under aerobic conditions.

Phylogenetic Analysis

A large continuous sequence (approximately 1,400 bases) of the 16S rRNA gene of five strains was determined bidirectionally using universal primers pA (5′- AGAGTTTGATCCTGGCT-CAG, positions 8–27, *Escherichia coli* numbering) and pH* (5′-AAGGAGGTGATCCAGCCGCA, positions 1541–1522, *E. coli* numbering) as described previously [10], and subjected to a comparative analysis. The identification of the phylogenetic relatives and calculations of pair-wise 16S rRNA gene sequence similarities were achieved, using the EzTaxon-e server [15]. The 16S rRNA gene sequences of the type strains of all validly published species of the genus *Flavobacterium* were retrieved from

Table 1. Cellular fatty acid compositions of *Flavobacterium plurextorum* 1126-1H-08[T] and its closest phylogenetic neighbours.

Fatty acid	1	2	3	4	5	6	7	8
Saturated								
$C_{12:1}$	1	tr	–	–	–	–	tr	–
$C_{14:0}$	tr	–	tr	1.1	tr	tr	–	tr
$C_{15:0}$	15	13.5	11.9	5.5	5.6	6.9	20.6	15.7
$C_{16:0}$	2	1.6	1.1	2.3	2.8	2.2	tr	2.9
Hydroxy								
$C_{15:0}$ 2OH	–	–	1.1	–	tr	–	–	–
$C_{15:0}$ 3OH	3	3.3	1.9	–	–	–	1.8	–
iso-$C_{15:0}$ 3OH	6	7.8	6.9	7.7	8.6	5.8	7.1	5.8
$C_{16:0}$ 3OH	3	–	1.1	3.5	4.5	1.4	–	2.5
iso-$C_{16:0}$ 3OH	2	1.0	2.1	1.6	1.9	tr	2.1	1.5
iso-$C_{17:0}$ 3OH	5	8.2	7.3	5.9	10.3	5.1	7.0	
Branched								
$C_{14:0}$ aldehyde	1.0	–	–	–	–	–	–	tr
iso-$C_{15:0}$	19	26.1	14.6	28.2	23.5	28.0	24.8	25.5
anteiso-$C_{15:0}$	1.0	1.3	3.0	3.2	tr	4.3	2.5	1.9
iso-$C_{15:0}$ aldehyde	2.0	3.2	1.2	1.3	tr	1.3	2.3	2.0
iso-$C_{15:1}$ G	6.0	2.9	7.4	3.7	5.8	7.2	5.0	5.0
iso-$C_{16:0}$	1	–	1.1	1.0	tr	1.0	–	1.1
iso-$C_{16:1}$ H	tr	–	1.0	1.0	tr	–	–	tr
iso-$C_{17:1}$ ω9c	3	6.0	5.2	4.3	4.1	6.0	1.1	2.9
Unsaturated								
$C_{15:1}$ ω6c	9	12.3	10.1	4.1	2.9	5.5	12.2	7.6
$C_{16:1}$ ω7c	10	3.7	11.2	19.2	15.7	18.1	2.2	9.8
$C_{17:1}$ ω6c	6	5.9	6.4	3.5	2.5	3.2	6.2	2.4
$C_{17:1}$ ω8c	1	1.0	1.5	–	tr	tr	1.3	tr
Summed feature 1[a]	–	–	–	2.0	1.7	1.4	–	1.7
Unidentified fatty acid[b]								
ECL 11.541	2	1.4	tr	tr	tr	tr	1.2	1.1
ECL 12.555	1	–	tr	–	–	–	1.1	tr
ECL 14.809	1	–	–	–	–	–	–	–
ECL 16.580	–	–	tr	–	1.1	–	–	tr

Taxa: 1, *F. plurextorum* 1126-1H-08[T]; 2, *F. pectinovorum* CCUG 58916[T]; 3, *F. aquidurense* CCUG 59847[T]; 4, *F. frigidimaris* CCUG 59364[T]; 5, *F. hydatis* DSM 2063[T]; 6, *F. araucananum* CCUG 61031[T]; 7, *F. chungangense* CCUG 58910[T]; 8, *F. oncorhynchi* CECT 7678[T].
Values are percentages of total fatty acids; fatty acids representing less than 1% in all strains were omitted. tr = trace amount, i.e., <1%. - = not detected.
CFA values for type strains other than *F. plurextorum* 1126-1H-08[T] were taken from the CCUG culture collection (http://www.ccug.se/). Strains were cultivated on the same medium and growth conditions.
[a]Summed features represent groups of two or three fatty acids that cannot be separated by GLC with the MIDI system. Summed feature 1 comprised iso-$C_{17:1}$ I/$C_{16:0}$ DMA.
[b]ECL, equivalent chain length.

GenBank and aligned with the newly determined sequences using the program SeqTools [16]. Phylogenetic trees were constructed according to three different algorithms: neighbour-joining [17], using the programs SeqTools and TREEVIEW [18]; maximum-likelihood, using the PHYML software [19]; and maximum-parsimony, using the software package MEGA (Molecular Evolutionary Genetics Analysis) version 5.0 [20]. Genetic distances for the neighbour-joining and the maximum-likelihood algorithms were calculated by the Kimura two-parameter [21] and close-neighbour-interchange (search level = 2, random additions = 100) was applied in the maximum-parsimony analysis. The stability of the groupings was estimated by bootstrap analysis (1000 replications).

Genomic DNA G+C Content Determination and DNA-DNA Hybridizations

The G+C content of the genomic DNA of a representative strain (1126-1H-08[T]) was determined from the mid-point value (Tm) of the thermal denaturation profile [22], obtained with a Perkin-Elmer UV-Vis Lambda 20 spectrophotometer at 260 nm.

Genomic DNA-DNA hybridizations were carried out between strains 1126-1H-08[T], 986-08, 424-08, 1084B-08 and 51B-09, and between strain 1126-1H-08[T] and the type strains of the closest

Table 2. Characteristics that differentiate *Flavobacterium plurextorum* sp. nov. from closely related *Flavobacterium* species based in the 16S rRNA tree topology.

Characteristic	1	2	3	4	5	6	7	8
Growth on Marine agar	−	−	−	+	−	−	−	−
Growth at 30°C	+	+	+	−	+	+	+	+
Hydrolysis of:								
L- tyrosine	+	−	+	−	+	−	−	+
DNA	−	−	−	−	+	−	+	−
Urea	−	−	+	−	−	−	−	−
Nitrate reduction	+	+	−	−	+	+	+	+
Assimilation of:								
Arabinose	+	+	+	+	−	+	+	+
Mannitol	−	−	−	+	−	−	−	−
N-acetyl-glucosamine	+	+	+	+	+	+	-	+
Production of:								
Valine arylamidase	−	+	+	+	+	+	+	−
α-Glucosidase	+	−	−	+	+	+	−	+
β-Glucosidase	−	+	−	+	−	+	+	−
N-Acetyl-β-glucosaminidase	+	−	−	+	+	−	−	+

Taxa: 1, *F. plurextorum* 1126-1H-08[T]; 2, *F. pectinovorum* CCUG 58916[T]; 3, *F. aquidurense* CCUG 59847[T]; 4, *F. frigidimaris* CCUG 59364[T]; 5, *F. hydatis* DSM 2063[T]; 6, *F. araucananum* CCUG 61031[T]; 7, *F. chungangense* CCUG 58910[T]; 8, *F. oncorhynchi* CECT 7678[T].
Data are from this study.
+, positive reaction; −, negative reaction.

phylogenetically related species. DNA was extracted and purified by the method of Marmur [22]. Hybridization studies were carried out, using the membrane method of Johnson [23], described in detail by Arahal *et al.* [24]. The hybridization experiments were carried out under optimal conditions, at a temperature of 44°C, which is within the limits of validity for the membrane method [25]. The percentages of hybridization were calculated as described by Johnson [26]. Three independent determinations were carried out for each experiment and the results reported as mean values. The type strains of species *F. aquidurense* CCUG 59847[T], *F. araucananum* CCUG 61031[T], *F. hydatis* DSM 2063[T], *F. pectinovorum* CCUG 58916[T], *F. frigidimaris* CCUG 59364[T], *F. chungangense* CCUG 58910T and *F. oncorhynchi* CECT 7678[T] were included in this study.

Chemotaxonomic Characteristics

Respiratory quinones of strain 1126-1H-08[T] were extracted from 100 mg of freeze-dried cell material, using the two stage method described by Tindall [27,28], and further separated by thin layer chromatography on silica gel and analyzed, using HPLC, by the identification service of the DSMZ (Braunschweig, Germany).

For cell fatty acid-fatty acid methyl ester (CFA-FAME) analyses, strain 1126-1H-08[T] was grown on Columbia II agar base (BBL 4397596) with 5% horse blood, at 30°C for 30–48 h, under aerobic conditions. The CFA-FAME profile was determined using gas chromatography (Hewlett Packard HP 5890) and a standardized protocol similar to that of the MIDI Sherlock MIS system [29], described previously [10]. CFAs were identified and the relative amounts were expressed as percentages of the total fatty acids of the respective strains.

Morphological, Physiological and Biochemical Characteristics

The minimal standards for the description of new taxa in the family *Flavobacteriaceae* [30] were followed for the phenotypic characterization of the strains. Gram-staining was performed as described by Smibert & Krieg [31]. Oxidase activity was determined by monitoring the oxidation of tetramethyl-*p*-phenylenediamine on filter paper and catalase activity was determined, using 3% H_2O_2 solution [31]. Hydrolysis of L-tyrosine (0.5%, w/v), lecithin (5%, w/v) [31], esculin (0.01% esculin and 0.05% ferric citrate, w/v), gelatin (4%; w/v), starch (0.2%, w/v), and casein [50% skimmed milk (Difco), v/v] were tested using nutrient agar as basal medium [30]. DNase test agar (Difco) was used for the DNase assay. Hydrolysis of urea (1%, w/v) was tested as described by Bowman *et al.* [32]. Growth in brain heart infusion broth was assessed at 15, 25, 30, 37 and 42°C, with 3.0, 4.5 and 6.5% added NaCl, and under anaerobic (with 4–10% CO_2) and micro-aerobic (with 5–15% O_2 and 5–12% CO_2) conditions, using GasPak Plus and CampyPak Plus systems (BBL), respectively. Growth was tested on MacConkey (bioMérieux), nutrient (Difco) and trypticase-soy (bioMérieux) agar plates. The presence of gliding motility, using the hanging drop technique, and the production of flexirubin-type pigments and extracellular glycans were assessed, using the KOH and Congo red tests, respectively [1]. The strains were further biochemically characterized using the API 20NE and API ZYM systems (bioMérieux) according to the manufacturer's instructions, except that incubation temperature was 25°C. The type strains of species *F. aquidurense* CCUG 59847[T], *F. araucananum* CCUG 61031[T], *F. hydatis* DSM 2063[T], *F. pectinovorum* CCUG 58916[T], *F. frigidimaris* CCUG 59364[T], *F. chungangense* CCUG 58910[T] and *F. oncorhynchi* CECT 7678[T] were included in this study as references for the investigation of the phenotypic properties of the trout strains, using the same laboratory conditions.

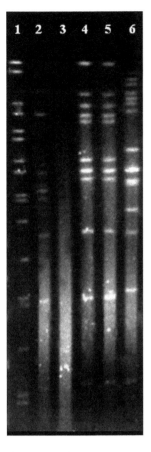

Figure 2. PFGE patterns generated after Bsp120I macrorestriction of *Flavobacterium plurextorum* sp. nov. Lane 1, DNA molecular size marker; Lanes 2 to 6, strains 1126-1H-08^T, 51B-09, 986-08, 1084B-08 and 424-08, respectively.

PGFE Typing

The five strains were characterized by pulsed-field gel electrophoresis (PFGE), after digestion of their genomic DNAs with the restriction enzymes *Bsp120*I and *Xho*I, according to the specifications of Chen *et al.* [33]. DNA fragments were resolved in a 1% agarose gel with a pulse-field gel electrophoresis apparatus, CHEF-DR III (Bio-Rad), at 6V/cm for 40 hours, with switching times ramped from 0.1 to 12 s at 14°C, with an angle of 120°. The gels were stained for 30 min with Syber-Safe and photographed under UV light (Gel-Doc, Bio-Rad). Strains differing in at least one band were considered different.

Results and Discussion

16S rRNA gene sequences were determined for the five trout strains, displaying 100% 16S rRNA sequence similarity among them. Sequence searches showed that the 16S rRNA gene sequence of the strains were most similar to those of species of the genus *Flavobacterium*, exhibiting the highest levels of similarity with the sequence of the type strains of *Flavobacterium oncorhynchi* CECT 7678^T and *Flavobacterium pectinovorum* DSM 6368^T (98.5% and 97.9% sequence similarity, respectively). In addition, strains exhibited 16S rRNA gene sequence similarities greater than 97.0% with other seventeen other *Flavobacterium* species. It is clear from the phylogenetic analysis (Fig. 1) that the trout strains held a clear affiliation to the genus *Flavobacterium* and represented a distinct sub-lineage clustering with a cluster of four species that

included *F. pectinovorum*, *F. chilense*, *F. oncorhynchi* and *F. hercynium*. However, their position within this sub-group was not supported by significant bootstrap values. The GenBank accession numbers for the 16S rRNA gene sequences of five strains sequenced in this study are shown in Fig. 1.

Genomic DNA–DNA hybridizations between the trout strains yielded binding values of 87 to 100%. *Flavobacterium* species with 16S rRNA gene sequence similarities to the sequences of the trout strains lower than 98.0% correlated with levels of genomic DNA-DNA relatedness always lower than 70% [9–11,34–36]. For that reason, DNA-DNA hybridizations were carried out only between strain 1126-1H-08^T and the type strains of the phylogenetically closest related species; *i.e.*, those species with 16S rRNA gene sequence similarities greater than 97.5%. The levels of DNA-DNA relatedness for strain 1126-1H-08^T with respect to *F. aquidurense* CCUG 59847^T, *F. araucananum* CCUG 61031^T, *F. hydatis* DSM 2063^T, *F. pectinovorum* CCUG 58916^T, *F. frigidimaris* CCUG 59364^T, *F. chungangense* CCUG 58910^T and *F. oncorhynchi* CECT 7678^T ranged between 21 and 48%. These values were below the 70% cut-off point for species delineation [37,38] and clearly confirmed that the trout strains belong to a distinct genomic species of the genus *Flavobacterium*. The DNA G+C content of strain 1126-1H-08^T was 33.2 mol%, a value consistent with those of the genus *Flavobacterium* [1,30].

Chemotaxonomic characteristics of strain 1126-1H-08^T were in accordance with those of members of the genus *Flavobacterium* [5,6]: the major quinone was MK-6 (95%) with minor amounts of MK-5 (5%). The predominant cell fatty acids of strain 1126-1H-08^T were iso-$C_{15:0}$ (19%) and $C_{15:0}$ (15%). Strain 1126-1H-08^T also contained moderate or small amounts of $C_{16:1}$ $\omega 7c$ (10%), $C_{15:1}$ $\omega 6c$ (9%), iso-$C_{15:0}$ 3-OH, $C_{17:1}$ $\omega 6c$, isoG-$C_{15:1}$ (6%/each), iso-$C_{17:0}$ 3-OH (5%), iso-$C_{17:1}$ $\omega 9c$, $C_{15:0}$ 3-OH, $C_{16:0}$ 3-OH (3%/each), isoaldehyde-$C_{15:0}$, $C_{16:0}$, iso-$C_{16:0}$ 3-OH, unknown fatty acids with an equivalent chain length of 11.5 (2%/each) and $C_{17:1}$ $\omega 8c$, iso-$C16:0$, $C_{12:1}$, aldehyde-$C_{14:0}$, anteiso-$C_{15:0}$ and unknown fatty acids with an equivalent chain lengths of 14.8 and 12.5 (1%/each) (Table 1).

The trout strains exhibited identical physiological and biochemical characteristics. Cells were Gram-negative rods, 0.7 μm wide and 3 μm long, non-endospore-forming, and non-gliding. Strains grew well under aerobic conditions and grew weakly under micro-aerobic conditions. Strains grew at 15–30°C with optimal growth at approximately 25°C, while no growth was observed at 37°C or 42°C. Growth occurred on trypticase-soy and nutrient agars but not on Marine agar after incubation at 25°C for 72 hours. Colonies were circular, yellow-pigmented, smooth and entire on TGE agar after 72 hours incubation at 25°C. Colonies are non-hemolytic on Columbia agar after 72 hours incubation at 25°C. Diffusible flexirubin-type pigments were produced and congo red was not absorbed by colonies. Growth did not occur in brain heart infusion broth containing 3, 4.5 and 6.5% NaCl. Catalase and oxidase were produced and nitrate and nitrite were reduced. Starch and tyrosine were degraded but DNA, gelatin, casein or agarose were not. A brown pigment was not produced on tyrosine agar. Aesculin was hydrolyzed but not urea, lecithin and arginine. Indole and H_2S were not produced. Acid was not produced from D-glucose. Arabinose, D-glucose, mannose, N-acetyl-glucosamine, and maltose were used as sole carbon and energy sources but not citrate, mannitol, gluconate, caprate, adipate, and malate. Activities for alkaline phosphatase, leucine arylamidase, N-acetyl-β-glucosaminidase, α-glucosidase, acid phosphatase, and naphthol-AS-BI-phophohydrolase were detected. Esterase C4, valine arylamidase, β-galactosidase, ester lipase C8, lipase C14, cystine arylamidase, α-chymotrypsin, trypsin, αgalactosidase, β-glucuronidase, β-glucosidase, α-mannosidase and α-fucosidase were not detected.

The phenotypic characteristics that differentiated the trout strains from phylogenetically related species are shown in Table 2. The new species also can be also differentiated from the clinically relevant fish pathogens *F. columnare*, *F. psycrophilum* and *F. branchiophilum*, by the inability of these three species to grow in trypticase-soy agar and to hydrolyze aesculin [4]. Other species isolated from diseased fish such as *F. hydatis*, *F. jonshoniae* and *F. succinicans* are motile (gliding), degrade DNA and produce acid from carbohydrates [4], while the new species exhibited opposite results for those tests. Moreover, the new species can be readily differentiated from *F. chilense* and *F. araucananum* because the latter species are motile (gliding), grow in 3% NaCl and assimilate mannitol [9] and from *F. oncorhynchi* which produces β-galactosidase while the new species give opposite results for this test [10].

After PFGE typing, the trout strains were characterized by 3 different restriction profiles with the enzymes *Bsp120*I (Fig. 2) and *Xho*I (not shown). Strains 986-08 and 1084B-08 exhibited indistinguishable restriction profiles with both enzymes and strain 51B-09 could not be characterized because its DNA systematically was autodegraded.

Flavobacteria are known to belong to the microbiota of fish and fish eggs [4,5]. Therefore, although two strains were isolated from internal organs, the other three were recovered from gills and eggs which suggest that the new species could be saprophytic or commensal and able to colonize fish, and produce disease under stressful conditions or other predisposing circumstances such as coinfections with other bacteria or viruses, poor farming conditions or environmental disorders [4,39]. This assumption should be confirmed by experimental infection trials. Nevertheless, the formal description of *Flavobacterium plurextorum* and the availability of tests to facilitate its identification from other *Flavobacterium* species associated with fish disease or isolated from diseased fish will aid laboratories in its recognition and identification in the future, and to improve the knowledge of its distribution and possible association with disease.

Conclusion

The phylogenetic, genotypic and phenotypic results of the present polyphasic study demonstrated that the new strains isolated from rainbow trout represented a novel species of the genus *Flavobacterium*, for which the name *Flavobacterium plurextorum* sp. nov. is proposed (plu.rex.to'rum. L. comp. pl. plures, more, several, many; L. pl. n. exta -orum, entrails; N.L. gen. pl. n. plurextorum, of several internal organs). Detailed description of the morphological, physiological and biochemical characteristics of this species were indicated above. The type strain is 1126-1H-08^T (= CECT 7844^T = CCUG 60112^T).

Acknowledgments

The authors thank Professor J. P. Euzéby of the Ecole Nationale Vétérinaire in Toulouse for advice concerning the Latin species name and A. Casamayor (VISAVET) for technical assistance in PFGE analysis and Kent Molin (CCUG) for the analyses of CFAs.

Author Contributions

Conceived and designed the experiments: JFF-G AIV LD. Performed the experiments: LZ CS-P. Analyzed the data: ERBM AV AIV. Wrote the paper: LZ JFF-G AIV. Obtained clinical specimens: MAP. Critical revision and final approval: ERBM AV JFF-G.

References

1. Bernardet JF, Segers P, Vancanneyt M, Berthe F, Kersters K, et al. (1996) Cutting a Gordian knot: emended classification and description of the genus *Flavobacterium* emended description of the family *Flavobacteriaceae* and proposal of *Flavobacterium hydatis* nom nov (basonym *Cytophaga aquatilis* Strohl and Tait 1978). Int J Syst Bacteriol 46: 128–148.

2. Bernardet JF, Nakagawa Y (2006) An introduction to the family *Flavobacteriaceae*. In: Dworkin M, Falkow S, Rosenberg E, Schleifer KH, Stackebrandt E, editors. The Prokaryotes: a Handbook on the Biology of Bacteria, 3rd ed, vol, 7, New York: Springer. 455–480.

3. Euzéby JP (1997) List of Bacterial Names with Standing in Nomenclature: a folder available on the Internet. Int J Syst Bacteriol 47 590–592. Available: http://wwwbacterionet Accessed (April 13, 2013).

4. Bernardet JF, Bowman JP (2006) The genus *Flavobacterium*. In: Dworkin M, Falkow S, Rosenberg E, Schleifer KH, Stackebrandt E, editors. The Prokaryotes: a Handbook on the Biology of Bacteria, 3rd ed. Vol, 7. New York: Springer. 481–531.

5. Bernardet JF, Bowman JP (2011) Genus I. *Flavobacterium* Bergey, Harrison, Breed, Hammer and Huntoon 1923, 97AL emend. Bernardet, Segers, Vancanneyt, Berthe, Kersters and Vandamme 1996, 139. In: Krieg NR, Staley JT, Brown DR, Hedlund BP, Paster BJ, et al. editors. Bergey's Manual of Systematic Bacteriology, 2nd ed. Vol, 4. New York; Springer. 112–155.

6. Roberts RJ (2012) The Bacteriology of Teleosts. In Fish Pathology, 4 ed. UK: Wiley-Blackwell. 339–382.

7. Starliper CE, Schill WB (2012) Flavobacterial diseases: columniaris disease, coldwater disease and bacterila gill disease. In: Woo TTK, Bruno DW, editors. Fish Diseases and Disorders, 2nd ed, Vol. 3. UK: CAB International. 606–631.

8. Flemming L, Rawlings D, Chenia H (2007) Phenotypic and molecular characterization of fish-borne *Flavobacterium johnsoniae*-like isolates from aquaculture systems in South Africa. Res Microbiol 158: 18–30.

9. Kämpfer P, Lodders N, Martin K, Avendaño-Herrera R (2012) *Flavobacterium chilense* sp nov and *Flavobacterium araucananum* sp. nov. two novel species isolated from farmed salmonid in Chile. Int J Syst Evol Microbiol 62: 1402–1408.

10. Zamora L, Fernández-Garayzábal JF, Svensson-Stadler LA, Palacios MA, Domínguez L, et al. (2012) *Flavobacterium oncorhynchi* sp. nov. a new species isolated from rainbow trout (*Oncorhynchus mykiss*). Syst Applied Microbiol 35: 86–91.

11. Chen WM, Huang WC, Young CC, Sheu SY (2013) *Flavobacterium tilapiae* sp. nov. isolated from a freshwater pond and emended descriptions of *Flavobacterium defluvii* and *Flavobacterium johnsoniae*. Int J Syst Evol Microbiol 63: 827–834.

12. Sheu SY, Lin YS, Chen WM. (2012) *Flavobacterium squillarum* sp. nov., isolated from a freshwater shrimp culture pond and emended descriptions of *Flavobacterium haoranii*, *F. cauense*, *F. terrae* and *F. aquatile*. Int J Syst Evol Microbiol (in press) doi: 10.1099/ijs.0.046425-0.

13. Sheu SY, Chiu TF, Young CC, Arun AB, Chen WM (2011) *Flavbacterium macrobrachii* sp. nov. isolated from a freshwater shrimp culture pond. Int J Syst Evol Microbiol. 61: 1402–1407.

14. Ilardi P, Avendaño-Herrera R (2008) Isolation of *Flavobacterium*-like bacteria from diseased salmonids cultured in Chile. Bull Eur Assoc Fish Pathol 28: 176–185.

15. Kim OS, Cho YJ, Lee K, Yoon SH, Kim M, et al. (2012) Introducing EzTaxon-e: a prokaryotic 16S rRNA gene sequence database with phylotypes that represent uncultured species. Int J Syst Evol Microbiol 62: 716–721. Available: http://eztaxon-e.ezbiocloud.net Accessed May 29, 2013.

16. Rasmussen SW (2002) SEQtools a software package for analysis of nucleotide and protein sequences. Available: http://wwwseqtoolsdk Accessed May 27, 2013.

17. Saitou N, Nei M (1987) The neighbour-joining method: a new method for reconstructing phylogenetic trees. Mol Biol Evol 4: 406–425.

18. Page RDM (1996) TREEVIEW: an application to display phylogenetic trees on personal computers Comput Appl Biosci. 12: 357–358.

19. Guindon S, Gascuel O (2003) A simple fast and accurate algorithm to estimate large phylogenies by maximum likelihood. Syst Biol 52: 696–704.

20. Tamura K, Peterson D, Peterson N, Stecher G, Nei M, et al. (2011) MEGA5: Molecular evolutionary genetics analysis using maximum likelihood, evolutionary distance, and maximum parsimony methods. Mol Biol Evol 28: 2731–2739.

21. Kimura M (1980) A simple method for estimating evolutionary rates of base substitutions through comparative studies of nucleotide sequences. J Mol Evol 16: 111–120.

22. Marmur J (1961) A procedure for the isolation of deoxyribonucleic acid from microorganisms. J Mol Biol 3: 208–219.

23. Johnson JL (1994) Similarity analysis of DNAs. In: Gerhardt P, Murray RGE, Wood WA, Krieg NR, editors. Methods for General and Molecular Bacteriology. Washington DC: American Society for Microbiology. 655–681.

24. Arahal DR, García MT, Vargas C, Canovas D, Nieto JJ, et al. (2001) *Chromohalobacter salexigens* sp. nov. a moderately halophilic species that includes *Halomonas elongata* DSM 3043 and ATCC 33174. Int J Syst Evol Microbiol 51: 1457–1462.

25. De Ley J, Tijtgat R (1970) Evaluation of membrane filter methods for DNA-DNA hybridization. A Van Leeuw J Microb 36: 461–474.

26. Johnson JL (1994) Similarity analysis of DNAs. In: Gerhardt P, Murray RGE, Wood WA, Krieg NR, editors. Methods for General and Molecular Bacteriology, Washington DC: American Society for Microbiology. 655–681.

27. Tindall BJ (1990) A comparative study of the lipid composition of *Halobacterium saccharovorum* from various sources. Syst Appl Microbiol 13: 128–130.

28. Tindall BJ (1990) Lipid composition *of Halobacterium lacusprofundi*. FEMS Microbiol Letts 66: 199–202.

29. Sasser M (2001) Identification of bacteria by gas chromatography of cellular fatty acids MIDI. Available: http://wwwmicrobialidcom/PDF/TechNote_101pdf Accessed May 27, 2013.

30. Bernardet JF, Nakagawa Y, Holmes B (2002) Proposed minimal standards for describing new taxa of the family *Flavobacteriaceae* and emended description of the family. Int J Syst Evol Microbiol 52: 1049–1070.

31. Smibert RM, Krieg NR (1994) Phenotypic Characterization. In: Gerhardt P, Murray RGE, Wood WA, Krieg NR, editors. Methods for General and Molecular Bacteriology, Washington DC: American Society for Microbiology. 607–653.

32. Bowman JP, Cavanagh J, Austin JJ, Sanderson K (1996) Novel *Psychrobacter* species from Antarctic ornithogenic soils. Int J Syst Bacteriol 46: 841–848.

33. Chen YC, Davis MA, Lapatra SE, Cain KD, Snekvik KR, et al. (2008) Genetic diversity of *Flavobacterium psychrophilum* recovered from commercially raised rainbow trout *Oncorhynchus mykiss* (Walbaum) and spawning coho salmon *O. kisutch* (Walbaum). J Fish Dis 31: 765–73.

34. Lim CS, Oh YS, Lee JK, Park AR, Yoo JS, et al. (2011) *Flavobacterium chungbukense* sp. nov. isolated from soil. Int J Syst Evol Microbiol 61: 2734–2739.

35. Xu M, Xin Y, Tian J, Dong K, Yu Y, et al. (2011) *Flavobacterium sinopsychrotolerans* sp. nov. isolated from a glacier. Int J Syst Evol Microbiol 61: 20–24.

36. Yoon JH, Park S, Kang SJ, Oh SJ, Myung SC, et al. (2011) *Flavobacterium ponti* sp. nov.isolated from seawater. Int J Syst Evol Microbiol 61: 81–85.

37. Wayne LG, Brenner DJ, Colwell RR, Grimont PAD, Kandler O, et al. (1987) International Committee on Systematic Bacteriology Report of the ad hoc Committee on Reconciliation of Approaches to Bacterial Systematics. Int J Syst Bacteriol 37: 463–464.

38. Stackkebrandt E, Goebel BM (1994) Taxonomic note: a place for DNA-DNA reassociation and 16S rRNA sequence analysis in the present species definition in bacteriology. Int J Syst Bacteriol 44: 846–849.

39. Georgiadis MP, Gardner IA, Hedrick RP (2001) The role of epidemiology in the prevention, diagnosis, and control of infectious diseases of fish. Prev Vet Med 48: 287–302.

10

An Important Natural Genetic Resource of *Oreochromis niloticus* (Linnaeus, 1758) Threatened by Aquaculture Activities in Loboi Drainage, Kenya

Titus Chemandwa Ndiwa[1,2], Dorothy Wanja Nyingi[1]*, Jean-François Agnese[2]

1 Kenya Wetlands Biodiversity Research Group, Ichthyology Section, National Museums of Kenya, Nairobi, Kenya, **2** Département Conservation et Domestication, UMR IRD 226 CNRS 5554, Institut des Science de l'Evolution, Université de Montpellier 2, Montpellier, France

Abstract

The need to improve food security in Africa through culture of tilapias has led to transfer of different species from their natural ranges causing negative impacts on wild fish genetic resources. Loboi swamp in Kenya is fed by three hot springs: Lake Bogoria Hotel, Chelaba and Turtle Springs, hosting natural populations of *Oreochromis niloticus*. The present study aimed at better genetic characterization of these threatened populations. Partial mtDNA sequences of the D-loop region and variations at 16 microsatellite loci were assessed in the three hot spring populations and compared with three other natural populations of *O. niloticus* in the region. Results obtained indicated that the hot spring populations had mitochondrial and nuclear genetic variability similar to or higher than the large closely related populations. This may be attributed to the perennial nature of the hot springs, which do not depend on rainfall but rather receive permanent water supply from deep aquifers. The study also revealed that gene flow between the three different hot spring populations was sufficiently low thus allowing their differentiation. This differentiation was unexpected considering the very close proximity of the springs to each other. It is possible that the swamp creates a barrier to free movement of fish from one spring to the other thereby diminishing gene flow. Finally, the most surprising and worrying results were that the three hot spring populations are introgressed by mtDNA genes of *O. leucostictus*, while microsatellite analysis suggested that some nuclear genes may also have crossed the species barrier. It is very likely that the recent intensification of aquaculture activities in the Loboi drainage may be responsible for these introgressions. Taking into account the importance of these new genetic resources, protection and management actions of the Loboi swamp should be accorded top priority to prevent the loss of these spring populations.

Editor: Valerio Ketmaier, Institute of Biochemistry and Biology, Germany

Funding: TCN received a personal grant from IRD (Institut de Recherche pour le Développement) and MAE (French Ministry of Foreign Affairs). Field work was funded by Kenya Wetlands Research and National Museums of Kenya (NMK). Laboratory work was funded 50% by IRD and 50% by NMK. The funders had no role in study design, data collection and analysis, decision to publish, or preparation of the manuscript.

Competing Interests: The authors have declared that no competing interests exist.

* Email: dorothynyingi@yahoo.com

Introduction

Tilapias are some of the most important species for fisheries and aquaculture in Africa. This has contributed to their massive transfer not only within Africa, but to other countries around the world [1,2,3]. These transfers are a major concern because of the invasive nature of tilapia species and their ability to hybridize with local species [3,4,5]. Introgression of alien genes can leads to disruption of specific allele combinations responsible for adaptation of the populations to their environments hence reducing their fitness [6,7].

Within tilapias, *Oreochromis niloticus* (Linnaeus, 1758), the Nile tilapia, is the most economically important species, with a wide natural distribution in Africa. Its natural range covers the entire Nilo-Sudanian province (Senegal to Nile), Ethiopian Rift-Valley province, Kivu province, North Tanganyika Province (Ruzizi) and the Northern part of the East African Rift-Valley [8,9]. Currently, Nile tilapia is cultured in more than 100 countries and its

production estimated at 2,790,350 metric tonnes in 2011, and valued at 4.52 billion USD [10]. This species is favoured by fish farmers due to its fast growth rate and its ability to tolerate a wide range of environmental conditions [11]. Resistance to diseases and good consumer acceptance has further promoted its culture worldwide [12].

Seven subspecies of Nile tilapia have been described based on morphological characteristics [8]: *O. n. niloticus* from West-Africa and River Nile, *O. n. baringoensis* from Lake Baringo, *O. n. sugutae* from River Suguta (Kenya), *O. n. eduardianus* from Lakes Edward, Albert, George and Tanganyika, *O. n. vulcani* from Lake Turkana, *O. n. cancellatus* from River Awash, and Lake Tana and *O. n. filoa* from hot springs in the Awash system. Recently, Nyingi *et al.* [5] discovered a natural population from Lake Bogoria Hotel Spring (Kenya). This spring that drains into Loboi swamp is characterized by an elevated water temperature (36°C) and pH ranging from 6.4–6.9 [13]. This population of *O. niloticus* was characterised by 10 private microsatellite alleles and five private mtDNA haplotypes. Both extents of mitochondrial and microsatellite differentiations were in the range of those observed among other naturally occurring discrete populations in the region (Lakes Baringo and Turkana, and River Suguta). These observations indicated that the Lake Bogoria Hotel Spring population was not as earlier hypothesised, introduced from other neighbouring East African natural populations, but represents a previously unknown natural population.

This population offers new opportunities for aquaculture due to various adaptations including its ability to survive in relatively high temperatures (approx. 36°C). The fish may therefore have developed hypoxic resistance mechanisms since dissolved oxygen levels are generally low in warm waters. In addition, special adaptations may also be present that regulate sex determination mechanisms known to involve temperature [14,15]. This population may in this regard potentially offer a model for the study of sex determination in tilapine fishes.

In a first study [5] samples from a single hot spring in the Loboi drainage, the Lake Bogoria Hotel Spring were analysed. At least two other springs with similar ecological conditions exist, each hosting O. niloticus population, the Chelaba Spring that is located close to the Lake Bogoria Hotel Spring and the Turtle Spring located at the border of the papyrus marsh of Loboi Swamp. In order to protect these populations from various anthropogenic and environmental threats, there is an urgent need to characterise their natural genetic diversity as a first step of a future action plan for sustainable management. The Loboi Swamp itself has receded in size by over 60% over a short period of 30 years due to expansion of irrigation via a ditch constructed in 1970 [13,16] and conversion of wetlands for agriculture. Another anthropogenic threat is due to the rapid expansion of aquaculture activities in the Rift Valley region enhancing fish transfer from one drainage system to another, and allowing mixing between populations and or species. It has been demonstrated [17], using mtDNA (D-loop) and microsatellite studies, that the O. niloticus population from Lake Baringo has been introgressed by O. leucostictus from Lake Naivasha. However, even though mtDNA introgression was clearly established, no nuclear introgression was apparent. These introgressions have been attributed to possible introduction of O. leucostictus into Lake Baringo to boost tilapia fisheries caused by the decline of O. niloticus population by unsustainable high fishing pressure. Another possible explanation for this may have been the expansion of aquaculture in the region, which has been a continuous threat to wild populations of the Nile tilapia in Kenya. The human population of Lakes Baringo and Bogoria, mainly composed of small-scale crop farmers, have begun to diversify their livelihoods by construction of earthen ponds along the rivers and streams in the region [17,18]. Fish farmers are now breeding different tilapia species with fingerlings from a diversity of sources within East Africa. Most of these ponds are not isolated from streams and wetlands, thus farmed fish can easily escape and hybridize with autochthonous O. niloticus.

In order to characterize the three hot spring populations of the Loboi Swamp, we carried out a study of their genetic variation in mtDNA (sequence of partial D-loop region) and microsatellites together with other O. niloticus populations from the region: Lake Albert, Lake Turkana, Lake Baringo, River Suguta and O. leucostictus from Lake Naivasha.

Materials and Methods

Sampling design

Oreochromis niloticus is not an endangered or protected species in Kenya. Seven populations (Figure 1, Table 1) were studied and compared, three from the hot springs of the Loboi drainage (Turtle Spring, Chelaba Spring and Lake Bogoria Hotel Spring) (Figures S1–S3), while other populations were from Kenyan Lakes or Rivers (Baringo, Turkana, Suguta). Authorization to collect and sacrifice fish within the protected areas in Lake Turkana was provided by the means of a permit (number KWS/BRM/5001). All fishing and processing of tissue and whole fish specimens was carried out under existing collaborative arrangements with the Kenya Wildlife Service (KWS) and the National Museums of Kenya (NMK) for purposes of biodiversity research. For the other locations outside protected areas, no special authorization was necessary.

As there is no Institutional Animal Care and Use Committee or equivalent animal ethic committee in Kenya, no formal approval was obtained for this work. To minimize suffering of individuals studied, specimens were captured using seine nets and immediately anaesthetized and killed using an overdose of MS-222 (Tricaine-S, Western Chemicals Inc). After death, a fragment of muscle or fin tissue was taken from each specimen and immediately preserved in 95% ethanol for later DNA analyses. Voucher whole fish specimens were fixed in 4% formalin and later preserved in 70% ethanol, and are presently curated at the National Museums of Kenya (NMK) in Nairobi. Other voucher specimens whose sequences were obtained from Genbank are curated at the Musée Royal de l'Afrique Centrale (MRAC) in Tervuren, Belgium or at the Institut des Sciences de l'Evolution de Montpellier (ISEM), France (Table 1).

DNA extraction, sequencing and microsatellite analysis

Approximately 50 mg of the sample tissue was sheared into fine pieces before being digested at 55°C overnight using 10 µl proteinase K (10 mM/ml) in 190 ml of an extraction buffer solution (1 M Tris, 0.5 M NaCl$_2$, 1% SDS). DNA was then extracted from each sample following the protocol described for genomic DNA extraction for PCR-based techniques [19]. The extracted DNA was suspended in sterile double distilled water and stored in −20°C until PCR amplification. A 450 bp fragment in the 5′ region of the D-loop was amplified in each sample using two primers: 5′-ACCCCTAGCTCCCAAAGCTA-3′ (forward) and 5′- CCTGAAGTAGGACCAGATG-3′ (reverse). Amplifications were performed in a final volume of 50 µl containing 0.25 mM MgCl$_2$, 0.2 mM of each dNTP, 1 µM of each primer, 5 µl of 10x buffer and 10 units of Taq polymerase (Promega). PCR reaction conditions were: 3 minutes pre-heating at 93°C followed by 40 cycles of 30 seconds at 93°C, 30 seconds at 62°C and 1 minute at 72°C, with a final elongation phase of 72°C for five minutes.

A total of 16 microsatellite loci were analysed in this study: PRL1AC and PRL1GT [20], UNH104 (GenBank reference number G12257), UNH115 (G12268), UNH129 (G12282), UNH142 (G12294), UNH146 (G12298), UNH154 (G12306), UNH162 (G12314), UNH189 (G12341), UNH211 (G12362), UNH860 (G68195), UNH874 (G68202), UNH887 (G68210), UNH995 (G68274) and UNH1003 (G68280). Amplifications were carried out using Qiagen Multiplex PCR Kit and following manufacturer's protocol. Reaction conditions were: 5 minutes at 95°C pre-heating followed by 30 cycles of 30 seconds at 95°C, 90 seconds at 60°C (except for UNH115, UNH154, UNH860 and UNH1003 for which the temperature was 50°C) and 1 minute at 72°C, with a final elongation phase of 5 minutes at 72°C. 3 µl of PCR products were diluted in 12 µl of HiDi formamide and 0.2 µl of 500 LIZ size standard added. Electrophoresis was thereafter carried out in an ABI 3730 XL automated sequencer. GeneMapper software was used to analyse the electrophoregrams and allele sizes.

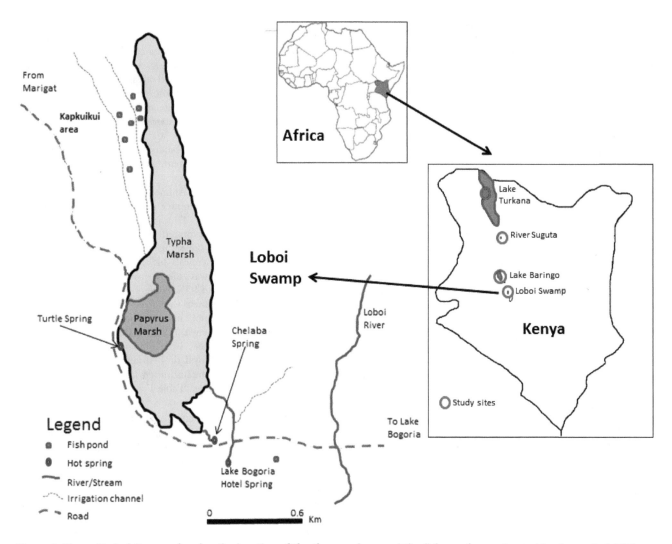

Figure 1. Map of Loboi Swamp showing the location of the three springs and the fish ponds constructed by the end of 2011.

Data Analysis

Sequence analyses

Sequences were aligned manually using BioEdit 5.09 [21]. Additional sequences obtained from Genbank were also included in the analysis (Table 1). Genetic diversity was calculated using DnaSP 5 [22]. Aligned sequences were analysed using Maximum Likelihood (ML) and Distance Method (DM) using MEGA (Molecular Evolutionary Genetics Analysis) version 5.1 [23]. Prior to analysis, an evolutionary model for ML was selected by MEGA 5.1 [24] using the Bayesian information criterion (BIC) [25]. Models with the lowest BIC scores are considered to describe the substitution pattern the best. Pair-wise sequence divergences between unique mtDNA haplotypes were calculated using the Kimura two-parameter model [26], followed by Neighbor Joining methods [27] to construct trees. Supports for inferred clades were obtained through the non-parametric bootstrap [28] with 2000 replicates for both methods.

Genetic diversity estimates were computed using DnaSP version 5.10.01 [29] and involved analysis of π, the average number of nucleotide differences per site between two sequences [30], and k, the average number of nucleotide differences between sequences [31].

Genotype analyses

Microsatellite data were checked for scoring errors due to stuttering or large allele dropout and presence of null alleles, using MICRO-CHECKER software [32]. Intra-population genetic variability was measured by estimating observed heterozygosity (H_O), expected heterozygosity (H_E) and the inbreeding coefficient (F_{IS}) using GENEPOP version 3.4 [33]. F_{ST} and associated probabilities were estimated using a Markov chain method following [34]. The length of the Markov chain involved a burn in period of 1000 iterations and 100 batches of 1000 iterations thereafter.

A Factorial Correspondence Analysis (FCA) was carried out with GENETIX Version 4.05 [35], in order to investigate the relationships between individuals. This type of analysis can explain a maximal amount of genetic variation using a minimal number of factors and can provide the means for visualizing the genetic relationships between populations or species. We first carried out an analysis of all the specimens studied from both species, *O. niloticus* and *O. leucostictus*, in order to visualize differentiation at the species level. Thereafter, individuals from Lake Baringo and the three hot springs were compared to individuals of *O. leucostictus* in order to observe possible introgression. Finally,

Table 1. Original Genbank references and corresponding voucher specimen numbers of the samples studied.

Population	Coordinates	Species	Genbank No.	Voucher Specimens
Loboi swamp				
Bogoria Hotel Spring	0°21'44"N, 36°03'04"E	*O. niloticus*	FJ440588-440603	(NMK)/1798/1-6
		O. niloticus	KJ746025-746032	(NMK) FW/36549/1-18
Loboi swamp				
Chelaba Spring	0°21'30"N, 36°02'58"E	*O. niloticus*	KJ746033-746040	(NMK) FW/2997/1-31
Loboi swamp				
Turtle Spring	0°21'44"N, 36°02'41"E	*O. niloticus*	KJ746051-746058	(NMK) FW/2996/1-25
Lake Baringo				
Kambi samaki	0°36'54"N, 36°01'59"E	*O. niloticus*	KJ746041-44	(NMK).FW/2660/1-31
Robertson camp	0°36'43"N, 36°01'31"E	*O. niloticus*	EF016697-016708	MRAC 95-027-P-0074-0084
			FJ440604-440607	
River Suguta				
Kapedo	1°10'44"N, 36°06'25"E	*O. niloticus*	EF016709-016714	(ISEM) JFA-5287-5313
Lake Turkana				
North Island	4°05'50"N, 36°02'46"E	*O. niloticus*	EF016680-016696	(NMK)1639/1-11,1641/1-13
Fergusson Golf	3°30'46"N, 35°54'53"E	*O. niloticus*	KJ746045-746050	(ISEM) JFA-2010-LT1-30
Lake Albert				
Butiaba	1°48'47"N, 31°16'04"E	*O. niloticus*	FJ440577-FJ440587	(ISEM) JFA-1922-1932
River Senegal				
Saint Louis	16°03'47"N, 16°28'33"W	*O. niloticus*	EF016715-016723	(ISEM) JFA-5683-5692
Lake Naivasha	0°46'18"N, 36°20'51"E	*O. leucostictus*	EF016702	(ISEM) JFA-5346-5375

In bold, specimens studied or sequences obtained during the present work. ISEM Institut des Sciences de l'Evolution de Montpellier, France; MRAC, Musée Royal de l'Afrique Centrale de Tervuren, Belgium; NMK, National Museums of Kenya.

genotypes of individuals from the hot springs and the neighbouring Lake Baringo were compared to establish their relationship.

A Bayesian cluster approach as implemented in STRUCTURE version 2.3.4 [36] was used to assess genetic admixture between *O. leucostictus* and *O. niloticus* at microsatellite loci assuming two populations ($K = 2$). The admixture model with correlated allele frequencies was chosen [36,37]. Three different runs to estimate $q(i)$, the fraction of the genome of an individual i of a given species (*O. niloticus* or *O. leucostictus*), inherited from the other species, three different runs were done for 500,000 generations after discarding 200,000 generations as burn-in. The 90% probability intervals around the different $q(i)$ were calculated.

To confirm occurrence of nuclear introgression, we used Bayesian method implemented in NEWHYBRIDS 1.1 [38]. This method is designed to identify new hybrids between populations or species, and unlike STRUCTURE, NEWHYBRIDS model takes into account the predictable patterns of gene inheritance in hybrids [38,39]. An analysis was carried out for 100,000 iterations of Markov Chain Monte Carlo (MCMC) after 100,000 burn-in steps. Affinity of an individual to respective genotype class was assessed by posterior probability values.

Results

mtDNA differentiation

Forty-two new partial D-Loop sequences were obtained during this study (GenBank accession numbers are given in Tab. II). Thirty different haplotypes were identified using these and an additional 58 sequences from GenBank (Tab. II). All populations studied were polymorphic except for River Senegal (*O. niloticus*)

and Lake Naivasha (*O. leucostictus*), where a single haplotype was present in 20 and five specimens examined respectively. Sample size (n), number of observed haplotypes (Hob), number of polymorphic sites (p), average number of nucleotide differences (k) and nucleotide diversity (π) are presented in Table 2. Out of the 30 haplotypes detected, 22 were unique to specific sample localities: 14 were observed only in Lake Turkana (total number of haplotypes sequenced $n = 23$), two in Lake Albert ($n = 20$), one in River Suguta ($n = 9$), five in Lake Bogoria Hotel Spring ($n = 24$). Furthermore, two other haplotypes were only found in hot springs populations.

Surprisingly, the haplotype of *O. leucostictus* from Lake Naivasha, which had been previously described to occur in *O. niloticus* from Lake Baringo population [17] was also detected in all three hot spring populations: one in Turtle Spring, two in Chelaba Spring and 1 in Lake Bogoria Hotel Spring.

According to the different values of the corrected Akaike Information Criterion (AIC) obtained with MEGA 5.1, the optimal model of sequence evolution was HKY model [40]. This model was used in the ML analysis. Phylogenetic relationships among all haplotypes observed based on ML or DM methods were congruent. Fig. 2 presents a consensus NJ tree where bootstrap values obtained with ML and DM have been indicated. [4,17] reported that *O. niloticus* from West Africa (from River Senegal to the Nile) has been naturally introgressed with mtDNA from *O. aureus*. The haplotype observed in the River Senegal sample corresponds to this alien haplotype. Consequently, the network has been rooted using the single haplotype from River Senegal.

All other haplotypes observed, except one clustered in a single highly supported group (bootstrap value = 97% for ML and MD

Table 2. mtDNA (partial D-loop sequences) variability observed in of 8 populations of *O. niloticus*.

Population	n	Hob	p	k	π
Bogoria Hotel Spring	24 (23)	10 (9)	51 (32)	8.39 (6.41)	0.024 (0.018)
Chelaba Spring	8 (6)	4 (3)	45 (25)	18.57 (9.13)	0.053 (0.026)
Turtle Spring	8 (7)	5 (4)	46 (26)	18.07 (13.33)	0.051 (0.038)
Lake Baringo	15 (8)	4 (3)	37 (12)	18.06 (6.00)	0.051 (0.017)
Lake Turkana	23	15	19	2.53	0.007
River Suguta	9	3	6	1.33	0.004
Lake Albert	20	4	3	0.30	0.001
River Senegal	18	1	0	0.00	0.000

n, number of sequences; *Hob*, number of different haplotypes observed; *P*, number of polymorphic sites; *k*, average number of nucleotide difference; π, nucleotide diversity. In bracket, results without taking into account introgressed specimens.

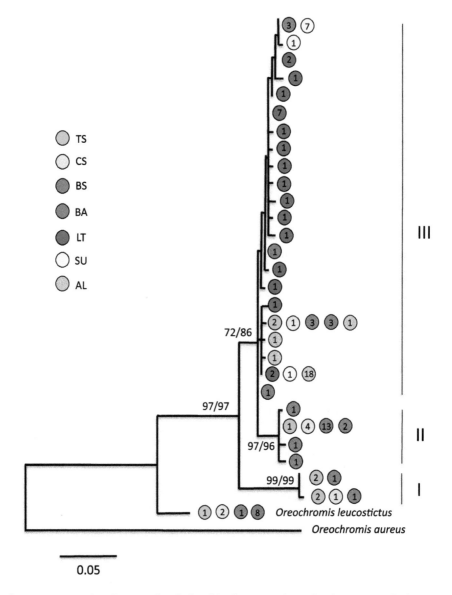

Figure 2. Phylogenetic tree representing the genetic relationships between the 30 haplotypes studied. The network was rooted using haplotype observed in the River Senegal population naturally introgressed by *O. aureus. O. niloticus* haplotypes have been clustered in three groups (I, II and III). Numbers above branches indicate bootstrap values (in percentage) based on 2000 replicates for ML and DM methods. The bar represents 0.05 units divergence, TS, Turtle spring, CS, Chelaba Spring, BS, Lake Bogoria Hotel Spring, BA, Lake Baringo, LT, Lake Turkana, SU, Suguta River and AL, Lake Albert. Values inside the circles represent number of haplotype.

Table 3. Heterozygosity and coefficient of inbreeding in six populations of O. niloticus and O. leucostictus from Lake Naivasha.

Species		Oreochromis niloticus						O. leucostictus
Population		Bogoria	Chelaba	Turtle	Lake Baringo	River Suguta	Lake Turkana	Lake Naivasha
Locus		Spring	Spring	Spring	Baringo	Suguta	Turkana	Naivasha
	N	31	31	25	30	21	29	20
PRL1AC	H_e	0.6683	0.7472	0.8523	0.8839	0.7971	0.9031	0.5639
	H_o	0.4000	0.5333	0.6818	0.7333	0.8571	0.8571	0.5000
	F_{is}	0.4156	0.3017	0.2222	0.1867	-0.0511	0.0690	0.1300
	P	**0.0000**	**0.0172**	0.3320	0.095	0.3632	0.1056	0.2257
	A	10	11	9	18	10	18	4
PRL1GT	H_e	0.5899	0.7733	0.8504	0.8761	0.7088	0.8456	0.6142
	H_o	0.5000	0.6667	0.7600	0.8000	0.6500	0.7600	0.3214
	F_{is}	0.1701	0.1545	0.1264	0.1037	0.1083	0.1214	0.4906
	P	0.0601	0.1910	0.5222	**0.0000**	0.5784	0.2748	**0.0000**
	A	10	10	10	13	5	16	4
UNH211	H_e	0.8798	0.8250	0.8620	0.9310	0.7275	0.8644	0.8272
	H_o	0.7419	0.7333	0.9583	0.8966	0.7000	0.7931	0.7857
	F_{is}	0.1727	0.1278	-0.0907	0.0546	0.0634	0.0999	0.0682
	P	**0.0046**	**0.0020**	0.2035	0.5198	0.5585	0.0596	0.1343
	A	12	11	12	18	6	12	8
UNH1003	H_e	0.7756	0.7513	0.7456	0.8195	0.7450	0.7111	0.3428
	H_o	0.4333	0.6452	0.6000	0.7153	0.6500	0.5862	0.2667
	F_{is}	0.4548	0.1573	0.2148	0.1463	0.1527	0.1925	0.2381
	P	**0.0000**	0.0902	0.0785	0.2722	0.3487	0.2712	0.1424
	A	10	10	8	11	6	12	4
UNH104	H_e	0.6239	0.7456	0.7694	0.7564	0.7742	0.8977	0.7487
	H_o	0.600	0.7000	0.8696	0.8571	0.7895	0.9655	0.8571
	F_{is}	0.0552	0.0780	-0.1083	-0.1153	0.0074	-0.0580	-0.1270
	P	0.9434	0.1105	0.9019	0.7534	0.8804	0.6980	0.9064
	A	4	7	8	9	7	14	6
UNH115	H_e	0.7846	0.7456	0.6354	0.7878	0.7506	0.6327	0.3200
	H_o	0.8065	0.8065	0.6667	0.6897	0.6667	0.5714	0.2000
	F_{is}	-0.0115	-0.0653	-0.0279	0.1418	0.1358	0.1333	0.3895
	P	0.9078	0.3620	0.4527	**0.0006**	**0.0250**	0.3973	0.0598
	A	8	6	7	7	8	7	2
UNH129	H_e	0.4938	0.5128	0.5408	0.6161	0.5839	0.5344	0.6650
	H_o	0.5161	0.3333	0.5714	0.6000	0.7143	0.5000	0.6667

Table 3. Cont.

Species		*Oreochromis niloticus*						*O. leucostictus*
Population		Bogoria Spring	Chelaba Spring	Turtle Spring	Lake Baringo	River Suguta	Lake Turkana	Lake Naivasha
Locus								
	N	31	31	25	30	21	29	20
	F_{is}	-0.0289	0.3647	-0.0323	0.0431	-0.2000	0.0825	0.0144
	P	0.6644	**0.0429**	0.7965	0.6357	0.5035	0.2638	0.9113
	A	3	3	4	5	3	4	3
UNH142	H_e	0.5994	0.6906	0.7360	0.8033	0.7022	0.8011	0.2494
	H_o	0.5161	0.8000	0.8400	0.8333	0.7778	0.7037	0.2857
	F_{is}	0.1549	-0.1419	-0.1213	-0.0204	-0.0794	0.1401	-0.1279
	P	0.1515	0.5706	0.8908	0.1061	0.2617	0.1590	1.0000
	A	5	6	5	8	5	11	3
UNH146	H_e	0.7594	0.7423	0.7736	0.8733	0.6020	0.5511	0.5594
	H_o	0.7667	0.6071	0.7600	0.6667	0.6190	0.4138	0.5667
	F_{is}	0.0074	0.1997	0.0380	0.2525	-0.0039	0.2656	0.0040
	P	0.7886	**0.0496**	0.4238	**0.0102**	0.7029	0.0540	0.3045
	A	10	7	7	10	3	4	4
UNH154	H_e	0.8713	0.8663	0.8624	0.9019	0.8491	0.9477	0.7300
	H_o	0.7692	0.8065	0.6800	0.7931	0.3077	0.8929	0.8000
	F_{is}	0.1364	0.0854	0.2309	0.1379	0.6608	0.0760	-0.0791
	P	0.5460	0.3935	**0.0003**	0.0568	**0.0000**	**0.0416**	0.9411
	A	15	11	10	10	10	20	7
UNH162	H_e	0.7361	0.7283	0.8628	0.8756	0.6429	0.8668	0.7128
	H_o	0.7000	0.6000	0.8095	1.0000	0.6190	0.9655	0.6667
	F_{is}	0.660	0.1926	0.0860	-0.1255	0.0614	-0.0965	0.0816
	P	0.1675	0.1581	0.8164	0.1468	0.8427	0.1697	0.8282
	A	9	7	10	12	5	14	5
UNH189	H_e	0.8538	0.7078	0.8376	0.8389	0.4206	0.8984	0.4923
	H_o	0.9032	0.7000	0.8400	0.7931	0.1905	0.8750	0.5714
	F_{is}	-0.0415	0.0279	0.0175	0.0721	0.5640	0.0473	-0.1429
	P	0.6913	0.3419	0.1503	0.2293	**0.0078**	0.2097	0.3020
	A	15	9	11	12	3	16	5
UNH860	H_e	0.7877	0.8341	0.7584	0.8029	0.1863	0.9251	0.7483
	H_o	0.6452	0.7586	0.8400	0.8929	0.2000	0.8276	0.7333
	F_{is}	0.1968	0.1079	-0.0874	-0.0940	-0.0483	0.1227	0.0370
	P	**0.0201**	0.4876	0.8627	0.7953	1.0000	0.1618	0.6471

Table 3. Cont.

Species		Oreochromis niloticus						O. leucostictus
Population		Bogoria	Chelaba	Turtle	Lake Baringo	River Suguta	Lake Turkana	Lake Naivasha
Locus		Spring	Spring	Spring				
	N	31	31	25	30	21	29	20
UNH874	A	11	10	10	11	4	19	8
	H_e	0.7683	0.6906	0.8084	0.8339	0.7687	0.8992	0.6653
	H_o	0.6333	0.7667	0.8095	0.8000	0.9048	0.8571	0.7241
	F_{is}	0.1921	−0.0934	0.0230	0.0576	−0.1533	0.0650	−0.0710
	P	**0.0082**	0.8139	0.5181	**0.0260**	0.3119	0.0607	0.1354
UNH887	A	10	9	9	9	9	17	4
	H_e	0.6550	0.7028	0.7760	0.8133	0.6066	0.7478	0.5938
	H_o	0.6129	0.6667	0.8400	0.8333	0.6667	0.7308	0.6786
	F_{is}	0.0807	0.0683	−0.0622	−0.0076	−0.0749	0.0423	−0.1250
	P	**0.0201**	0.4876	0.8627	0.7953	1.0000	0.1618	0.6471
UNH995	A	6	9	8	10	5	8	4
	H_e	0.6972	0.7494	0.8176	0.7527	0.7637	0.9001	0.7783
	H_o	0.5161	0.5667	0.7600	0.7931	0.7500	0.9310	0.7000
	F_{is}	0.2749	0.2598	0.0907	−0.0362	0.0436	−0.0168	0.1174
	P	**0.0006**	**0.0039**	**0.0220**	0.9307	0.8199	0.3499	0.1225
	A	5	8	10	8	5	14	9
Global mean	H_e	0.7215	0.7383	0.7805	0.8229	0.6643	0.8080	0.6007
	H_o	0.6288	0.6682	0.7679	0.7935	0.6819	0.7607	0.5828
	F_{is}	0.1444	0.1111	0.0375	0.0536	0.0628	0.0765	0.0481
	A	143	134	180	138	93	217	80

N, sample size; H_e, expected heterozygosity; H_o, observed heterozygosity; F_{is}, inbreeding coefficient; P, associated F_{is} probability and A, number of alleles.

Table 4. Pairwise F_{ST} estimates between six *O. niloticus* populations and one *O. leucostictus* population (Lake Naivasha).

Population	Lake Bogoria Hotel Spring	Chelaba Spring	Turtle Spring	Lake Baringo	River Suguta	Lake Turkana
Chelaba Spring	0.0177					
Turtle Spring	0.0689	0.0749				
Lake Baringo	0.0373	0.0408	0.0253			
River Suguta	0.2123	0.2201	0.1822	0.1368		
Lake Turkana	0.1645	0.1642	0.1334	0.1085	0.1366	
Lake Naivasha	0.2893	0.2653	0.2769	0.2591	0.3522	0.2728

Every F_{ST} were statistically highly significant.

methods). The outlier haplotype present in Lake Baringo, the three hot springs populations and *O. leucostictus* from Lake Naivasha, belongs to *O. leucostictus* as demonstrated earlier [17]. This haplotype was previously observed in the *O. niloticus* population from Lake Baringo, and considered introgressed, but was not observed in the Lake Bogoria Hotel Spring [5], as is the case in the present study. These results gave evidence that

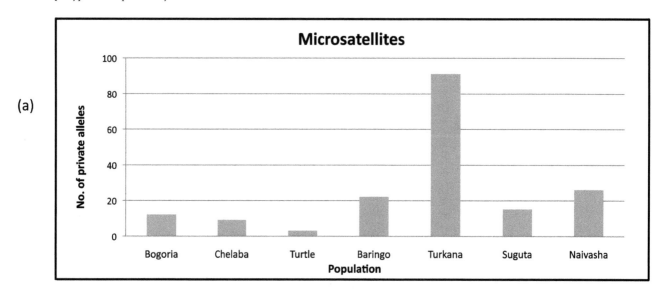

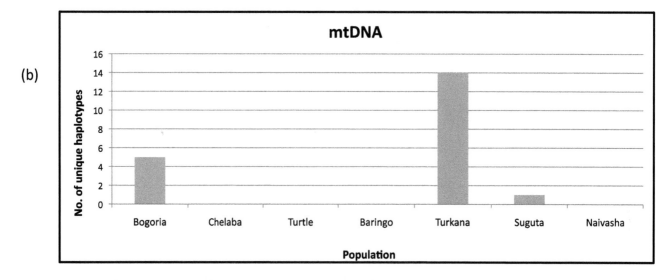

Figure 3. Number of private microsatellite alleles and private mtDNA haplotypes present in six populations of *O. niloticus* and one population of *O. leucostictus* (mtDNA not included for Naivasha population).

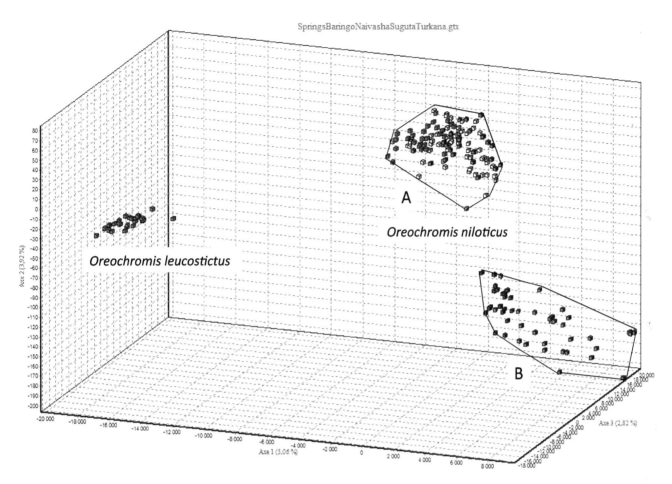

SpringsBaringoNaivashaSugutaTurkana.gtx

Figure 4. Factorial Correspondence Analysis (GENETIX 4.05) based on individual microsatellite genotypes at 16 loci showing differentiation of the seven populations into two distinct groups corresponding to *Oreochromis leucostictus* **and** *O. niloticus.* *O. niloticus* populations further clustered into two groups; individuals from Lake Baringo Hotel Spring, Chelaba Spring, Turtle Spring and Lake Baringo (A), and Lake Turkana and River Suguta (B).

introgression by *O. leucostictus* mtDNA has also occurred in all the hot spring populations of the Loboi swamp.

The 28 remaining haplotypes are divided in three groups. The first one (I), supported by high bootstrap values (99% for ML and MD), is the sister group of the two others (II and III) and is composed of two different haplotypes present only in the hot spring populations. The four haplotypes of group II (bootstrap 97% for ML and 96% for DM) are found only in the three hot spring populations and in Lake Baringo. The other 22 haplotypes composed group III, which was supported by important bootstrap values: 72% for ML and 86% for DM. All the populations studied possess at least one haplotype in this group.

Microsatellite analysis

A total of 197 individuals from seven localities were genotyped at 16 microsatellite loci. All loci were polymorphic in all populations with the number of alleles ranging from nine to 44. Locus UNH154 was the most polymorphic with 44 alleles, while UNH129 was the least polymorphic with nine alleles. Lake Turkana population recorded the highest mean number of alleles (12.9±1.3, 217 allele observed; X±SE where SE is standard error), while Lake Naivasha population (*O. leucostictus*) had the least (5.0±0.5, 80 allele observed). The three hot spring populations had mean number of alleles ranging from 8.4±0.6, (Chelaba Spring), 8.6±0.5 (Turtle Spring) to 8.9±0.9 (Lake

Bogoria Hotel Spring), and total number of alleles ranging from 134, 138 to143 at Chelaba, Turtle and Lake Bogoria Hotel Springs, respectively. Distribution of alleles within the populations is shown in Table 3.

Using Micro-Checker, the presence of null alleles could not be rejected in the following loci and populations: PRL1AC in all three springs and Lake Baringo populations, UNH154 in Turtle Spring and River Suguta populations, UNH995 in Lake Bogoria Hotel and Chelaba Springs, PRL1GT and UNH115 in Lake Naivasha population, UNH1003 in Lake Bogoria Hotel population, UNH146 in Lake Baringo population, and UNH189 and UNH860 in River Suguta and Lake Turkana populations. No scoring error was detected in the data due to either stuttering or large allele dropout.

All populations had high observed heterozygosities ranging from 0.5828 in Lake Naivasha population to 0.7935 in Lake Baringo. F_{IS} values indicated heterozygote deficiencies in all populations studied. A total of 23 (24.15%) F_{IS} values out of the 105 calculated indicated significant heterozygote deficiencies (Table 3).

Pairwise F_{ST} values among the seven populations ranged from 0.0177 to 0.3522 (Table 4) and were all highly significant (P<0.001), indicating that all the samples can be considered genetically differentiated. F_{ST} values between Turtle Spring population and the other two hot spring populations (Lake

HotspringsBaringoNaivasha.gtx

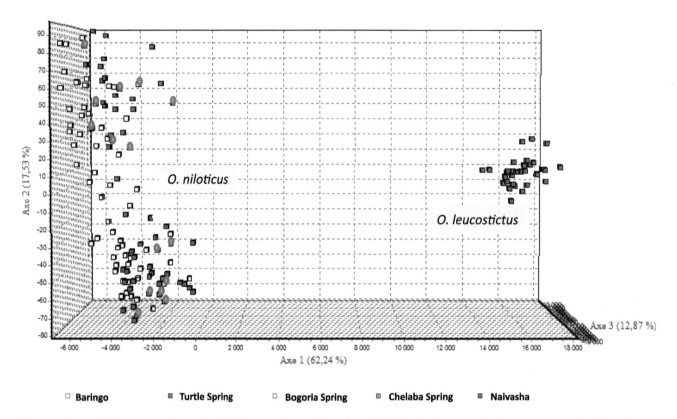

○ Baringo   ■ Turtle Spring   □ Bogoria Spring   ■ Chelaba Spring   ■ Naivasha

Figure 5. Factorial Correspondence Analysis (GENETIX 4.05) based on individual microsatellite genotypes at 16 loci showing no intermediate genotypes between mtDNA introgressed populations, the three hot springs and Lake Baringo populations, and *O. leucostictus* from Lake Naivasha. Individuals of *O. niloticus* that have *O. leucostictus* mtDNA are highlighted in red. TS, Turtle Spring; BA, Lake Baringo; BS, Bogoria Spring; OL, *Oreochromis leucostictus* from Lake Naivasha. Note: only two introgressed individuals from Lake Baringo are shown instead of eight (Fig. 1) because sequences of the six omitted individuals were obtained from Genbank hence not genotyped.

Bogoria Hotel and Chelaba Springs), were relatively higher (0.0689 and 0.0749, respectively) than the F_{ST} observed between the latter two populations (0.0177). As expected, the highest F_{ST} values were observed for interspecific comparisons between *O. niloticus* and *O. leucostictus* populations, and ranged between 0.2653 at Chelaba Spring to 0.3522 at River Suguta.

All populations had private microsatellite alleles as shown in Figure 3. The highest number of private alleles was observed in Lake Turkana population (91). Lake Bogoria Hotel Spring population had the highest number of private alleles (12) within the hot spring populations, while Chelaba Spring and Turtle Spring had nine and three private alleles, respectively.

FCA of all the genotypes observed (Figure 4) highlighted the differences between the two species *O. niloticus* and *O. leucostictus*. Based on the first axis (5.06% of total genetic variations), two main clusters corresponding to the two different species were identified. Within the *O. niloticus* cluster, two distinct groups were observed which correspond on one hand to populations from Lake Bogoria Hotel Spring, Chelaba Spring, Turtle Spring and Lake Baringo (A), and on the other hand to Lake Turkana and River Suguta populations (B).

Taking into account that some populations have been introgressed by mtDNA from *O. leucostictus*, we carried out another FCA (Figure 5) of all three introgressed populations (the three hot springs and Lake Baringo) and *O. leucostictus* from Lake Naivasha. This enabled a clear separation between populations

from *O. niloticus* and the one from *O. leucostictus* based on the first axis (62.24% of the total genetic variation). The fact that mtDNA of the introgressed individuals did not appear related to *O. leucostictus*, and the absence of any intermediate individuals between these two groups suggested that there has been no nuclear introgression between the two species.

Finally, an FCA analysis including the three hot spring populations and the Lake Baringo populations was run (Figure 6). The individuals were separated into four different clusters corresponding to the four populations. The axis 1 (43.34% of the genetic variation) discriminated individuals from Lake Bogoria Hotel and Chelaba Springs from Turtle Spring and Lake Baringo. The second axis (32.05% of the genetic variation) discriminated on one hand Chelaba Spring population from Lake Bogoria Hotel Spring population and on the other hand, Turtle Spring population from Lake Baringo population. It is noticeable that Lake Bogoria Hotel Spring and Chelaba Spring populations which are geographically closer to each other than to Turtle Spring are also genetically close, leading to a genetic isolation by distance pattern.

Further analysis of genotypes from introgressed populations was carried out using STRUCTURE in order to determine the proportions of nuclear genetic admixtures between *O. niloticus* and *O. leucostictus*. No F1 hybrid genotypes were detected from the analysis; this means there were no individuals with Q values around 0.5 (Figure 7). All individuals of *O. leucostictus* had

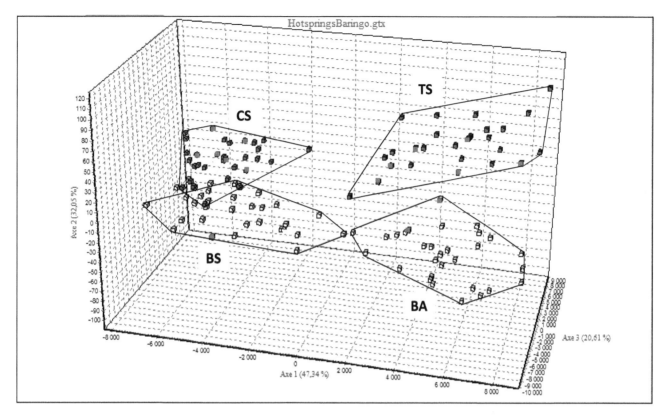

Figure 6. Factorial Correspondence Analysis (GENETIX 4.05) based on individual microsatellite genotypes at 16 loci showing clear separation of the four populations, the three hot spring populations and the lake Baringo population, using the two first axis. CS, Chelaba Spring; BS, Bogoria Hotel Spring; TS, Turtle Spring; BA, Lake Baringo.

Q values ≤0.003 (estimated fraction proportion of their genome inherited from any other species, in this case, *O. niloticus*) and could be considered as pure (not introgressed). All individuals of *O. niloticus* from Lake Turkana and River Suguta, and most of those from the hot springs and Lake Baringo could also be considered as pure (were characterized by Q values ≤0.01). Eleven individuals had Q values ≥ 0.01 (five from Chelaba Spring, three from Turtle Spring, two from Lake Bogoria Hotel Spring and one from Lake Baringo) and this number rises to 30 when considering

the 90% probability intervals (13, 6, 6, 5, respectively). Only one individual from Chelaba Spring had a Q value >0.1 and ten when considering the 90% intervals (4, 3, 2, 1 respectively).

When analysing individual genotypes with NEWHYBRIDS, all specimens of *O. leucostictus* had a posterior probability of being pure *O. leucostictus* greater or equal to 0.9989. In *O. niloticus* populations, seven specimens had a posterior probability of being pure *O. niloticus* inferior to 0.9574, and could be considered as possible hybrids. For all the concerned specimens, the probability

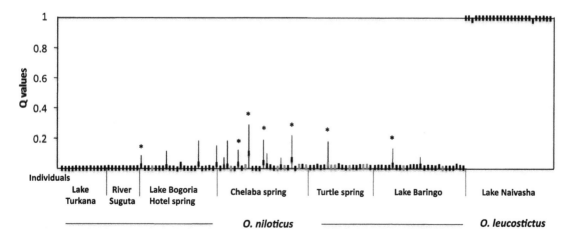

Figure 7. Distribution of specimen membership Coefficient (Q±90% probability intervals) based on genotypes of *O. niloticus* and *O. leucostictus* in the hot springs and lakes Baringo and Naivasha, identified through microsatellite analysis using STRUCTURE. Each sample along the x axis represents an individual.

to be F2 hybrid backcrossed with *O. niloticus* was higher than 0.0416 (Table S1).

It is important to take into account that all the specimens identified by NEWHYBRIDS as potential F2 backcrossed hybrids have been considered as hybrids according to their *Q* values obtained with STRUCTURE (Fig. 7).

Only one specimen from Turtle Spring population was considered as a hybrid by all three methods of analysis used; mtDNA haplotype identification and nuclear genotypes composition analysed by STRUCTURE and NEWHYBRIDS (Table S1).

Discussion

In a survey of *O. niloticus* populations in East Africa, the occurrence of a new (previously unknown) native population of *O. niloticus* in Loboi Swamp (Lake Bogoria Hotel Spring), Kenya has been documented [5]. This population was initially assumed to have been introduced from other localities within the region. However, its native status was confirmed due to its significant and unique genetic variability with a large number of private microsatellite alleles and mtDNA haplotypes (partial D-loop). The presence of tilapia in two other hot springs in the Loboi Swamp namely Turtle and Chelaba Springs, prompted us to investigate these fish. The present study clearly establishes that these two hot springs also host native fish closely related to those from Lake Bogoria Hotel Spring and that all three hot spring populations are in addition also closely related to the Lake Baringo population of the Nile tilapia. They share the most common haplotype in group II, and one haplotype in group III (Figure 2). Nevertheless, the three hot spring populations are also characterized by the private haplotypes of group I.

Lake Baringo is considered a modern representation of a much larger middle to upper Pleistocene Lake Kamasia [41], whose shores extended to Kapthurin River to the west and Bogoria River (which Loboi system is part of) headwaters to the south. This may explain the close relationship between the hot springs and the Lake Baringo, as observed in our study.

The analysis of partial mtDNA sequences of the D-loop region showed that the haplotype of *O. leucostictus* from Lake Naivasha previously confirmed to be present in Lake Baringo [17], is also present in the three *O. niloticus* populations of the springs. The microsatellite study and visual observations of fish specimens during their capture confirmed that the spring populations were composed of a single species, *O. niloticus*. It is therefore evident that some of the fish in each of the hot springs have been introgressed by *O. leucostictus* mitochondrial genome.

The first studies of the Loboi swamp spring population [5,17] detected introgressed specimens in Lake Baringo by *O. leucostictus* but none from the Lake Bogoria Hotel Spring. However, it is possible that the authors failed to detect any introgression of the Lake Bogoria Hotel Spring population due to the low number of hot spring specimens available for their study (16 specimens).

If these introgressions originated from aquaculture practices *i.e.* escapees of *O. leucostictus* specimens or from *O. niloticus* introgressed beforehand in aquaculture farms within the Loboi swamp drainage, as it was hypothesized, then it is possible that these introgressions may have occurred after 2007.

Indeed, fish-farming activities within the region has been enhanced as a result of funding by Kenyan government through the Economic Stimulus Programme (ESP) introduced in 2009. The aim of the ESP project was to construct 200 fishponds worth 10.6 million euros in 140 constituencies by June 2013, with each constituency receiving 70,000 € [42]. Eight fishponds have been constructed between 2012 and 2013, in close proximity to the

Loboi Swamp (within a range of 230 – 700 metres), and some stocked with Nile tilapia fingerlings from unspecified sources. Other fishponds in the Baringo-Bogoria catchment are located along the main rivers and streams in the region. Fingerlings used to stock the ponds within the region are produced by Omega Fish Farms located at Ol Kokwe Island on Lake Baringo. The farm was established to produce fingerlings for the local fish farmers [43]. Broodstock for the establishment was obtained from the Lake Baringo, hence may have acted as a source of introgressions observed in the Loboi Swamp fish.

At a first glance, it appears that these introgressions only concern mitochondrial DNA as earlier concluded in the study of Lake Baringo population [17]. Indeed, the results of the Factorial Correspondence Analysis (Figure 5) based on individuals microsatellites genotypes (16 loci) showed a clear separation between the three hot spring populations on one hand and on the other hand, the Lake Baringo, and population of *O. leucostictus* from Lake Naivasha. In case of large amount of introgression, individuals with intermediate genotypes were expected.

Recently STRUCTURE has been used [44,45] to quantify in more details the proportion of individual's genome originating from a different hybridizing species in tilapia from Lake Victoria and in Cichlids from Lake Tanganyika, respectively. Results obtained by these authors concern tilapia species from the Lake Victoria region: *O. niloticus* and *O. esculentus*. These two species are suspected to hybridize even though no mitochondrial introgression has so far been observed in samples (30 individuals of *O. niloticus* from Lakes Victoria, Kanyaboli, Namboyo and Sare, and 30 individuals of *O. leucostictus* from Lakes Kanyaboli and Namboyo). Results of this microsatellite study (eight loci) allowed the separation of both species using a Factorial Correspondence Analysis. However, using STRUCTURE, eight individuals (six *O. esculentus* and two *O. niloticus*) appeared to have 90% probability intervals that extend more than 30% out of a pure species *Q* value which suggested that these individuals had an important degree of genetic introgression.

In the present study, all *O. leucostictus* had *Q* values ≤0.003 and may be considered as not introgressed by *O. niloticus* genes (Figure 7). At least ten *O. niloticus* specimens (representing 11% of the individuals studied in the three hot springs and Lake Baringo populations) when considering the 90% probability intervals had 10% of their genes or more that may be considered to have originated from *O. leucostictus* (three in Lake Bogoria Hotel Spring, five in Chelaba Spring, one in Turtle Spring and one in Lake Baringo).

Results obtained with NEWHYBRIDS are congruent with these findings and indicated that seven individuals (as observed using STRUCTURE) may have been introgressed. Taking into account that *O. niloticus* populations are genetically more variable than the *O. leucostictus* population, it is possible that this hybridization pattern was induced by a high genetic variance of the *O. niloticus* samples from the hot springs and Lake Baringo rather than introgression with *O. leucostictus*.

Nevertheless, by assessing the specimens supposed to be hybrids, all seven specimens determined by NEWHYBRIDS were also pointed out by STRUCTURE. It is also worth noting that even if there is a poor correspondence between specimens with mtDNA introgression and specimens designated by the two Bayesian programs as hybrids (only one specimen is concerned), with the FCA analysis, all the mtDNA introgressed individuals appeared within their respective *O. niloticus* cluster, but closer to the *O. leucostictus* cluster (Fig. 5). Lastly, specimens from Lake Turkana population which is the most genetically variable, is clearly composed of non-hybridized individuals (Fig. 7). This suggests that hybrids identified by both STRUCTURE and NEWHYBRIDS are not artefacts due to genetic variance in the *O. niloticus* populations.

All of these observations are congruent and confirm with more detail the previous conclusions [17] pointing out the nuclear introgression of the *O. niloticus* populations from the hot springs of Loboi Swamp and Lake Baringo by nuclear *O. leucostictus* genes.

Even if the extent of the hot spring biotopes are small, and consequently the size of the effective populations, native mtDNA diversity was higher in the hot springs populations than in Lake Baringo and River Suguta populations. Nine different haplotypes were found in the hot springs (36 pure individuals analysed) while three were found in Lake Baringo (eight pure individuals analysed) or River Suguta (nine pure individuals analysed). Differences in samples sizes may partially account for the observed differences but when one considers Chelaba Spring on its own with four haplotypes for six individuals or Turtle Spring with five haplotypes for seven individuals, one can conclude that haplotype diversity in the hot spring is high.

This does not seem to be congruent with the very small size of the biotopes. The hot springs are located within a small swamp of about 1.5 km². The swamp itself is not expected to be a favourable biotope for fish survival as it is completely covered with macrophytes of which the most dominant are *Typha domingensis* and *Cyperus papyrus* [46]. This unexpected high mtDNA diversity can be explained if we consider that the characteristics of the hot springs are not directly related to pluvial water. The hot springs originate from faults and fractures that allow water from deep aquifers to rise to the surface [13,47,48] hence maintaining their perennial existence throughout the year, regardless of climatic changes. The high mtDNA diversity may have been favoured by the stable environmental conditions and consequently decreased levels of environmental stress in the hot springs.

Moreover, gene flow between the three hot spring populations is not enough to prevent their differentiation, as shown by F_{ST} values that were all statistically significant between populations (Table 4), and the FCA analysis (Figure 6). Even if there are only three populations, it seems that the genetic differentiation is congruent with the genetic differentiation through isolation by distance. This was also unexpected considering that the springs are very close to each other. When they join the swamp, Lake Bogoria Hotel and Chelaba Springs are separated only by a few hundred metres, and Turtle Spring is separated from Lake Bogoria Hotel and Chelaba Springs by approximately one kilometre.

To explain this amount of observed genetic differentiation between the springs, the first hypothesis is to consider that the swamp may create a barrier to free movement of fish from one spring to another thereby diminishing gene flow.

Both water temperature and dissolved oxygen concentration within the swamp are lower than within the springs mainly due to the shading effect provided by the dense vegetation and organic decomposition from the decaying matter, respectively. This situation has been described in a study of cichlid fishes from Lake Victoria [49,50]. It has been demonstrated that differences in eco-physiological properties of water can limit gene flow as a result of adaptation or physiological avoidance. Our observations are also congruent with the apparent lack of gene flow between two populations of both *O. esculentus* and *O. niloticus* from different satellite lakes of Lake Victoria separated by a wetland and papyrus swamps [44]. However, if the swamp represents an impenetrable barrier to free movement of fish so that F_{ST} values are statistically significant, it is difficult to explain how *O. leucostictus* or introgressed *O. niloticus* could have crossed the swamp, from fish ponds connected to the swamp drainage, and subsequently

hybridized with native fish of the hot springs. It is still possible, though very unlikely, that someone may have directly introduced fish into each hot spring. Whereas this hypothesis cannot be rejected without consideration, there is simply no objective reason to carry out such fish translocation. Even though the exact way by which fishes have been introgressed is still unknown, the reality of the introgression has clearly been demonstrated and that aquaculture activities played a role in this process.

It was also apparent that the three populations of *O. niloticus* in the hot springs are distinct from other related populations by the presences of private alleles and haplotypes. This differentiation could be the consequence of their local adaptations to the extreme thermal conditions of their habitat. Studies indicate that exposure of *O. niloticus* to water temperatures of more than 36°C at an early stage of their development (10 days post fertilization) influences sex ratios towards males [14,41]. These characteristics are of utmost importance to aquaculture, as they can be utilised to inexpensively produce all-male cultures. Tilapia farming mainly relies on male mono-sex culture in order to avoid overcrowding in the ponds due to breeding, and to benefit from fast growth of males [51,52,53]. Hence the rich genetic diversity and adaptations to masculinizing effect of water temperature makes these populations an important aquaculture resource, and an important natural model for understanding sex determination in Nile tilapias.

These populations and the genetic resources they represent need to be conserved. The conservation and management of Loboi swamp system should thus be accorded top priority in order to safeguard its critical genetic resource that has great potential in development of tilapia aquaculture.

Supporting Information

Figure S1   Photograph showing Lake Bogoria Hotel Hot Spring.

Figure S2   Photograph showing Chelaba Hot Spring.

Figure S3   Photograph showing Turtle Hot Spring.

Table S1   Species specific mtDNA identification, membership probability (STRUTURE) and Posterior probability (NewHybrids) associated with the introgressed specimens. Post. Prob, posterior probability; On, *Oreochromis niloticus*, Ol, *Oreochromis leucostictus*, CS, Chelaba Spring; BS, Bogoria Hotel Spring; TS, Turtle Spring; BA, Lake Baringo.

Acknowledgments

Molecular genetic data used in this work were produced through molecular genetic analysis technical facilities of "Centre Méditerraéen de l'Environnement et de la Biodiversité" (CeMEB). We are grateful to Frédérique Cerqueira, Erick Desmarais (ISEM) and Joseph Gathua (NMK) and for their technical support in the laboratory and field collections during the study. We are grateful to Dr. Judith Nyunja (KWS) for her support during sample collections in Lake Turkana. Finally, we are grateful to C. Firmat for useful comments on the first draft of this manuscript.

Author Contributions

Conceived and designed the experiments: TCN DWN JFA. Performed the experiments: TCN DWN JFA. Analyzed the data: TCN DWN JFA. Contributed reagents/materials/analysis tools: TCN DWN JFA. Contributed to the writing of the manuscript: TCN DWN JFA.

References

1. Shelton WL (2002) Tilapia culture in the 21st century. In: Guerrero III. Guerrero-del CastilloRD, M.R. (Eds.), Tilapia Farming in the 21st Century, Proc. Int. Forum on Tilapia Farming in the 21st Century, February 2002, Los Baños, Philippines, pp. 1–20.

2. Romana-Eguia MRR, Ikeda M, Basiao ZU, Taniguchi N (2004) Genetic diversity in farmed Asian Nile and red hybrid tilapia stocks evaluated from microsatellite and mitochondrial DNA analysis. Aquaculture 236, 131–150.

3. Eknath AE, Hulata G (2009) Use and exchange of genetic resources of Nile tilapia (Oreochromis niloticus). Reviews in Aquaculture 1: 197–213.

4. Rognon X, Guyomard R (2003) Large extent of mitochondrial DNA transfer from Oreochromis aureus to O. niloticus in West Africa. Molecular Ecology 12: 435–445.

5. Nyingi D, De-Vos L, Aman R, Agnèse J-F (2009) Genetic characterization of an unknown and endangered native population of the Nile tilapia Oreochromis niloticus (Linnaeus, 1758) (Cichlidae; Teleostei) in the Loboi Swamp (Kenya). Aquaculture 297: 57–63.

6. Ryman N (1991) Conservation genetics considerations in fishery management. Journal of Fish Biology 39 (21): 1–224.

7. Lind CE, Brummett RE, Ponzoni RW (2012) Exploitation and conservation of fish genetic resources in Africa: issues and priorities for aquaculture development and research. Reviews in Aquaculture 4: 125–141.

8. Trewavas E (1983) Tilapiine fishes of the genera Sarotherodon Oreochromis and Danakilia, British Museum (Natural History), London. 583p.

9. Bezault E, Balaresque P, Toguyeni A, Fermon E, Araki H, et al. (2011) Spatial and temporal variation in population structure of wold Nile tilapia (Oreochromis niloticus) across Africa. Genetics, 12: 102.

10. FAO website. Fisheries and aquaculture information and statistics service. Available: http://www.fao.org/figis/servlet/SQServlet?ds=Capture&k1=SPECIES&K1v=1&k1s=3217&outtype=html. Accessed 2014 Aug 19.

11. Grammer GL, Slack WT, Peterson MS, Dugo MA (2012) Nile tilapia Oreochromis niloticus (Linnaeus, 1758) establishment in temperate Mississippi, USA: multi-year survival confirmed by otolith ages. Aquatic Invasions 7: 367–376.

12. Chakraborty SB, Banerjee S (2009) Culture of monosex Nile tilapia under different traditional and non-traditional methods in India. World Journal of Fish and Marine Sciences, 1 (3): 212–217.

13. Ashley GM, Maitima Mworia J, Muasya AM, Owen RB, Driese SG, et al. (2004) Sedimentation and recent history of a freshwater wetland in a semi-arid environment: Loboi Swamp, Kenya, East Africa. Sedimentology 51, 1301–1321.

14. Baroiller JF, D'Cotta H (2001) Environment and sex determination in farmed fish. Comparative Biochemistry and Physiology Part C 130: 399–409.

15. Tessema M, Muller-Belecke A, Horstgen-Schwark G (2006) Effects of rearing temperatures on the sex ratios of Oreochromis niloticus populations. Aquaculture, 258 (1–4): 270–277.

16. Owen RB, Renaut RW, Hover VC, Ashley GM, Muasya AM (2004) Swamps, springs, and diatoms: wetlands regions of the semi-arid Bogoria–Baringo Rift, Kenya. Hydrobiologia, 518: 59–78.

17. Nyingi DW, Agnèse J-F (2007) Recent introgressive hybridization revealed by exclusive mtDNA transfer from Oreochromis leucostictus (Trewavas, 1933) to Oreochromis niloticus (Linnaeus, 1758) in Lake Baringo, Kenya. J. Fish BioLake 70 (Supplement A): 148–154.

18. Mumba M (2008) Adapting to climate change: lessons from Lake Bogoria catchement, Kenya. Programme report No. 3/08.

19. Aljanabi SM, Martinez I (1997) Universal and rapid salt-extraction of high quality genomic DNA for PCR-based techniques. Nucleic Acids Research 25(22): 4692–4693.

20. Agnese J-F, Gourene B, Nyingi DW (2009) Functional microsatellite and selective sweep in the black-chinned tilapia Sarotherodon melanotheron (Teleostei, Cichlidae) natural populations. Marine Genomics, 1: 103–107.

21. Hall TA (1999) BioEdit: a user friendly biological sequence alignment editor and analysis program for windows 95/98/NT. Nucleic Acids Symposium Series No. 41: 95–98.

22. Librado P, Rozas J (2009) DnaSP v5: A software for comprehensive analysis of DNA polymorphism data. Bioinformatics 25: 1451–1452.

23. Kumar S, Tamura K, Nei M (2004) MEGA3: Integrated software for molecular evolutionary genetic analysis and sequence alignment. Briefing in Bioinformatics, 5: 150–163.

24. Tamura K, Peterson D, Peterson N, Stecher G, Nei M, et al. (2011) MEGA5. Molecular evolutionary genetics analysis using maximum likelihood, evolutionary distance, and maximum parsimony methods. Molecular Biology and Evolution 28(10): 2731–2739.

25. Schwarz G (1978) Estimating the dimensions of a model. The Annals of statistics, 6(2): 461–464.

26. Kimura M (1980) A simple method for estimating evolutionary rate of base substituition through comparative studies of nucleotide sequences. Journal of Molecular Evolution 16: 111–120.

27. Saitou N, Nei M (1987) The neighbor-joining method: A new method for reconstructing phylogenetic trees. Molecular Biology and Evolution 4:406–425.

28. Felsenstein J (1985) Confidence limits on phylogenies: an approach using the bootsrap. Evolution, 39(4): 783–791.

29. Rozas J, Sanchez-DelBarrio JC, Messeguer X (2003) DnaSP, DNA polymorphism analyses by the coalescent and other methods. Bioinformatics 19, 2496–2497.

30. Nei M (1987) Molecular Evolutionary Genetics. Columbia University Press, New York, NY.

31. Tajima F (1983) Evolutionary relationships of DNA sequences in finite populations. Genetics 105: 437–460.

32. Van Oosterhout C, Hutchinson WF, Wills DPM, Shipley P (2004) Micro-Checker: software for identifying and correcting genotyping errors in microsatellite data. Molecular Ecology Notes, 4: 535–538.

33. Rousset F (2008) Genepop'007: a complete re-implementation of the genepop software for Windows and Linux. Molecular Ecology Resources, 8:103–106.

34. Weir BS, Cockerham CC (1984) Estimating F-Statistic for the analysis of population structure. Evolution 38(6): 1358–1370.

35. Belkhir K, Borsa P, Chikhi L, Raufaste N, Bonhomme F (2004) GENETIX 4.05, logiciel sous Windows TM pour la génétique des populations. Laboratoire Génome, Populations, Interactions, CNRS UMR 5000, Université de Montpellier II, Montpellier (France).

36. Pritchard JK, Stephens M, Donelly P (2000) Inference of population structure using multilocus genotype data. Genetics,155(2): 945–959.

37. Falush D, Stephens M, Pritchard JK (2003) Inference of population structure using multilocus genotype data: linked loci and correlated allele frequencies. Genetics 164 (4): 1567–1587.

38. Anderson EC, Thompson EA (2002) A model-based method for identifying species hybrids using multilocus genetic data. Genetics 160: 1217–1229.

39. Anderson EC (2008) Bayesian inference of species hybrids using multilocus dominant genetic markers. Phil Trans R Soc 363: 2841–2850.

40. Hasegawa M, Kishino H, Yano T (1985) Dating of human-ape splitting by a molecular clock of mitochondrial DNA. J. Mol. Evol. 22 (2): 160–174.

41. Fuch VE (1950) Pleistocene events in the Baringo Basin, Kenya Colony. Geol. Mag.Lond. 87: 149–174.

42. Nyingi DW, Agnèse J-F (2011) Tilapia in eastern Africa – a friend and foe. Species in the Spotlight. Darwall, W.R.T., Smith, K.G., Allen, D.J., Holland, R.A, Harrison, I.J., and Brooks, E.G.E. editors. The Diversity of Life in African Freshwaters: Under Water, Under Threat. An analysis of the status and distribution of freshwater species throughout mainland Africa. Cambridge, United Kingdom and Gland, Switzerland: IUCN. Pp. 76–78.

43. The Fish Site website. Developing a Tilapia breeding station in Kenya: Available: http://www.thefishsite.com/articles/1186/developing-a-tilapia-breeding-station-in-kenya. Accessed 2014 Aug 19.

44. Angienda PO, Lee HJ, Elmer KR, Abila R, Waindi EN, et al. (2010) Genetic structure and gene flow in an endangered native tilapia fish (Oreochromis esculentus) compared to invasive Nile tilapia (Oreochromis niloticus) in Yala swamp, East Africa. Conservation genetics, 12: 243–255.

45. Nevado B, Fazalova V, Backeljau T, Hanssens M, Verheyen E (2011) Repeated unidirectional introgression of Nuclear and Mitochondrial DNA between four congeneric Tanganyikan cichlids. Molecular Biology and Evolution 28(8): 2253–2267.

46. Muasya AM, Hover VC, Ashley GM, Owen RB, Goman MF, et al. (2004) Diversity and distribution of macrophytes in a freshwater wetland, Loboi swamp (Rift valley) Kenya. Journal of East African Natural History, 93: 39–47.

47. Deocampo DM (2002) Sedimentary processes and lithofacies in lake-margin groundwater-fed wetlands in East Africa. In: Ashley, G.M, Renaut, R.W. editors. Sedimentation in Continental Rifts, vol. 73. Special Publication-SEPM, Tulsa, OK, pp. 295–308.

48. Ashley GM, Goman M, Hover VC, Owen RB, Renaut RW, et al. (2002) Artesian blister wetlands, a perennial water resource in the semi-arid Rift-Valley of East Africa, Wetlands 22(4): 686–695.

49. Chapman LJ, Chapman CA, Nordlie FG, Rosenberger AE (2002) Physiological refugia: swamps, hypoxia tolerance and maintenance of fish diversity in the Lake Victoria region. Comp Biochem Physiol A 133:421–437.

50. Crispo E, Chapman LJ (2008) Population genetic structure across dissolved oxygen regimes in an African cichlid fish. Molecular Ecology 17:2134–2148.

51. Baroiller JF, D'Cotta H, Bezault E, Wessels S, Hoerstgen-Schwark G (2009) Tilapia sex determination: where temperature and genetics meet. Comparative Biochemistry and Physiology Part A, 153: 30–38.

52. Baroiller J-F, Jalabert B (1989) Contribution of research in reproductive physiology to the culture of tilapias. Aquatic Living Resources 2: 105–116.

53. Beardmore JA, Mair GC, Lewis RI (2001) Monosex male production in finfish as exemplified by tilapia: applications, problems, and prospects. Aquaculture 197: 283–301.

11

Three Decades of Farmed Escapees in the Wild: A Spatio-Temporal Analysis of Atlantic Salmon Population Genetic Structure throughout Norway

Kevin A. Glover[1]*, María Quintela[2], Vidar Wennevik[1], François Besnier[1], Anne G. E. Sørvik[1], Øystein Skaala[1]

1 Section of Population Genetics and Ecology, Institute of Marine Research, Bergen, Norway, 2 Dept of Animal Biology, Plant Biology and Ecology, University of A Coruña, Spain

Abstract

Each year, hundreds of thousands of domesticated farmed Atlantic salmon escape into the wild. In Norway, which is the world's largest commercial producer, many native Atlantic salmon populations have experienced large numbers of escapees on the spawning grounds for the past 15–30 years. In order to study the potential genetic impact, we conducted a spatio-temporal analysis of 3049 fish from 21 populations throughout Norway, sampled in the period 1970–2010. Based upon the analysis of 22 microsatellites, individual admixture, F_{ST} and increased allelic richness revealed temporal genetic changes in six of the populations. These changes were highly significant in four of them. For example, 76% and 100% of the fish comprising the contemporary samples for the rivers Vosso and Opo were excluded from their respective historical samples at $P = 0.001$. Based upon several genetic parameters, including simulations, genetic drift was excluded as the primary cause of the observed genetic changes. In the remaining 15 populations, some of which had also been exposed to high numbers of escapees, clear genetic changes were not detected. Significant population genetic structuring was observed among the 21 populations in the historical (global $F_{ST} = 0.038$) and contemporary data sets (global $F_{ST} = 0.030$), although significantly reduced with time ($P = 0.008$). This reduction was especially distinct when looking at the six populations displaying temporal changes (global F_{ST} dropped from 0.058 to 0.039, $P = 0.006$). We draw two main conclusions: 1. The majority of the historical population genetic structure throughout Norway still appears to be retained, suggesting a low to modest overall success of farmed escapees in the wild; 2. Genetic introgression of farmed escapees in native salmon populations has been strongly population-dependent, and it appears to be linked with the density of the native population.

Editor: Martin Krkosek, University of Otago, New Zealand

Funding: This study is financed by the Norwegian Minitry of Fisheries. The funders had no role in study design, data collection and analysis, decision to publish, or preparation of the manuscript.

Competing Interests: The authors have declared that no competing interests exist.

* E-mail: Kevin.glover@imr.no

Introduction

Delineation of historical genetic structure can provide an insight into how contemporary evolutionary relationships among populations have been shaped by demographic, environmental and anthropogenic factors. Understanding these processes and their potential interactions will assist in predicting how natural populations are likely to evolve in relation to present and future challenges.

Salmonid fishes provide excellent opportunities to study evolutionary relationships among populations in both time and space. They inhabit a variety of habitats and display phenotypic and life-history variation among populations [1], some of which reflect local adaptations [1–3]. Furthermore, salmonids tend to exhibit highly distinct population genetic structuring, also in anadromous forms where high fidelity to natal stream (homing) serves to limit gene flow [4]. The Atlantic salmon (*Salmo salar*) is no exception to these characteristics, and the analysis of molecular genetic markers has revealed highly significant population genetic structuring throughout its entire range [5–8].

The contemporary population genetic structure of Atlantic salmon can be ascribed to a hierarchical system, whereby the largest genetic differences are observed among fish from different continents and regions [9–14]. These differences are to a large degree thought to reflect the patterns of post-glacial colonization. Within regions, highly significant genetic differentiation has been observed among salmon originating from different rivers [11,15,16], and in some cases, also between tributaries within the same river system [15,17–19]. These differences, as revealed by molecular genetic markers, primarily reflect a combination of reproductive isolation and genetic drift, whereby demographics and landscape features play a modifying role [16,17,19]. Generally, where wild populations experience low human impacts, temporal genetic stability has been reported [20,21].

Atlantic salmon populations have been heavily exploited and influenced by a wide-range of anthropogenic factors over a long period of time [22]. Adding to the list of challenges since the 1970's, is the hundreds of thousands of domesticated salmon that escape from farms on a yearly basis, which display a wide range of interactions with wild conspecifics [23]. Although escapees display high mortality post-escapement [24,25], they have been recorded in rivers throughout the species' native range, such as England [26], Scotland [27,28], North America [29], and Norway [30]. Escapees have also been observed in rivers located in countries where salmon farming is not practiced [29].

Genetic changes in native Atlantic salmon populations as a result of introgression from farmed escapees have been observed in Ireland [31–34] and North America [35]. Looking beyond these studies that have been conducted in single rivers, an analysis of seven Norwegian Atlantic salmon populations revealed significant changes in several rivers that had displayed large numbers of farmed escapees on the spawning grounds [36]. However, although farmed escapees have been observed in natural populations for over three decades, and in many regions these numbers exceed wild spawner abundance, the impact this has had on population genetic structure remains elusive. It is therefore not surprising that there are global concerns regarding the genetic integrity of wild populations [23,37–41].

Norway is the world's largest commercial producer of Atlantic salmon, and is the country where the highest numbers of farmed escapees have been recorded on the spawning grounds. Therefore, Norway represents an ideal country in which to examine how genetic structure has changed both within and among native Atlantic salmon populations in response to widespread migration of farmed escapees onto the spawning grounds. Here, we have conducted a spatio-temporal genetic analysis in order to investigate the potential genetic impacts of farmed escapees on population structure throughout an entire country.

Materials and Methods

Study Design

Atlantic salmon farming in Norway is currently based upon rearing multiple domesticated strains and sub-strains that were initially founded on fish originating from over 40 Norwegian rivers in the 1970's [42]. Thus, while the allele frequencies of the farmed strains are generally distinct to each other due to founder effects [43], they overlap with the allele frequencies of Norwegian wild populations [43,44]. Over time, farmed escapees do not originate from a single farmed strain, but from multiple strains. The result of this is that the gene flow signal from escapees represents a dynamic mixture of allele frequencies. Thus, the detection of genetic changes in wild populations when gene flow comes from multiple farmed sources is far more complicated [45] than where a set of populations are supplemented by a single and readily defined hatchery source [32,46]. In the latter case, it is straight-forward to demonstrate that the allele frequencies in the recipient wild population converges with the allele frequencies with its donor. However, for the case of multiple farmed strains, the recipient wild population will not converge with any given farmed strain over time, and genetic introgression may be partially concealed [45].

Increasing the complexity of detecting genetic introgression of farmed escapees in wild Atlantic salmon populations is that the farmed strains (and therefore their allele frequencies) have, and continue to change significantly with time, i.e., some of the populations used at an earlier stage have been terminated or combined with other strains, while new sub-strains (e.g., in response to QTL selection) have been established. Consequently, it is not possible to accurately reconstruct the allele frequencies of the farmed escapees in Norway over the 15–30 year period in which this study is conducted. Nevertheless, despite the above challenges, modeling has demonstrated that gene flow from farmed escapees will lead to a reduction in genetic structure among wild populations [45,47]. This is because over time, wild

populations will be exposed to the average allele frequency from the major strains, and this will start to erode the existing allele frequency differences among wild populations. Furthermore, modeling has shown that genetic changes in wild populations as a result of farmed escapees spawning may be detected, although its likely to be underestimated [45].

As a consequence of the situation described above, the methodological approach implemented in this study is to look at both within and among-population genetic structure in the time-period where the numbers of escapees reported in Norwegian rivers has been highest (i.e., the last 15–30 years). Have native Norwegian salmon populations displayed temporal genetic changes in this period? And if so, can genetic drift be excluded as the primary driver of these temporal changes? Furthermore, where temporal genetic changes have been observed, have the populations become more similar or more differentiated to each other?

Biological Samples

Historical and contemporary samples of Atlantic salmon populations were collected from 21 rivers spanning the entire Norwegian coastline which extends over 2500 km (Fig. 1; Table 1, 2). Populations were chosen primarily due to the availability of archived scale samples which were essential to re-construct the historical baseline (pre- or early aquaculture industry), and, availability of contemporary samples (year 2000+).

Historical samples were exclusively represented by fish scales taken from adult spawners captured in their specific rivers by rod and line (Table 1). Intermediate (neither the oldest nor newest set of samples from any given river system), and contemporary samples, were mostly represented by scale samples taken from adult fish captured by rod and line fishing or various research projects. Therefore, no specific licenses were applied for nor required to collect these samples for this study. Prior to any genetic analysis, all scale samples were analysed for growth patterns in order to exclude any salmon that had directly escaped from fish farms [48]. For some of the intermediate and contemporary samples, adult spawners were not available (for example due to closure of rod and line fishery). Instead, samples of juvenile fish were included for these populations. The historical samples were not collected from the exact same time period (Table S1), however, this was factored into some of the analyses.

Some of the relevant available information for the populations included in this study, which can be found in Norwegian reports [49–52] have been placed into Table 2. Importantly, this information includes the frequency of farmed salmon that have been observed in these populations in the period 1989–2009. Observations of farmed escaped salmon in Norwegian populations are primarily recorded by two approaches. One of the methods is based upon the percent of farmed fish in the angling catch during the summer sports fishing season, while the other is based upon the percent of farmed fish observed during dedicated autumn (spawning site) surveys. As farmed salmon tend to migrate later than wild salmon into freshwater [30], the autumn surveys tend to show higher percentages of farmed fish. However, the surveys of farmed fish frequency in the autumn usually involve sample sizes smaller than the summer angling catch surveys, are conducted less frequently, and are conducted in fewer rivers [49]. Nevertheless, the potential for genetic interaction is more tightly linked to the frequency of escapees observed on the spawning sites during the autumn than found in the summer angling catches. Therefore, we have chosen to use both estimates in the present study. First we use the un-weighted mean percent of farmed fish observed in the spawning surveys (i.e., averaging the percent farmed fish observed

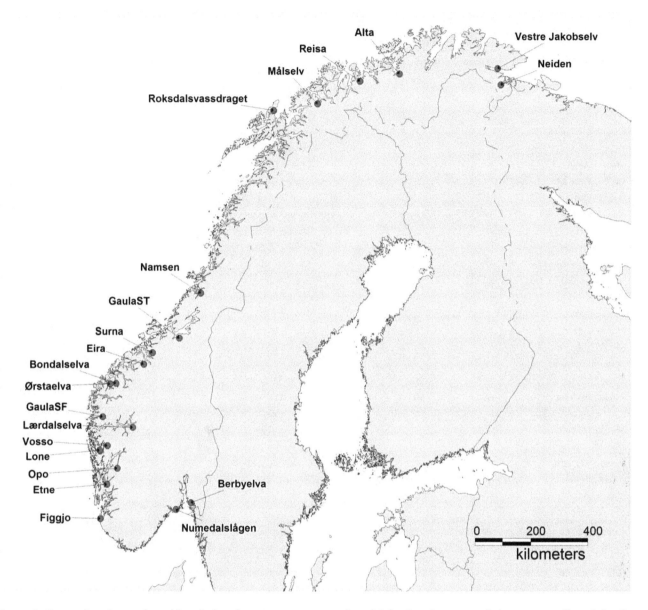

Figure 1. Norwegian rivers where historical and contemporary samples of Atlantic salmon populations were collected for the present study.

for the number of years in which they survey was conducted), in addition to using a weighted average based upon combining both summer sports fishing and autumn survey data that has been recently used to categorise over 100 Norwegian rivers in their degree of potential influence from farmed escaped salmon [52]. These estimates have then been compared with the temporal genetic changes observed for each river by regression analysis.

Samples of farmed salmon have been included for the analysis of admixture. These samples were selected from multiple data sets that have been analysed to identify the farms of origin for escapees as a DNA forensic service for the Norwegian ministry of fisheries in the period 2006 - present [53–57]. A total of nine farm samples, each of approximately 45 fish, were chosen based upon their large genetic differences to each other, and, in order to represent some of the genetic diversity found among salmon farms and farmed strains in Norway.

Genotyping

DNA extraction was performed in 96-well plates using the Qiagen DNeasy®96 Blood & Tissue Kit. Each DNA plate contained two or more negative controls.

The following twenty two microsatellite loci were used; *SSsp3016* (Genbank no. AY372820), *SSsp2210*, *SSspG7*, *SSsp2201*, *SSsp1605*, *SSsp2216* [58], *Ssa197*, *Ssa171*, *Ssa202* [59], *SsaD157*, *SsaD486*, *SsaD144* [60], *Ssa289*, *Ssa14* [61], *SsaF43* [62], *SsaOsl85* [63], *MHC I* [64] *MHC II* [65], *Ssa19NVH* (Genbank no. AF256670), *CA060208* [66], *SsalR002TKU* and *SsalR010TKU* [67]. Amplifications were conducted in four multiplex reactions (conditions available from the authors). PCR products were analysed on an ABI 3730 Genetic Analyser and sized by a 500LIZ^{TM} size-standard. Automatically binned alleles were manually checked by two researchers prior to exporting data for statistical analyses.

Table 1. Numbers and types of samples collected from 21 Atlantic salmon rivers.

Population	Sample size (n)	Sample type (NSR)	Population	Sample size (n)	Sample type (NSR)
Neiden H (1979–82)	79	SP (1)	GaulaSF H (1987–93)	40	SP (1)
Neiden I (1989–93)	43	SP	GaulaSF C (2006–08)	83	SP
Neiden C (2009)	93	SP	Lærdalselva H (1973)	95	SP (1)
V. Jakobselva H (1989–91)	96	SP (1)	Lærdalselva I (1996–97)	65	?
V. Jakobselva C (2007–08)	101	SP	Lærdalselva C (2005–08)	53	SP
Alta H (1988–90)	39	SP (1)	Vosso H (1980)	49	SP (1)
Alta C (2005–2007)	85	P	Vosso I1 (1993–96)	66	SP
Reisa H (1986–91)	48	SP (1)	Vosso I2 (2007–08)	48	SM
Reisa C (2006)	61	P	Vosso C (2008)	42	SP
Målselva H (1986–88)	47	SP (1)	Loneelva H (1986–93)	60	SP (0)
Målselva C (2008)	30	P	Loneelva C (2001–07)	52	SP
Roksdalsvassdraget H (1987–93)	37	SP (1)	Opo H (1971–73)	54	SP (0)
Roksdalsvassdraget C (2008)	94	SP	Opo I (2000)	46	P
Namsen H (1977)	92	SP (1)	Opo C (2010)	60	P
Namsen I (2000)	58	SP	Etne H (1983)	88	SP (1)
Namsen C (2008)	102	SP	Etne I (1997–98)	76	P
GaulaST H (1986–94)	48	SP (1)	Etne C (2006–2008)	88	SP
GaulaST C (2006–08)	106	SP	Figgjo H (1972–75)	57	SP (1)
Surna H (1986–89)	30	SP (1)	Figgjo I (1987–90)	41	SP
Surna C (2005–08)	52	SP	Figgjo C (2006)	72	SP
Eira H (1986–94)	34	SP (0)	Numedalslågen H (1989–93)	43	SP (1)
Eira C (2005–2008)	50	SP	Numedalslågen C (2007–08)	72	SP
Bondalselva H (1986–88)	44	SP (0)	Berbyelva H (1988–93)	46	SP (1)
Bondalselva C (2007)	16	P	Berbyelva C (2007–08)	94	SP
Ørsta H (1986–89)	40	SP (1)			
Ørsta C (2006–08)	34	SP			

Population = name of river with postscript letter H = historical sample, I = intermediate sample, C = contemporary sample. Life stage sampled = SP = spawners, E = eggs, A = alevins, F = fry, P = parr, SM = smolt, NSR = National Salmon River (river protected by extra legislation from government): 1 = yes, 0 = no.

Microsatellites are known to be prone to genotyping errors [68,69], even under strict protocols [70]. Eighteen of the microsatellite markers implemented here are routinely genotyped at IMR, and have revealed low error rates [55]. Within the present data set, some samples were re-analysed in order to increase the genotyping coverage and provide an ad-hoc quantification of genotyping quality.

Statistical Analyses

For most of the statistical analyses conducted, samples were grouped into historical, intermediate and contemporary data sets. Other sub-sets of the data set were analysed for specific tests (i.e., including reduced sets of populations and markers). These variations are identified in the results. Bonferroni adjustment of the significance level for multiple testing was not presented. Instead, statistical significance was tested at α 0.05 and a more stringent level of α 0.001.

The genotype distribution of each locus in each population was compared with the expected Hardy-Weinberg distribution using the program GenePop [71] as was the linkage disequilibrium. Both were examined using the following Markov chain parameters: 10000 steps of dememorisation, 1000 batches and 10000 iterations per batch. Relative genetic variation in each population was assessed using allele frequency data from which observed

heterozygosity Ho, expected heterozygosity He, allelic richness, F_{IS} and pairwise F_{ST} were calculated using MSA 4.05 [72].

In order to test whether the global F_{ST} among historical populations was significantly larger than the global F_{ST} among contemporary populations, a bootstrap test based on 10 000 re-sampled datasets was computed. For each resample, the global F_{ST} in historical and contemporary data was calculated based on a random sample of 30% of the individuals from each population and 30% of the markers (7 out of 22). After re-sampling, the distribution of the 10 000 differences between historical and contemporary F_{ST} was used to test the alternative hypothesis (H$_1$: F_{ST} historical > F_{ST} contemporary) against the null hypothesis (H$_0$: F_{ST} historical $\leq$ F_{ST} contemporary).

The program Geneclass 2.0 [73] was used to perform genetic assignment. First, the program was used to conduct self-assignment among the 21 populations in the historical and contemporary data sets. Thereafter, the historical genetic profile for each population was used as the baseline, while individual fish representing the contemporary sample for each population was assigned to their respective baseline population. Exclusion was assessed at a significance level of α 0.001 using all 22 loci, and the reduced set of 14 loci, with the Rannala & Mountain simulation method [74].

Table 2. Characteristics of the rivers including catch statistics and numbers of escapees.

Population	Farmed escapees in the river			River characteristics						
	Years counted	Unweighted mean* (Range)	Weighted mean**	Local stocking?	2010 catch (kg)	2010 catch (n)	1990 catch (kg)	1990 catch (n)	Anadromous area (km²)	Conservation attainment (2007–2010)
Neiden	1	12%	2%	No	4.907**	1390	7099	NA	21.4	98%
V. Jakobselv	18	30% (3–65)	20%	No	7.127	2283	1008	272	15.4	322%
Alta	15	6% (0–22)	5%	M(S)	15.865	3403	9959	1953	57.0	228%
Reisa	12	31% (0–100)	5%	L	7.280	1324	3044	585	53.0	177%
Målselv	15	16% (4–36)	8%	L	11.614	2362	4992	908	20.0	249%
Roksdalsvass.	19	7% (0–47)	3%	No	1.317	556	NA	NA	3.3	130%
Namsen	21	27% (10–59)	11%	L(A)	20.360	4818	32075	8019	190.7	188%
GaulaST	16	6% (0–22)	4%	M(A+E)	32.721	5690	25068	5334	93.6	224%
Surna	7	28% (0–56)	14%	H(S+F)	7.320	1364	7750	2348	35.1	136%
Eira	7	16% (0–44)	17%	H(S+P)	2206	549	580	NA	7.0	119%
Bondalselva	10	27% (0–83)	17%	L(A)	521	175	7500	2143	2.1	124%
Ørstaelva	15	41% (8–78)	22%	M(A)	1.375	502	4040	1616	4.9	60%
GaulaSF	13	31% (4–65)	17%	M(A+E)	891	300	2071	628	10.5	144%
Lærdalselva	4	2% (0–2)	4%	H(F)	Banned*	NA	4371	599	18.2	NA
Vosso	14	45% (0–71)	29%	H(S+P)	Banned***	NA	880	91	15.3	NA
Loneelva	16	8% (0–26)	7%	M(A+F)	244	107	363	214	0.4	133%
Opo	2	50% (0–100)	89%	L(F+S)	Banned***	NA	612	146	5.8	NA
Etne	19	57% (0–100)	35%	L(E+S)	Banned***	NA	7778	2431	3.7	156%
Figgjo	14	9% (0–28)	9%	L(A+E)	4393	1466	7326	3330	5.4	175%
Numedalslågen	15	7% (0–50)	5%	L(A)	7.729	1695	8791	2442	79.4	93%
Berbyelva	6	4% (0–11)	2%	L	1134	181	304	74	3.3	582%

Years counted = numbers of years in which farmed salmon were counted in the river, % of farmed salmon = the mean percent of farmed salmon observed in these populations based upon the unweighted mean = average percentage of farmed salmon in spawning population in the period 1989–2009 [50,51], weighed mean = weighted average percentage of farmed salmon in the population combining data from both sports-fishing and spawning population samples [52]; range for the unweighted mean refers to the lowest and maximum percentages of farmed salmon observed in the spawning populations (this also includes recordings with very low numbers of observations in some years [49]). Local stocking history and river catch in 2010 statistics Norway www.ssb.no, and 1990 [125];
Na = not available.
* = treated against *Gyrodactylus salaris*;
** = Norwegian zone;
*** = population collapse or strongly reduced;
smolt and parr stocking activity: <5000 : Low; 5–15000: Medium; >15000: High (eggs, alevins and fry converted to smolt numbers by calculating 10% survival); anadromous area available to smolts [49], and conservation attainment which is the average attainment of the conservation limit for each specific river as defined by the numbers of female salmon left in the river after fishing mortality in relation to the number of eggs required to achieve the rivers estimated carrying capacity [49].

In order to investigate the potential relationship between geographic and genetic distance (F_{ST}) in the historic and contemporary data sets, Mantel tests were conducted with the software PASSaGE [75] and significance was tested after 10 000 permutations. Genetic differentiation among populations was estimated by the Analysis of Molecular Variance, AMOVA [76] implemented in the program Arlequin [77], and significance was based upon 10 000 permutations.

A growing number of statistical approaches are available to identify putative non-neutral loci [78]. First, we used a hierarchical Bayesian method [79] as implemented in BayeScan software [80]. Secondly, we used the Fdist approach [81], implemented in LOSITAN [82] selection detection workbench for codominant markers. As a result, a subset of fourteen neutral microsatellite loci was obtained. Full details and results of these analyses are available in Text S1.

To investigate population structure we identified genetic clusters in the total and neutral dataset with the Bayesian model-based clustering algorithms implemented in STRUCTURE v. 2.3.3 [83–

85] under a model assuming admixture and correlated allele frequencies without using population information. Five to ten runs with a burn-in period of 50000–100000 replications and a run length of 500000–1000000 Markov chain Monte Carlo (MCMC) iterations were performed for a variable number of clusters (see footnotes of corresponding barplots for more detailed information). We then applied an ad hoc summary statistic ΔK which is based on the rate of change of the 'estimated likelihood' between successive K values [86]. When needed, runs of the selected K were averaged with CLUMPP version 1.1.1 [87] using the LargeKGreedy algorithm and the G' pairwise matrix similarity statistics and results were visualized as a barplot. Admixture analyses were conducted both with wild salmon and with a combination of wild and farmed salmon (see results).

Genetic drift may be considered as a random evolutionary process whereby a population's allele frequency at one or more loci can change through time. This process is especially influential in small populations [88,89]. Thus, in order to evaluate whether any of the populations included in the present study were very small and likely to be strongly influenced by genetic drift, the effective population size (Ne) was computed in each river. This was conducted separately for both the historical and contemporary samples, using the one sample linkage disequilibrium method implemented in the program LDNE [90]. Furthermore, in order to investigate the plausibility that genetic drift could have been the primary driver of the temporal genetic changes observed in some of the populations studied (see results), we simulated genetic drift on these historical populations. For these computations, a methodological approach inspired by an available software for simulating genetic drift [91] was implemented in R (R Development Core Team). Starting from the observed historical sample, additional generations were simulated by gene dropping, so that every additional generation were obtained from the previous one assuming random mating, equal sex proportions, no migration, selection nor migration. Drift was thus assumed to be the only evolutionary force acting upon the populations and markers were unlinked. In order to investigate how Ne influences genetic drift over multiple loci simultaneously, these simulations were conducted 1000 times for each population assuming Ne of 25, 50, 75, 100, 200, 300, 400 and 500, and setting a non-overlapping generation interval to 5 years. The number of generations in which drift was simulated was thereafter a function of the number of years between the historical sample and the corresponding contemporary one, divided by 5, and then rounded up to the nearest whole generation. The genetic distance (F_{ST}) between the observed historical genetic profile for that population, and the 1000 simulated contemporary populations at each level of Ne, were then compared to the genetic distance that was actually observed between the historical and contemporary sample. The probability that the observed pair-wise F_{ST} was greater than the genetic drift simulated F_{ST} was thereafter computed. As in [91], this was achieved by comparing the proportion of the observed F_{ST} values exceeding the genetic-drift simulated F_{ST} values for that population. These simulations were also used to look at global F_{ST} values, and evaluate allelic richness in the presence of genetic drift.

Results

Genotyping Quality

The final data set consisted of 3049 salmon displaying a mean genotyping coverage of 96.1%. Coverage ranged from 87.1% for the marker Ssa$D157$, to 99.4% for the marker Ssa$F43$. When genotyping success was broken down into the historical and contemporary data sets, coverage was 94.8% and 97.9% respectively.

From 9314 alleles scored independently on two occasions, a mean genotyping error rate (defined here as inconsistent scoring between two independent runs of the same sample) of 0.1% was computed. The absolute number of alleles scored twice/errors observed = 7506/7, 806/1, and 1002/2 for the historical, intermediate and contemporary samples respectively. This is consistent with previous estimates for these [55] and other genetic markers [70,92] in this laboratory. Allelic distribution in the historical and contemporary data sets (pooled populations) did not reveal a disproportionate loss of the large alleles in the historical samples (Table S2).

HWE, LD and Potential Neutrality of Markers

Analysis of HWE and LD can identify technical issues (marker robustness and genetic linkage between loci) and biological processes (mixing of populations and population disturbance through introgression). At the significance level of α 0.05, a total of 32 (7.1%), 5 (2.9%) and 32 (7.2%) loci by sample combinations

displayed significant deviations from HWE in the historical, intermediate, and contemporary samples respectively (Table 3; Table S3–supporting information). At α 0.001, the number of deviations dropped to 2, 1, and 1 in the three data sets respectively. No more than 4 of the 21 populations deviated for any given locus in any of the three data sets demonstrating once again that the markers were of high technical quality. Excluding the historical sample for Vestre Jacobselv, where 9 loci departed from equilibrium at α 0.05 (one of which remained significant at α 0.001), deviations from HWE were distributed among the rivers, with most displaying deviations in 0–3 loci at α 0.05 (Table 3; Table S3).

When computed for all combinations of pairs of loci, within each population separately, LD was detected 309 (6.4%) and 35 (0.7%) times among the historical samples, 122 (6.6%) and 12 (0.6%) times in the intermediate samples, and 422 (8.7%) and 25 (0.5%) times in the contemporary samples at α 0.05, and α 0.001, respectively. Deviations were distributed evenly among the different combinations of pairs of loci, but unevenly distributed among the samples (Table 3). For example, in the historical samples, Vestre Jacobselv displayed 85 pair-wise LD combinations among loci (28% of all LD observed in the historical samples). Together, HWE and LD suggest some form of disturbance in the Vestre Jacobselv in the historical sample. Within the contemporary samples, three populations (Rokdalsvassdraget, Reisa and Opo) accounted for 44% of the pair-wise LD combinations observed.

All loci displayed statistically significant global F_{ST} estimates in the historical and contemporary data sets (Table S3). Samples corresponding to the historical data set identified three loci under possible directional selection (MHC2, Ssa$F43$, Ssa289) and five under possible stabilizing selection (SS$sp2216$, Ssa197, Ssa$D157$, Ssa$D144$, SS$sp2201$), whereas the contemporary data set showed the same loci under possible directional selection but only two of the former ones under possible stabilizing selection (Ssa$D157$, SS$sp2201$) (Text S1). Subsequently, analyses have been conducted on data sets comprised of the full (all 22 loci) and the neutral (14 loci only) markers.

Temporal Genetic Variation within Populations

The number of alleles observed among populations, and between temporal samples within populations varied greatly (Table S3). Differences in sample size were accounted for by computing allelic richness A_R. Looking specifically at temporal variation of A_R within populations, most showed a very slight increase with time, however, the populations Vosso, Opo and Loneelva increased by 18–27 (Table 3).

When considering data from the set of 22 loci, and the 14 neutral ones separately, statistically significant temporal genetic change, as measured by F_{ST}, was detected in 6 of the 21 populations (Table 3). Populations displaying LD, or distinctly increased A_R in the contemporary samples, were all among those displaying temporal genetic changes. In three of the populations the F_{ST} estimates between historical and contemporary samples exceed 0.01 (i.e., Opo, Vosso and Loneelva). The change in A_R from the historical to the contemporary samples was significantly higher ($P = 0.003$; non-parametric Mann-Whitney test) in the six populations showing temporal genetic changes (mean increase per population $= 15.8$), than in the six ones displaying the strongest temporal stability (mean increase per population $= 2.6$).

No statistically significant correlation was observed between the frequency of farmed escapees observed in a given population in the period 1989–2009 based upon the un-weighted mean from the autumn spawning surveys (see Table 2), and pair-wise F_{ST}

Table 3. Effective population size, within-sample genetic diversity estimates, and temporal genetic stability between historical and contemporary samples within 21 Atlantic salmon rivers located throughout Norway. For full data, including locus specific statistics see Table S2.

Rivers	Within-sample diversity								Temporal stability				
	Historical				Contemporary				F_{ST} historical vs. contemporary		Exclusion from hist. <0.001		Temporal change?
	LD	HW	A_R	Ne (95% CI)	LD	HW	A_R	Ne (95% CI)	22 loci	14 loci	22 loci	14 loci	
Neiden	22	0	201	430 (296–760)	7	1	203	Inf (3179-Inf)	0.0009	0.0011	6%	3%	No
V. Jakobselv	85	9	190	79 (71–91)	32	0	200	169 (148–196)	0.0064**	0.0076**	16%	7%	Yes
Alta	5	2	187	Inf (990-Inf)	13	1	190	4860 (856-Inf)	−0.0002	0.0010	2%	1%	No
Reisa	11	2	185	272 (180–533)	61	1	179	80 (69–94)	0.0041*	0.0020	15%	10%	No
Målselv	10	2	199	Inf (−1361-Inf)	3	0	207	411332# (322-Inf)	−0.0026	−0.0011	13%	7%	No
Roksdalsvass.	9	0	205	516 (241-Inf)	66	2	206	384 (291–554)	0.0014	0.0023	20%	12%	No
Namsen	10	0	208	3526 (835-Inf)	14	1	209	914 (549–2550)	0.0013*	−0.0012	9%	3%	No
GaulaST	4	0	206	Inf (2162-Inf)	10	1	208	24753 (1358-Inf)	0.0012	0.0018	12%	14%	No
Surna	9	0	203	1530# (252-Inf)	11	1	216	Inf (965-Inf)	0.0025	0.0035	34%	17%	No
Eira	11	2	209	378 (196–3201)	11	0	211	498 (293–1519)	0.0005	0.0000	14%	10%	No
Bondalselva	9	0	209	1283 (418-Inf)	12	3	NC.	34# (26–47)	0.0043	0.0017	6%	0%	No.
Ørstaelva	6	1	214	3678 (450-Inf)	17	0	210	400 (202–6501)	0.0003	−0.0013	0%	0%	No
GaulaSF	7	3	211	1193 (371-Inf)	19	2	205	439 (311–727)	0.0001	0.0008	17%	1%	No
Lærdalselva	8	1	193	Inf (−506-Inf)	13	2	200	333 (216–698)	0.0015	0.0010	15%	6%	No
Vosso	14	1	175	Inf (−304-Inf)	8	4	202	189 (138–294)	0.0179**	0.0213**	76%	67%	Yes
Loneelva	17	5	176	984 (348-Inf)	8	2	200	241 (172–390)	0.0120**	0.0116**	52%	29%	Yes
Opo	10	1	166	Inf# (−14-Inf)	58	1	184	68 (60–76)	0.0258**	0.0279**	100%	90%	Yes
Etne	25	1	209	752 (439–2405)	12	3	209	917 (507–4135)	0.0006	0.0000	5%	5%	No
Figgjo	9	1	204	Inf (−1638-Inf)	14	2	210	Inf (1070-Inf)	0.0048**	0.0058**	38%	4%	Yes
Numedalslågen	9	1	194	Inf (1194-Inf)	14	1	210	653 (383–2050)	0.0032*	0.0051*	29%	18%	No
Berbyelva	19	0	156	81 (67–101)	19	4	166	245 (194–327)	0.0053**	0.0071**	16%	7%	Yes

Within samples: LD = observed number of deviations from linkage disequilibrium (231 pair-wise tests per population, 211 tests for Opo) at α 0.05, HW = observed deviations from Hardy Weinberg Equilibrium (22 tests per population, 21 tests for Opo) at α 0.05, A_R = allelic richness computed using re-sample size of 25 (note Opo samples only computed with 21 loci therefore not directly comparable to other populations), Ne = effective population size as computed from LD method in LDNE [90] Inf = Infinity suggesting that the population is "relatively large" (i.e., >200) [93], # = harmonic mean sample size less than 30 and therefore estimated Ne not to be trusted. Between temporal samples: * = F_{ST} significant at α 0.05, ** = F_{ST} significant at α 0.001 (and following Bonferroni), NC = not computed, Exclusion from hist. = percentage of fish from the contemporary population that are excluded from the historical population profile in the program Geneclass at a cut off of α 0.001, temporal change ? = whether significant temporal genetic change is reported within rivers at α 0.001 based upon pair-wise F_{ST} for both sets of microsatellites.

between the historical and contemporary samples for the same population ($R^2 = 0.18$, $P = 0.052$) (Fig. 2a). When using the weighted mean number of escapees reported in a combination of the summer sports-fishing catch and the autumn spawning counts for each population [52], the correlation with pair-wise F_{ST} was statistically strong ($R^2 = 0.56$, $P<0.0001$) (Fig. 2b). However, when the river Opo was excluded (this river displayed by both the highest percentages of escapees and greatest temporal genetic change) the correlation was not significant ($R^2 = 0.09$, $P = 0.20$) (Fig. 2c). The lack of a clear relationship between percentage of farmed fish (by either of the two estimations) and observed genetic changes is readily illustrated by the fact that two of the populations (e.g., Opo and Vosso) displayed high numbers of escapees on the spawning grounds and large temporal genetic changes, while other populations (e.g., Ørsta and Etne) also displayed high numbers of escapees but did not reveal genetic change with time. Furthermore, several other rivers had been exposed to >10% escapees in the period 1989–2009 without displaying statistically significant temporal genetic changes (Table 2, 3, Fig. 2).

Individual admixture analysis was also applied to evaluate within-population temporal stability, using historical, intermediate (when available) and the contemporary samples both for the total and neutral sets of microsatellites. The assessment of ΔK in single-population assignment analyses revealed that the most likely number of clusters ranged between two and three (Fig. 3; Fig. S1), although in one population, Berbyelva, this was ≥ 4 [86]. Admixture analysis supported the results of temporal change from the F-statistics. Thus, populations such as Opo, Vosso, Loneelva and Vestre Jakobselv, which showed temporal genetic changes in F_{ST}, also showed evident signs of admixture (Fig. 3).

The percentage of fish from each contemporary sample that was excluded from its historical population sample when conducting genetic assignment ranged from 0–100% when using all 22 loci, and 0–90% when using the reduced set of neutral loci (Table 3). There was a strong correlation between percentage of fish that were excluded from their respective historical populations, and the pair-wise F_{ST} values ($R^2 = 0.86$ $P<0.0001$). For example, the populations Opo, Vosso and Loneelva displayed the highest pair-wise F_{ST} values between historical and contemporary samples

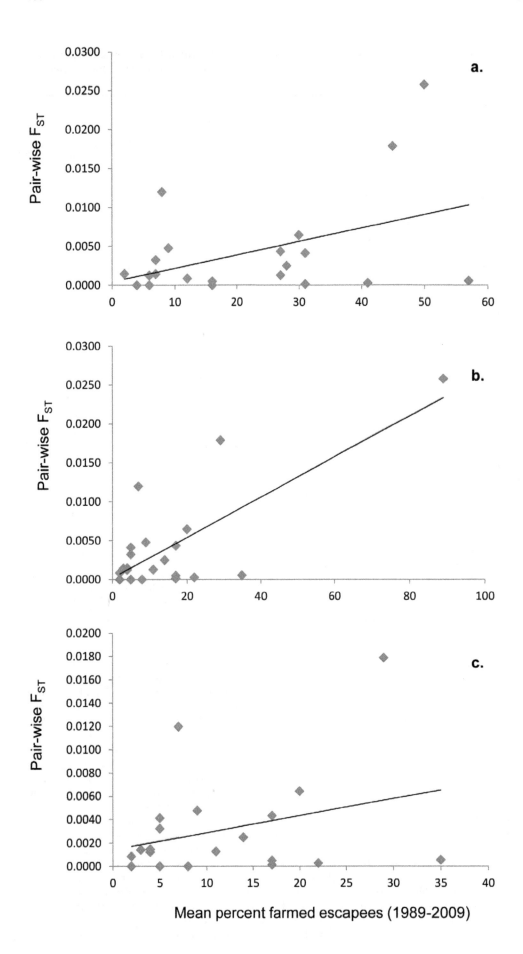

Figure 2. Relationship between average numbers of escapees observed in each population in the period 1989–2009, and the observed within-river temporal genetic changes as computed by pair-wise F_{ST} between the historical and contemporary sample. Graph a = relationship when using an un-weighted mean of the farmed escapees recorded in the autumn survey data ($R^2 = 0.18$, $P = 0.052$), graph b = relationship when using a weighted mean based upon a mixture of summer sports fishing and autumn survey data ($R^2 = 0.56$, $P < 0.0001$) [52], and c = same as b with the population Opo excluded ($R^2 = 0.09$, $P = 0.20$).

(0.028, 0.021, 0.012 respectively) in addition to the highest exclusion rates (100%, 76%, and 52% respectively). While other assignment methods implemented in the program Geneclass gave different absolute exclusion percentages, the above trend remained.

Spatio-temporal Genetic Variation

Global F_{ST} among the 21 historical samples was significantly larger than among the 21 contemporary ones (Table 4). Significantly, the reduction in global F_{ST} with time was observed in 21 of the 22 loci (Fig. 4, Table S3). This trend was also reflected in the self-assignment analyses conducted in Geneclass which showed a drop from 61.6% of fish correctly assigned to their source populations in the historical data set, to 57.6% in the contemporary. Finally, the AMOVA analysis revealed that the amount of genetic variation observed among populations dropped from 4.1% in the historical data set to 2.9% in the contemporary one.

The historical data set was drawn from a wider time-interval than the contemporary one (Table S1). Therefore, in order to test whether this was spuriously responsible for the drop in global F_{ST} between the two data sets, a reduced historical data set was established from 12 populations where samples were available from the interval 1986–1994. Likewise, a temporal reduction in global F_{ST} was still observed for the 12 populations (Table 4).

Looking specifically at the six populations displaying temporal genetic changes, global F_{ST} decreased from 0.058 among the historical samples, to 0.039 among the contemporary ones. In contrast, global F_{ST} estimated among the six populations that showed the highest level of within-river temporal stability did not display any change between the historical (0.026) and contemporary (0.027) data sets. Inspection of the pair-wise F_{ST} values among the six populations displaying within-population changes showed that all of them contributed to the distinct temporal decrease in global F_{ST} (Table 5, 6).

Using data from all 22 markers, a significant relationship between geographic and genetic distance was observed for the total set of populations both in the historical ($R^2 = 0.365$, $P < 0.0001$) and in the contemporary samples ($R^2 = 0.377$, $P < 0.0001$). When looking specifically at the six populations not displaying temporal genetic change, a strong relationship was found in the historical ($R^2 = 0.758$, $P = 0.0011$), and contemporary data sets ($R^2 = 0.668$, $P = 0.0013$). When examining the six populations displaying temporal genetic change, the relationship between genetic and geographic distance was not statistically significant in either the historical ($R^2 = 0.279$, $P = 0.1013$) nor the contemporary data sets ($R^2 = 0.221$, $P = 0.1411$).

Admixture analyses conducted on the 21 populations provided the strongest support for $K = 2$, both when considering the probability of the data [P(D)] and the ad hoc statistic ΔK, for historical and contemporary samples when using the 22 loci (Fig. 5) and the 14 neutral loci (Fig. S2). In both cases, the five northernmost populations formed a very distinct separate cluster. Following a hierarchical approach, we split the data set into the corresponding five and sixteen populations respectively and conducted the assignment analyses separately. Looking at the full set of markers, the five northernmost populations yielded K3 in the historic dataset and K4 in the contemporary one. Visual

inspection of either K3 or K4 for the northern populations revealed increased admixture in several of the rivers over time. This was most apparent for the rivers Vestre Jakobselv, and interestingly, Målselva, the latter of which did not display temporal genetic change as computed by F_{ST}, nor by single-river admixture analysis (Fig. S1). Turning to the remaining sixteen populations, both the historical and contemporary data sets revealed K = 3 as the most likely number of clusters. The southernmost population, Berbyelva was the most distinct (especially in the contemporary data set), and therefore, admixture analyses were also computed with this population excluded. Changes in genetic structure between the historical and contemporary data sets across these sixteen populations were subtle, and not as distinct as for changes within populations (Fig. 3; Fig. S1).

In order to investigate whether the inclusion of farmed salmon would improve the power to detect temporal genetic changes in population genetic structure (either within or among populations), samples from nine genetically distinct farm sources were included in the admixture analyses. Runs were conducted for K = 12 and K = 13 as the analyses included salmon from 9 distinct farm samples, and, that K for the northern and southern clusters had already been estimated at 3 or 4. Both sets of analyses were conducted with and without a prior for the farm samples (which made no difference to the result). As expected, samples from the farms were confirmed to be highly distinct to each other, whereas wild populations were strongly admixed in both the historical and contemporary samples (Fig. S3). Thus, inclusion of farmed fish did not reveal additional temporal genetic changes not already detected.

Effective Population Size and Simulations of Genetic Drift

In most of the historical and contemporary samples representing each population, the computed effective population size (Ne) was larger than 200 (Table 3). Confidence intervals associated with these estimates were large, often reaching infinity in the upper bound (Table 3, Table S4). Several of the samples also showed negative values, both in the upper and lower bound. Negative values occur when the variance observed can be ascribed entirely to sampling error alone, and suggests that these samples displayed relatively high Ne (i.e., >200) [93].

Simulations of genetic drift were conducted for the six populations identified as displaying statistically significant temporal genetic changes. These simulations were conducted in order to evaluate the possibility that genetic drift could have caused the observed changes given the number of generations that have occurred between the historical and contemporary samples.

Unsurprisingly, the mean pair-wise F_{ST} between the historical sample and the simulated contemporary population was strongly influence by Ne (Fig. 6); small Ne leading to large F_{ST}. For five of the six populations, a value of Ne of 100 was sufficient to exclude genetic drift as the primary driver of the observed temporal genetic changes ($P < 0.001$). In these cases the pair-wise F_{ST} that was observed between the historical and contemporary sample was greater than the pair-wise F_{ST} between the historical sample and the simulated population in all the replicates (i.e., $P < 0.001$ for 1000 replicates). In the river Figgjo, a value of Ne of 300 or more would be required to achieve the same level of significance

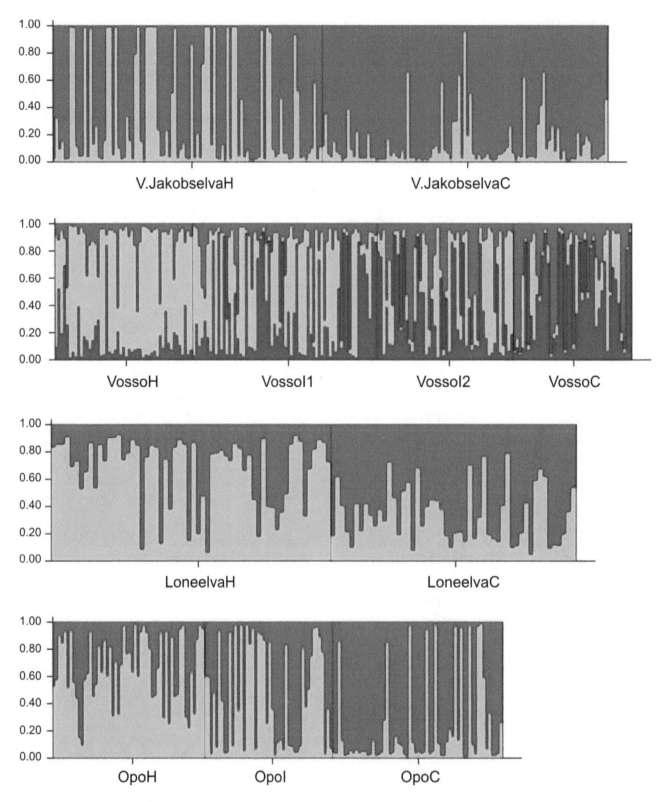

Figure 3. Bayesian clustering of historical (H), intermediate (I) and contemporary (C) samples representing the four rivers displaying the largest temporal genetic changes at 22 microsatellite loci. For the river Vosso, a total of four samples were available. Thus, the two intermediate samples for this river include a suffix I1 and I2 (linking to these specific samples to Table 1). These analyses were conducted on each river separately. Inferred ancestry was computed using STRUCTURE v. 2.3.3 [83,84], under a model assuming admixture and correlated allele frequencies without using population information. Ten runs with a burn-in period consisting of 100000 replications and a run length of 1000000 Markov chain Monte Carlo (MCMC) iterations were performed for a number of clusters ranging from K 1 to 5. Then an ad hoc summary statistic ΔK [86] was used to calculate the number of clusters (K) that best fitted the data for each river separately. For full computation details and results for all populations using both 22 and 14 markers see Fig. S1 (supporting information).

Table 4. Summary of global F_{ST} estimates, and, P values indicating whether the global F_{ST} estimates are significantly different between the historical and contemporary samples.

| | COMPARISON F_{ST} BETWEEN GROUPS (Historical *vs.* contemporary) | | | | | |
| | TOTAL LOCI | | | NEUTRAL LOCI | | |
	F_{ST} histor.	F_{ST} contemp.	P value	F_{ST} history.	F_{ST} contemp.	P value
All 21 populations	0.038	0.030	0.008	0.038	0.028	0.001
20 populations (excluding Opo)	0.038	0.030	0.010	0.034	0.026	0.006
12 populations in restricted data set*	0.039	0.032	0.078	0.032	0.025	0.042
6 populations displaying temporal changes	0.058	0.039	0.006	0.057	0.032	0.001
6 populations displaying the strongest temporal stability	0.027	0.028	0.550	0.027	0.026	0.470

All global F_{ST} estimates were significant at α 0.001.
*These 12 populations were selected due to narrow the historical temporal data-set to the period 1986–1994, Opo was excluded due to the fact that it was only genotyped for 21 of the 22 loci.

($P<0.001$). Comparing these genetic drift simulations with the computed Ne values (Table 3) revealed that genetic drift can be confidently excluded as the driver of the observed temporal genetic changes in the rivers Vosso, Loneelva and Figgjo. This is due to the fact that their Ne values ranged between several hundred and infinity in both the historical and contemporary samples (Table 3). For the rivers V. Jakobselv, Opo and Berbyelva, either the historical or contemporary sample displayed a Ne lower than 100 (79, 68 and 81 respectively). This is at the level of Ne where the potential for genetic drift to contribute to temporal genetic changes on the time-scale studied can be excluded at modest levels of statistical significance ($P = 0.04$, 0.01, and 0.01 for V. Jacokbselv, Opo and Berbyelva respectively for $Ne = 75$) (Fig. 6,

Table 7). Nevertheless, all of these three populations displayed Ne values >150 in one of the samples.

Strong genetic drift in small populations is not only expected to lead to within-population temporal instability, it is expected to simultaneously lead to increased inter-population differentiation (on average) when it is stronger than the influence of gene-flow [88,89]. The genetic drift based simulations reported above were also used to re-compute the global F_{ST} value between the six populations displaying statistically significant temporal genetic changes after having simulated genetic drift independently within each (Fig. 6). The "global" plot illustrates that as Ne decreases, and genetic drift becomes more pronounced within each population, the level of inter-population genetic differentiation increases rapidly. This is in stark contrast to the large and statistically

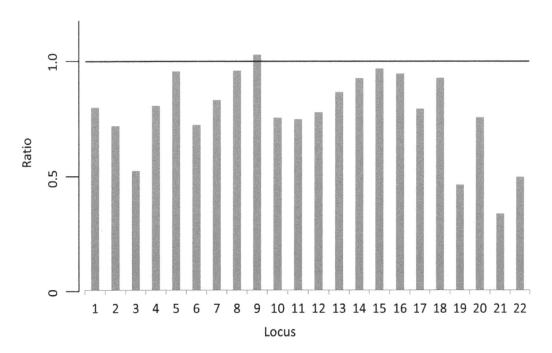

Figure 4. Ratio between global F_{ST} computed among the 21 contemporary samples divided by the global F_{ST} computed among the 21 historical samples for 22 microsatellite markers. Locus number is consequent with locus names and other locus-specific details available in Table S3.

Table 5. Pair-wise genetic distance as computed by F_{ST} among the 6 populations displaying within-river temporal genetic changes. Computed for historical (bottom left) and contemporary samples (top right), and based upon the analysis of 22 loci.

	V. Jakobselva	Loneelva	Vosso	Opo	Figgjo	Berbyelva
V. Jakobselva		0.035	0.026	0.031	0.035	0.074
Loneelva	0.056			0.013	0.017	0.020	0.063
Vosso	0.067	0.048		0.008	0.014	0.051
Opo	0.061	0.038	0.033		0.015	0.051
Figgjo	0.055	0.047	0.037	0.039		0.042
Berbyelva	0.086	0.086	0.078	0.069	0.053	

Computed for historical (bottom left) and contemporary samples (top right), and based upon the analysis of 22 loci.
All F_{ST} values significant at α 0.001 with the exception of those in bold.

Table 6. Pair-wise genetic distance as computed by F_{ST} among the 6 populations displaying the greatest within-river temporal stability.

	Alta	Målselva	Eira	Ørstaelva	GaulaSF	Etne
Alta		0.020	0.056	0.054	0.046	0.051
Målselva	0.021		0.029	0.024	0.023	0.026
Eira	0.053	0.031		0.012	0.015	0.012
Ørstaelva	0.051	0.029	0.009		**0.003**	**0.002**
GaulaSF	0.049	0.027	0.009	0.007		0.004
Etne	0.053	0.035	0.012	0.006	0.005	

Computed for historical (bottom left) and contemporary samples (top right), and based upon the analysis of 22 loci.
All F_{ST} values significant at α 0.001 with the exception of those in bold.

significant drop actually observed in the global F_{ST} among these six populations with time (Table 5).

Discussion

This study represents one of the largest temporal analyses of population genetic structure conducted thus far. Samples covering an entire country, and spanning up to four decades, have permitted the identification of genetic changes occurring both within and among 21 populations, through time. Two main conclusions can be drawn from these analyses. First, despite the fact that farmed escapees have been recorded on the spawning grounds for all of the populations studied, outnumbering wild conspecifics in some years in some of the populations, only weak to moderate changes in among-population genetic structure have been observed in the time-period studied, and in most rivers, statistically significant temporal genetic changes were not observed. This demonstrates that generally, farmed escaped salmon have had poor to moderate success in the wild. Second, not all populations were equally resilient. Genetic changes were observed in six of the populations (29% of those studied), and in four of them, the changes were highly significant. For example, 100%, 76% and 52% of the fish comprising the contemporary samples for Opo, Vosso and Loneelva were excluded from their respective historical baseline samples at $P = 0.001$ and when using data from all 22 loci. At the same time, genetic drift was excluded as the primary contributing factor. These changes have occurred during 15–30 years, equivalent to approximately 3–6 generations in native populations. Thus, these data demonstrate that farmed Atlantic salmon have successfully introgressed and caused genetic changes in some wild Norwegian populations.

A weak to moderate but statistically significant reduction in population genetic structure was observed among the 21 populations with time. This is consistent with an increase in gene flow, and has been previously reported in response to extensive supplementation and translocations of brown trout in Denmark [46], among stocks of pearl oyster (*Pinctada margaritifera cumingii*) throughout French Polynesia [94], and among brook charr (*Salvelinus fontinalis*) populations in Canadian lakes [95]. Importantly, a reduction in population genetic structure is a predicted response to widespread gene flow from farmed escapees, based upon simulations conducted with genetic data in Norway [45,47]. Nevertheless, although a decrease in population heterogeneity was observed with time, significant population genetic structure was still observed in the contemporary data set. Both the historical and contemporary datasets displayed a clear pattern of isolation by distance which is characteristic for Atlantic salmon [15,16]. In 15 of the 21 populations, temporal genetic changes were not detected despite the fact that all of them had experienced farmed escapees on the spawning grounds, and in some years, escapees had outnumbered wild spawners (Table 2). While it is possible that the set of markers implemented here may have failed to detect low-levels of introgression in some populations (see discussion below), it is concluded that the gene flow from farmed escapees into native populations throughout Norway, has been less than the numbers of escapees observed on the spawning grounds. We suggest that this is primarily due to the fact that farmed escapees display reduced spawning success [96–98], in addition to the fact that their offspring display lower survival in the wild when compared with native conspecifics [96,99,100].

Not all of the populations studied were equally resilient. Statistically significant temporal genetic changes were observed in six populations, and for some of these, the changes were very distinct and highly significant. For example, 100%, 75% and 52% of the contemporary samples from Opo, Vosso and Loneelva were excluded from their respective historical profiles. When focusing on the six populations displaying temporal changes, global F_{ST} nearly halved between the historical and contemporary data sets. From population genetics theory [88], classical experimental studies [89], and the simulations conducted within this study, genetic drift is expected to lead to greater differentiation among populations. This has been documented for example in the Spanish imperial eagle (*Aquila adalberti*) [101] and forest jaguars (*Panther onca*) [102] in response to habitat fragmentation, and among Atlantic salmon populations that have experienced significant population declines at the southernmost part of their natural distribution [103]. In addition, none of the six populations displaying temporal genetic changes had very low *Ne* estimates, and based upon simulations, genetic drift was conclusively excluded as the primary driver of the observed temporal genetic changes within most of these rivers. Furthermore, genetic drift was demonstrated to be incompatible with the observed drop in differentiation among these populations with time, and not least, cannot explain the increase in the number of alleles observed in all of these populations. Therefore, in consideration of the genetic data and simulations presented, characteristics of these populations, the high numbers of escapees observed on the spawning grounds (Table 2), and the fact that successful spawning of farmed escaped salmon has been documented in several Norwegian rivers

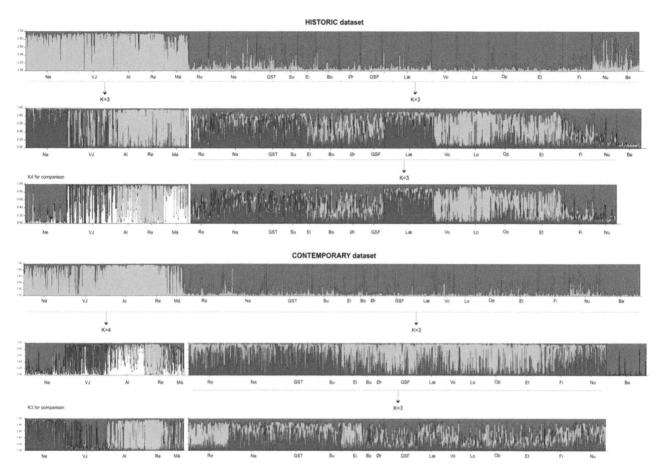

Figure 5. Hierarchical Bayesian clustering for the historical and contemporary data sets for 21 populations genotyped at 22 microsatellite loci. Inferred ancestry was computed using STRUCTURE v. 2.3.3 [83,84], under a model assuming admixture and correlated allele frequencies without using population information. Ten runs with a burn-in period consisting of 100000 replications and a run length of 1000000 Markov chain Monte Carlo (MCMC) iterations were performed for a number of clusters ranging from K 1 to 5. Then, the ad hoc summary statistic ΔK [86] was used to calculate the number of clusters (K) that best fitted the data. Populations are ordered North to South, thus corresponding with Tables 1 and 2. Barplots for K3 and K4 are presented for comparison between the historical and contemporary data sets (see results section). For full computation details and results using both 22 and 14 markers see Fig. S2 (supporting information).

Table 7. *P*-values testing whether the observed pair-wise F_{ST} between each population's historical and contemporary sample was significantly larger than the F_{ST} between each population's observed historical sample and 1000 computer simulated contemporary samples.

Ne	Population					
	V. Jakobselv	Vosso	Loneelva	Opo	Figgjo	Berbyelva
25	0.99	0.99	0.85	1.0	1.0	0.97
50	0.3	0.4	0.03	0.57	1.0	0.2
75	0.04	0.01	0.02	0.01	1.0	0.01
100	<0.001	<0.001	<0.001	<0.001	0.95	<0.001
200	<0.001	<0.001	<0.001	<0.001	0.03	<0.001
300	<0.001	<0.001	<0.001	<0.001	<0.001	<0.001
500	<0.001	<0.001	<0.001	<0.001	<0.001	<0.001

Simulations were based upon genetic drift at different *Ne*. Plots of observed and simulated F_{ST} values are presented in Fig. 6.

in the time period studied [104,105], it is concluded that genetic introgression of farmed escaped salmon represents the primary cause of the observed temporal genetic changes. Specifically in the case of the river Vosso, extensive spawning of farmed females has been documented by size and pigment measurements conducted on eggs deposited in the river, leading to the conclusion that the population in this river had been replaced by farmed escapees in the 1990's [105]. The results of that field experiment are highly consistent with both the timing and magnitude of genetic changes observed in the river Vosso in the present study. Nevertheless, it is worthy of note that the populations in Berbyelva and Figgjo both displayed relatively small temporal genetic changes. For these two populations, the influence of non-biological factors, for example sampling bias in the historical or contemporary samples, or unidentified natural or anthropogenic disturbances, may have had a proportionately high contribution to the observed changes.

No clear relationship between the reported frequency of farmed fish in each population, and the degree of within river genetic changes were revealed in this study. This was true when using both the unweighted mean percent of farmed fish observed in the autumn survey, and the weighted mean combining data from summer sports-fishing catches and autumn surveys [52] (in combination with removing the single river sample Opo which

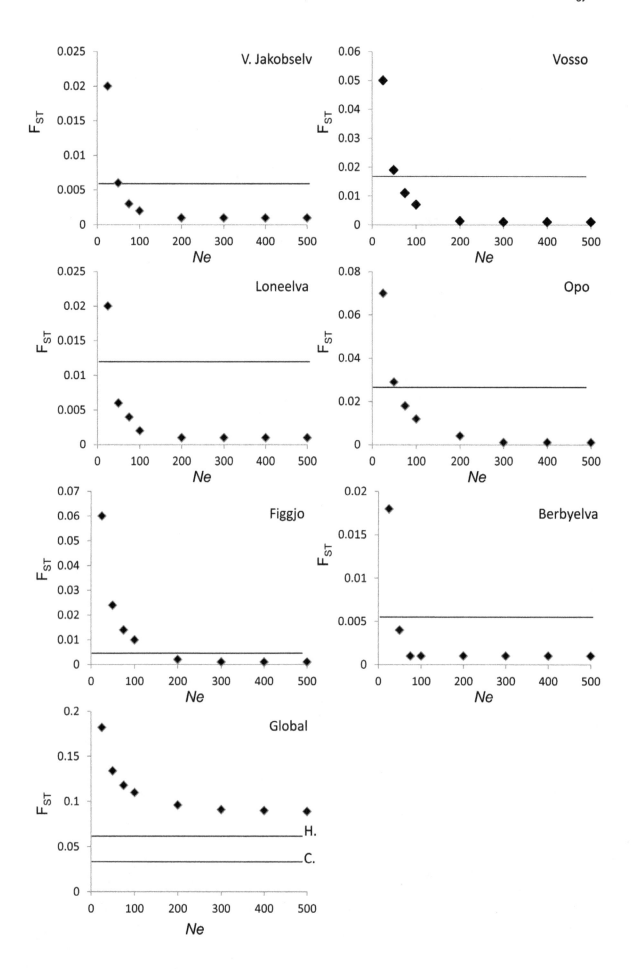

Figure 6. Simulations of genetic drift induced changes between the observed historical genetic profile and computed contemporary populations for each of the six populations displaying temporal genetic change. Black diamonds represent the mean F_{ST} between the historical population and the computed contemporary population based upon 1000 simulations of genetic drift with Ne set to 25, 50, 75, 100, 200, 300, 400 and 500. Horizontal black line for each plot represents the observed pair-wise F_{ST} between the historical and contemporary population (i.e., the values given in Table 2). "Global" plot represents the global F_{ST} computed among these six populations based upon the above mentioned simulations, while the horizontal black bar H = historical global F_{ST} observed among these populations, and C = contemporary global F_{ST} observed among these populations (i.e., the values given in Table 3). Statistical significance levels for these comparisons are presented in Table 5.

was solely responsible for the statistically significant relationship) (Fig. 2a, b, c). There are many potential explanations for this result. Firstly, it is important to consider the fact that the numbers of rivers investigated is only 21, limiting the ability to test for such a relationship in a statistically robust manner. Furthermore, and importantly, the data relating to the frequency of farmed fish in these populations (either the summer sports-fishing data or the autumn surveys) has limitations, such as missing counts in some years (Table 2), and the fact that the maturity status of these escapees is not often recorded. Nevertheless, the question still remains; why did some populations (e.g., Opo and Vosso) experiencing large numbers of domesticated escapees display very large temporal genetic changes, while other populations (e.g., Ørsta and Etne), also displaying high percentages of escapees, not reveal detectable temporal genetic changes? From both ecological and conservation viewpoints, these are vital questions in order to understand the evolutionary processes underlying the potential for natural populations to persist in the face of migration and potential gene flow from non-native sources. We suggest that there are both ecological and technical reasons for this. First we address the ecological reasons.

Farmed salmon are competitively inferior to wild salmon in spawning [96–98], and their relative spawning success is density-dependant [106]. Density-dependant spawning success has also been observed for hatchery reared salmon [107]. Together, these studies suggest that farmed escaped salmon will have a higher probability of introgression in native populations with low adult densities, than in populations with high adult densities. Once introgression has occurred, it is likely that the relative survival of the domesticated offspring and admixed individuals will be higher in rivers displaying low juvenile density and accordingly low intra-specific competition. This is because the offspring of domesticated and non-native conspecifics tend to display lower survival in the wild when compared to native fish [96,99,100]. This is consistent with the fact that successful introgression of hatchery reared brown trout in native Danish populations has been partially explained by low wild fish population density [46], and with a recent study that concluded that wild population density is the most important factor affecting the competitive balance between hatchery-reared and wild fish [108]. Furthermore, the two populations (Opo and Vosso) displaying the greatest genetic changes in the present study, have both experienced low numbers of adult spawners in the period where high numbers of escapees were reported. In contrast, two other populations (e.g., Ørsta and Etne) displaying relatively high numbers of wild adult spawners in the population, did not display temporal genetic changes, despite high numbers of escapees.

For several technical reasons, it is possible that the estimated level of within-population temporal genetic changes, as estimated by the 22 microsatellites implemented here, is lower than the *true* level of genetic introgression by farmed escapees. As detailed in the Materials and Methods, gene flow from farmed fish into wild populations may be concealed and thus underestimated [45]. Several of the populations studied here displayed close to significant temporal genetic changes in F_{ST}, relatively high exclusion rates from the historical population, and, some evidence

of linkage disequilibrium (Table 3). Furthermore, the ability to detect statistically significant temporal genetic changes is influenced by the ratio between sample and effective population size (S/Ne) [109]. Given that both factors varied among the samples and populations in this study (i.e., the contemporary sample for Bondalselva, which represented the smallest sample, was only N = 16), this may have limited the ability to detect temporal changes in some of the populations. It is possible however, that analysis of genetic markers putatively under domestication selection [44] may provide the ability to quantify introgression of escapees in rivers where this has occurred at a low level.

The effective population size (Ne) represents an important parameter in conservation genetics as it provides information about the potential for genetic drift, inbreeding and natural selection to act upon populations. A range of methods for computing Ne are available, and may be broadly split into temporal [109–112] and one-sample [90,113–115] based approaches. Here, we applied a one-sample based method [90] that utilizes a bias correction [116]. This provided us with the ability to compute Ne for both the historical and contemporary samples separately, in order to estimate whether these were small populations likely to be under the influence of genetic drift. All methods of computing Ne include underlying assumptions that are rarely fulfilled in the populations in which they are implemented. For example, linkage disequilibrium, which is the primary parameter used to estimate Ne in single-sample methods, can be caused by several factors not related to Ne, such as immigration and overlapping generations. Both of these two underlying assumptions were violated by the populations in the present study, although the LD method implemented by [90] has been demonstrated to be robust to equilibrium migration [117]. Thus, while the Ne estimations presented here should be treated with some caution, they nevertheless provide indications regarding each population's effective size, and thus potential for genetic drift.

The genetic changes observed here occurred over a period of 15–30 years, which is equivalent to approximately 3–6 generations for these wild populations. This time-scale is consistent with predictions from models of gene flow based upon experimental data in which it has been suggested that under high intrusion scenarios, it will be difficult to obtain broodstock from the original population after just a few generations [118]. This correlates strongly with the results of our genetic assignment tests, where over half of the contemporary populations for Opo, Vosso and Lonelva could be excluded from their historical population profiles at $P = 0.001$ (Table 3). Given that farmed salmon continue to escape into the natural environment, it is likely that the number of populations where introgression is observed, and the magnitude of introgression within each population, will increase with time. Several of the salmonid species in the Pacific are monitored, and in some circumstances, actively managed using genetic based methods [119]. Furthermore, there are a range of advantages in using genetic methods to monitoring populations for conservation and management [120]. Here, it is suggested that if farmed salmon continue to escape into the wild, a monitoring program to assess genetic stability in native salmon populations will be necessary in order to produce science-based management strategies in the future.

Salmonid fish populations are often regarded as locally adapted to their native environments [1–3], and supplementation with hatchery produced or non-native conspecifics is potentially negative to wild populations [121]. Farmed salmon have been selected for a range of economically important traits for approximately ten generations, and as a result, they display genetic differences to wild salmon. For example, farmed salmon grow significantly faster [122], transcribe genes differently [123], exhibit reduced anti-predator responses [124], and display lower fitness in the natural environment [96,99,100]. Nevertheless, analysis of neutral, or nearly neutral genetic markers as has been conducted here, can only describe changes in population genetic structure due to gene flow. While this represents a necessary step towards understanding the level of genetic-impact that farmed escaped fish may cause in native populations, such data cannot directly infer biological consequences in recipient wild populations. Ultimately, a major question will be how allele frequencies in genes causatively linked to adaptive traits have changed in these populations.

Supporting Information

Figure S1   Bayesian clustering of historical (H), intermediate (I) and contemporary (C) samples for 21 Atlantic salmon rivers separately.

Figure S2   Hierarchical Bayesian clustering of the 21 rivers in the historical and contemporary data sets.

Figure S3   Bayesian clustering of the 21 rivers in the historical and contemporary data sets when combined together with data from 9 distinct farm sources.

Table S1   Years in which samples were taken for the historical, intermediate, and contemporary data sets.

Table S2   Allele frequencies observed in the historical and contemporary data sets for 22 microsatellite markers.

Table S3   Locus by sample summary statistics for the historical, intermediate and contemporary samples collected from 21 Norwegian rivers.

Table S4   Effective population size for samples in the historical and contemporary data sets as computed by the LD method as implemented in LDNE [90].

Text S1   Description of methods and results for identification of markers putatively under selection.

Acknowledgments

A large number of people have been involved in collecting the samples in this study, and their important role is gratefully acknowledged. Of particular note is the Norwegian Gene bank run by the Norwegian Directorate for Nature management (DN), the Norwegian Institute for Nature Research (NINA), Erlend Waatevik, Eero Niemelä, Reidar Borgstrøm and Svein Jakob Saltveit who provided access to many of the historical scales for this study. In addition, fishermen, river owners and local volunteers are acknowledged for assisting the collection of contemporary samples. We would like to thank Robin Waples, Michael M. Hansen, Per E. Jorde, Terje Svåsand, Ove T. Skilbrei, John B. Taggart and anonymous referees for constructive advice on earlier drafts of this paper.

Author Contributions

Conceived and designed the experiments: kag øs vw. Performed the experiments: kag øs vw ages. Analyzed the data: kag vw mq fb. Contributed reagents/materials/analysis tools: kag vw mq fb ages. Wrote the paper: kag mq vw fb ages øs.

References

1. Garcia de Leaniz C, Fleming IA, Einum S, Verspoor E, Jordan WC, et al. (2007) A critical review of adaptive genetic variation in Atlantic salmon: implications for conservation. Biological Reviews 82: 173–211.
2. Taylor EB (1991) A review of local adaptation in salmonidae, with particular reference to Pacific and Atlantic salmon. Aquaculture 98: 185–207.
3. Fraser DJ, Weir LK, Bernatchez L, Hansen MM, Taylor EB (2011) Extent and scale of local adaptation in salmonid fishes: review and meta-analysis. Heredity 106: 404–420.
4. Stabell OB (1984) Homing and olfaction in salmonids - a critical review with special reference to the Atlantic salmon. Biological Reviews of the Cambridge Philosophical Society 59: 333–388.
5. Ståhl G (1987) Genetic population strcuture of Atlantic salmon. In: Ryman N, Utter F, editors. Population genetics and fishery management: University of Washington press, Seattle. 121–140.
6. Verspoor E, Beardmore JA, Consuegra S, De Leaniz CG, Hindar K, et al. (2005) Population structure in the Atlantic salmon: insights from 40 years of research into genetic protein variation. Journal of Fish Biology 67: 3–54.
7. Dionne M, Caron F, Dodson JJ, Bernatchez L (2008) Landscape genetics and hierarchical genetic structure in Atlantic salmon: the interaction of gene flow and local adaptation. Molecular Ecology 17: 2382–2396.
8. Griffiths AM, Machado-Schiaffino G, Dillane E, Coughlan J, Horreo JL, et al. (2010) Genetic stock identification of Atlantic salmon (Salmo salar) populations in the southern part of the European range. Bmc Genetics 11.
9. Taggart JB, Verspoor E, Galvin PT, Moran P, Ferguson A (1995) A minisatellite DNA marker for discriminating between European and North American Atlantic salmon (Salmo salar). Canadian Journal of Fisheries and Aquatic Sciences 52: 2305–2311.
10. Gilbey J, Knox D, O'Sullivan M, Verspoor E (2005) Novel DNA markers for rapid, accurate, and cost-effective discrimination of the continental origin of Atlantic salmon (Salmo salar L.). Ices Journal of Marine Science 62: 1609–1616.
11. Tonteri A, Veselov AJ, Zubchenko AV, Lumme J, Primmer CR (2009) Microsatellites reveal clear genetic boundaries among Atlantic salmon (Salmo salar) populations from the Barents and White seas, northwest Russia. Canadian Journal of Fisheries and Aquatic Sciences 66: 717–735.
12. King TL, Kalinowski ST, Schill WB, Spidle AP, Lubinski BA (2001) Population structure of Atlantic salmon (Salmo salar L.): a range-wide perspective from microsatellite DNA variation. Molecular Ecology 10: 807–821.
13. Lubieniecki KP, Jones SL, Davidson EA, Park J, Koop BF, et al. (2010) Comparative genomic analysis of Atlantic salmon, Salmo salar, from Europe and North America. Bmc Genetics 11.
14. Wennevik V, Skaala O, Titov SF, Studyonov I, Naevdal G (2004) Microsatellite variation in populations of Atlantic salmon from North Europe. Environmental Biology of Fishes 69: 143–152.
15. Dillane E, Cross MC, McGinnity P, Coughlan JP, Galvin PT, et al. (2007) Spatial and temporal patterns in microsatellite DNA variation of wild Atlantic salmon, Salmo salar, in Irish rivers. Fisheries Management and Ecology 14: 209–219.
16. Perrier C, Guyomard R, Bagliniere JL, Evanno G (2011) Determinants of hierarchical genetic structure in Atlantic salmon populations: environmental factors vs. anthropogenic influences. Molecular Ecology 20: 4231–4245.
17. Dillane E, McGinnity P, Coughlan JP, Cross MC, de Eyto E, et al. (2008) Demographics and landscape features determine intrariver population structure in Atlantic salmon (Salmo salar L.): the case of the River Moy in Ireland. Molecular Ecology 17: 4786–4800.
18. Dionne M, Caron F, Dodson JJ, Bernatchez L (2009) Comparative survey of within-river genetic structure in Atlantic salmon; relevance for management and conservation. Conservation Genetics 10: 869–879.
19. Vaha JP, Erkinaro J, Niemela E, Primmer CR (2007) Life-history and habitat features influence the within-river genetic structure of Atlantic salmon. Molecular Ecology 16: 2638–2654.

20. Palstra FP, Ruzzante DE (2010) A temporal perspective on population structure and gene flow in Atlantic salmon (Salmo salar) in Newfoundland, Canada. Canadian Journal of Fisheries and Aquatic Sciences 67: 225–242.

21. Tessier N, Bernatchez L (1999) Stability of population structure and genetic diversity across generations assessed by microsatellites among sympatric populations of landlocked Atlantic salmon (Salmo salar L.). Molecular Ecology 8: 169–179.

22. Parrish DL, Behnke RJ, Gephard SR, McCormick SD, Reeves GH (1998) Why aren't there more Atlantic salmon (Salmo salar)? Canadian Journal of Fisheries and Aquatic Sciences 55: 281–287.

23. Jonsson B, Jonsson N (2006) Cultured Atlantic salmon in nature: a review of their ecology and interaction with wild fish. Ices Journal of Marine Science 63: 1162–1181.

24. Hansen LP (2006) Migration and survival of farmed Atlantic salmon (Salmo salar L.) released from two Norwegian fish farms. Ices Journal of Marine Science 63: 1211–1217.

25. Whoriskey FG, Brooking P, Doucette G, Tinker S, Carr JW (2006) Movements and survival of sonically tagged farmed Atlantic salmon released in Cobscook Bay, Maine, USA. Ices Journal of Marine Science 63: 1218–1223.

26. Milner NJ, Evans R (2003) The incidence of escaped Irish farmed salmon in English and Welsh rivers. Fisheries Management and Ecology 10: 403–406.

27. Walker AM, Beveridge MCM, Crozier W, O Maoileidigh N, Milner N (2006) Monitoring the incidence of escaped farmed Atlantic salmon, Salmo salar L., in rivers and fisheries of the United Kingdom and Ireland: current progress and recommendations for future programmes. Ices Journal of Marine Science 63: 1201–1210.

28. Butler JRA, Cunningham PD, Starr K (2005) The prevalence of escaped farmed salmon, Salmo salar L., in the River Ewe, western Scotland, with notes on their ages, weights and spawning distribution. Fisheries Management and Ecology 12: 149–159.

29. Morris MRJ, Fraser DJ, Heggelin AJ, Whoriskey FG, Carr JW, et al. (2008) Prevalence and recurrence of escaped farmed Atlantic salmon (Salmo salar) in eastern North American rivers. Canadian Journal of Fisheries and Aquatic Sciences 65: 2807–2826.

30. Fiske P, Lund RA, Hansen LP (2006) Relationships between the frequency of farmed Atlantic salmon, Salmo salar L., in wild salmon populations and fish farming activity in Norway, 1989–2004. Ices Journal of Marine Science 63: 1182–1189.

31. Crozier WW (2000) Escaped farmed salmon, Salmo salar L., in the Glenarm River, Northern Ireland: genetic status of the wild population 7 years on. Fisheries Management and Ecology 7: 437–446.

32. Crozier WW (1993) Evidence of genetic interaction between escaped farmed salmon and wild Atlantic salmon (Salmo salar L) in a Northern Irish river. Aquaculture 113: 19–29.

33. Clifford SL, McGinnity P, Ferguson A (1998) Genetic changes in Atlantic salmon (Salmo salar) populations of northwest Irish rivers resulting from escapes of adult farm salmon. Canadian Journal of Fisheries and Aquatic Sciences 55: 358–363.

34. Clifford SL, McGinnity P, Ferguson A (1998) Genetic changes in an Atlantic salmon population resulting from escaped juvenile farm salmon. Journal of Fish Biology 52: 118–127.

35. Bourret V, O'Reilly PT, Carr JW, Berg PR, Bernatchez L (2011) Temporal change in genetic integrity suggests loss of local adaptation in a wild Atlantic salmon (Salmo salar) population following introgression by farmed escapees. Heredity 106: 500–510.

36. Skaala O, Wennevik V, Glover KA (2006) Evidence of temporal genetic change in wild Atlantic salmon, Salmo salar L., populations affected by farm escapes. Ices Journal of Marine Science 63: 1224–1233.

37. Heggberget TG, Johnsen BO, Hindar K, Jonsson B, Hansen LP, et al. (1993) Interactions between wild and cultured Atlantic salmon - a review of the Norwegian experiance. Fisheries Research 18: 123–146.

38. Hindar K, Ryman N, Utter F (1991) Genetic effects of cultured fish on natural fish populations. Canadian Journal of Fisheries and Aquatic Sciences 48: 945–957.

39. Araki H, Berejikian BA, Ford MJ, Blouin MS (2008) Fitness of hatchery-reared salmonids in the wild. Evolutionary Applications 1: 342–355.

40. Ferguson A, Fleming IA, Hindar K, Skaala Ø, McGinnity P, et al. (2007) Farm escapees. In: Verspoor E, Stradmeyer L, Nielsen JL, editors. The Atlantic salmon Genetics, Conservation and Management. Oxford, UK: Blackwell. 357–398.

41. Naylor R, Hindar K, Fleming IA, Goldburg R, Williams S, et al. (2005) Fugitive salmon: Assessing the risks of escaped fish from net-pen aquaculture. Bioscience 55: 427–437.

42. Gjedrem T, Gjoen HM, Gjerde B (1991) Genetic-Origin of Norwegian Farmed Atlantic Salmon. Aquaculture 98: 41–50.

43. Skaala O, Hoyheim B, Glover K, Dahle G (2004) Microsatellite analysis in domesticated and wild Atlantic salmon (Salmo salar L.): allelic diversity and identification of individuals. Aquaculture 240: 131–143.

44. Karlsson S, Moen T, Lien S, Glover KA, Hindar K (2011) Generic genetic differences between farmed and wild Atlantic salmon identified from a 7K SNP-chip. Molecular Ecology Resources 11: 247–253.

45. Besnier F, Glover KA, Skaala O (2011) Investigating genetic change in wild populations: modelling gene flow from farm escapees. Aquaculture Environment Interactions 2: 75–86.

46. Hansen MM, Fraser DJ, Meier K, Mensberg KLD (2009) Sixty years of anthropogenic pressure: a spatio-temporal genetic analysis of brown trout populations subject to stocking and population declines. Molecular Ecology 18: 2549–2562.

47. Mork J (1991) One-generation effects of farmed fish immigration on the genetic differentiation of wild Atlantic salmon in Norway Aquaculture 98: 267–276.

48. Lund RA, Hansen LP (1991) Identification of wild and reared Atlantic salmon, Salmo salaar L., using scale characters. Aquaculture and Fisheries Management 22: 499–508.

49. (2011) Status for norske laksebestander i 2011. Rapport fra vitenskapelig råd for lakseforvaltning nr 3, 285s (In Norwegian).

50. (2010) Vedleggsrapport merd vurdering av måloppnåelse for de enkelte bestandene. Rapport fra Vitenskapelig råd for Lakseforvaltning nr 2b (In Norwegian).

51. Fiske P, Lund RA, Østborg GM, Fløystad L (2001) Rømt oppdrettslaks i sjø- og elvefisket i årene 1989–2000. NINA oppdresgsmelding 704: 1–26 (In Norwegian).

52. Diserud OH, Fiske P, Hindar K (2012) Forslag til kategorisering av laksebestander som er påvirket av rømt oppdrettslaks. NINA Rapport 782 32s (In Norwegian).

53. Glover KA (2010) Forensic identification of fish farm escapees: the Norwegian experience. Aquaculture Environment Interactions 1: 1–10.

54. Glover KA, Skilbrei OT, Skaala O (2008) Genetic assignment identifies farm of origin for Atlantic salmon Salmo salar escapees in a Norwegian fjord. Ices Journal of Marine Science 65: 912–920.

55. Glover KA, Hansen MM, Lien S, Als TD, Hoyheim B, et al. (2010) A comparison of SNP and STR loci for delineating population structure and performing individual genetic assignment. Bmc Genetics 11: 12.

56. Glover KA, Hansen MM, Skaala O (2009) Identifying the source of farmed escaped Atlantic salmon (Salmo salar): Bayesian clustering analysis increases accuracy of assignment. Aquaculture 290: 37–46.

57. Glover KA, Skaala O, Sovik AGE, Helle TA (2011) Genetic differentiation among Atlantic salmon reared in sea-cages reveals a non-random distribution of genetic material from a breeding programme to commercial production. Aquaculture Research 42: 1323–1331.

58. Paterson S, Piertney SB, Knox D, Gilbey J, Verspoor E (2004) Characterization and PCR multiplexing of novel highly variable tetranucleotide Atlantic salmon (Salmo salar L.) microsatellites. Molecular Ecology Notes 4: 160–162.

59. O'Reilly PT, Hamilton LC, McConnell SK, Wright JM (1996) Rapid analysis of genetic variation in Atlantic salmon (Salmo salar) by PCR multiplexing of dinucleotide and tetranucleotide microsatellites. Canadian Journal of Fisheries and Aquatic Sciences 53: 2292–2298.

60. King TL, Eackles MS, Letcher BH (2005) Microsatellite DNA markers for the study of Atlantic salmon (Salmo salar) kinship, population structure, and mixed-fishery analyses. Molecular Ecology Notes 5: 130–132.

61. McConnell SK, Oreilly P, Hamilton L, Wright JN, Bentzen P (1995) Polymorphic microsatellite loci from Atlantic salmon (Salmo salar) - genetic differentiation of North-American and European populations. Canadian Journal of Fisheries and Aquatic Sciences 52: 1863–1872.

62. Sanchez JA, Clabby C, Ramos D, Blanco G, Flavin F, et al. (1996) Protein and microsatellite single locus variability in Salmo salar L (Atlantic salmon). Heredity 77: 423–432.

63. Slettan A, Olsaker I, Lie O (1995) Atlantic salmon, Salmo salar, microsatellites at the SsOSL25, SsOSL85, SsOSL311, SsOSL417 loci. Animal Genetics 26: 281–282.

64. Grimholt U, Drablos F, Jorgensen SM, Hoyheim B, Stet RJM (2002) The major histocompatibility class I locus in Atlantic salmon (Salmo salar L.): polymorphism, linkage analysis and protein modelling. Immunogenetics 54: 570–581.

65. Stet RJM, de Vries B, Mudde K, Hermsen T, van Heerwaarden J, et al. (2002) Unique haplotypes of co-segregating major histocompatibility class II A and class II B alleles in Atlantic salmon (Salmo salar) give rise to diverse class II genotypes. Immunogenetics 54: 320–331.

66. Vasemagi A, Nilsson J, Primmer CR (2005) Seventy-five EST-linked Atlantic salmon (Salmo salar L.) microsatellite markers and their cross-amplification in five salmonid species. Molecular Ecology Notes 5: 282–288.

67. Tonteri A, Vasemagi A, Lumme J, Primmer CR (2008) Use of differential expression data for identification of novel immune relevant expressed sequence tag-linked microsatellite markers in Atlantic salmon (Salmo salar L.). Molecular Ecology Resources 8: 1486–1490.

68. Hoffman JI, Amos W (2005) Microsatellite genotyping errors: detection approaches, common sources and consequences for paternal exclusion. Molecular Ecology 14: 599–612.

69. Pompanon F, Bonin A, Bellemain E, Taberlet P (2005) Genotyping errors: Causes, consequences and solutions. Nature Reviews Genetics 6: 847–859.

70. Haaland ØA, Glover KA, Seliussen BB, Skaug HJ (2011) Genotyping errors in a calibrated DNA -register: implications for identification of individuals. BMC Genetics 12: 36.

71. Raymond M, Rousset F (1995) GENEPOP (VERSION-1.2) - Population-genetics software for exact tests and ecumenicism. Journal of Heredity 86: 248–249.

72. Dieringer D, Schlotterer C (2003) MICROSATELLITE ANALYSER (MSA): a platform independent analysis tool for large microsatellite data sets. Molecular Ecology Notes 3: 167–169.

73. Cornuet JM, Piry S, Luikart G, Estoup A, Solignac M (1999) New methods employing multilocus genotypes to select or exclude populations as origins of individuals. Genetics 153: 1989–2000.

74. Rannala B, Mountain JL (1997) Detecting immigration by using multilocus genotypes. Proceedings of the National Academy of Sciences of the United States of America 94: 9197–9201.

75. Rosenberg MS, Anderson CD (2011) PASSaGE: Pattern Analysis, Spatial Statistics and Geographic Exegesis. Version 2. Methods in Ecology and Evolution 2: 229–232.

76. Excoffier L, Smouse PE, Quattro JM (1992) Analysis of molecular variance inferred from matric distances among DNA haplotypes - Application to human mitochondrial DNA restrction data. Genetics 131: 479–491.

77. Excoffier L, Laval G, Schneider S (2005) Arlequin (version 3.0): An integrated software package for population genetics data analysis. Evolutionary Bioinformatics 1: 47–50.

78. Joost S, Kalbermatten M, Bonin A (2008) Spatial analysis method(SAM): a software tool combining molecular and environmental data to identify candidate loci for selection. Molecular Ecology Resources 8: 957–960.

79. Beaumont MA, Balding DJ (2004) Identifying adaptive genetic divergence among populations from genome scans. Molecular Ecology 13: 969–980.

80. Foll M, Gaggiotti O (2008) A Genome-Scan Method to Identify Selected Loci Appropriate for Both Dominant and Codominant Markers: A Bayesian Perspective. Genetics 180: 977–993.

81. Beaumont MA, Nichols RA (1996) Evaluating loci for use in the genetic analysis of population structure. Proceedings of the Royal Society of London Series B-Biological Sciences 263: 1619–1626.

82. Antao T, Lopes A, Lopes RJ, Beja-Pereira A, Luikart G (2008) LOSITAN: A workbench to detect molecular adaptation based on a Fst-outlier method. Bmc Bioinformatics 9.

83. Falush D, Stephens M, Pritchard JK (2003) Inference of population structure using multilocus genotype data: Linked loci and correlated allele frequencies. Genetics 164: 1567–1587.

84. Pritchard JK, Stephens M, Donnelly P (2000) Inference of population structure using multilocus genotype data. Genetics 155: 945–959.

85. Hubisz MJ, Falush D, Stephens M, Pritchard JK (2009) Inferring weak population structure with the assistance of sample group information. Molecular Ecology Resources 9: 1322–1332.

86. Evanno G, Regnaut S, Goudet J (2005) Detecting the number of clusters of individuals using the software STRUCTURE: a simulation study. Molecular Ecology 14: 2611–2620.

87. Jakobsson M, Rosenberg NA (2007) CLUMPP: a cluster matching and permutation program for dealing with label switching and multimodality in analysis of population structure. Bioinformatics 23: 1801–1806.

88. Nei M, Maruyama T, Chakraborty R (1975) Bottleneck effect and genetic-variability in populations. Evolution 29: 1–10.

89. Buri P (1956) Gene-frequency in small populations of mutant dropsophila. Evolution 10: 367–402.

90. Waples RS, Do C (2008) LDNE: a program for estimating effective population size from data on linkage disequilibrium. Molecular Ecology Resources 8: 753–756.

91. Sandoval-Castellanos E (2010) Testing temporal changes in allele frequencies: a simulation approach. Genetics Research 92: 309–320.

92. Glover KA, Haag T, Oien N, Walloe L, Lindblom L, et al. (2012) The Norwegian minke whale DNA register: a database monitoring commercial harvest and trade of whale products. Fish and Fisheries Early online DOI: 10.1111/j.1467-2979.2011.00447.x.

93. Waples RS, Do C (2010) Linkage disequilibrium estimates of contemporary N-e using highly variable genetic markers: a largely untapped resource for applied conservation and evolution. Evolutionary Applications 3: 244–262.

94. Arnaud-Haond S, Vonau V, Bonhomme F, Boudry P, Blanc F, et al. (2004) Spatio-temporal variation in the genetic composition of wild populations of pearl oyster (*Pinctada margaritifera cumingii*) in French Polynesia following 10 years of juvenile translocation. Molecular Ecology 13: 2001–2007.

95. Marie AD, Bernatchez L, Garant D (2010) Loss of genetic integrity correlates with stocking intensity in brook charr (*Salvelinus fontinalis*). Molecular Ecology 19: 2025–2037.

96. Fleming IA, Hindar K, Mjolnerod IB, Jonsson B, Balstad T, et al. (2000) Lifetime success and interactions of farm salmon invading a native population. Proceedings of the Royal Society of London Series B-Biological Sciences 267: 1517–1523.

97. Fleming IA, Jonsson B, Gross MR, Lamberg A (1996) An experimental study of the reproductive behaviour and success of farmed and wild Atlantic salmon (*Salmo salar*). Journal of Applied Ecology 33: 893–905.

98. Weir LK, Hutchings JA, Fleming IA, Einum S (2004) Dominance relationships and behavioural correlates of individual spawning success in farmed and wild male Atlantic salmon, *Salmo salar*. Journal of Animal Ecology 73: 1069–1079.

99. McGinnity P, Prodohl P, Ferguson K, Hynes R, O'Maoileidigh N, et al. (2003) Fitness reduction and potential extinction of wild populations of Atlantic salmon, *Salmo salar*, as a result of interactions with escaped farm salmon. Proceedings of the Royal Society of London Series B-Biological Sciences 270: 2443–2450.

100. McGinnity P, Stone C, Taggart JB, Cooke D, Cotter D, et al. (1997) Genetic impact of escaped farmed Atlantic salmon (*Salmo salar* L.) on native populations: use of DNA profiling to assess freshwater performance of wild, farmed, and hybrid progeny in a natural river environment. Ices Journal of Marine Science 54: 998–1008.

101. Martinez-Cruz B, Godoy JA, Negro JJ (2007) Population fragmentation leads to spatial and temporal genetic structure in the endangered Spanish imperial eagle. Molecular Ecology 16: 477–486.

102. Haag T, Santos AS, Sana DA, Morato RG, Cullen L, et al. (2010) The effect of habitat fragmentation on the genetic structure of a top predator: loss of diversity and high differentiation among remnant populations of Atlantic Forest jaguars (*Panthera onca*). Molecular Ecology 19: 4906–4921.

103. Borrell YJ, Bernardo D, Blanco G, Vazquez E, Sanchez JA (2008) Spatial and temporal variation of genetic diversity and estimation of effective population sizes in Atlantic salmon (*Salmo salar* L.) populations from Asturias (Northern Spain) using microsatellites. Conservation Genetics 9: 807–819.

104. Lura H, Saegrov H (1991) Documentation of successful spawning of escaped farmed female Atlantic salmon, *Salmo salar*, in Norwegian rivers. Aquaculture 98: 151–159.

105. Saegrov H, Hindar K, Kalas S, Lura H (1997) Escaped farmed Atlantic salmon replace the original salmon stock in the River Vosso, western Norway. Ices Journal of Marine Science 54: 1166–1172.

106. Lura H (1995) Domesticated female Atlantic salmon in the wild: spawning success and contribution to local populations. DSc thesis, University of Bergen, Norway.

107. Fleming IA, Lamberg A, Jonsson B (1997) Effects of early experience on the reproductive performance of Atlantic salmon. Behavioral Ecology 8: 470–480.

108. Tatara CP, Berejikian BA (2012) Mechanisms influencing competition between hatchery and wild juvenile anadromous Pacific salmonids in fresh water and their relative competitive abilities. Environmental Biology of Fishes 94: 7–19.

109. Waples RS (1989) Temporal variation in allele frequencies - testing the right hyopthesis. Evolution 43: 1236–1251.

110. Wang JL, Whitlock MC (2003) Estimating effective population size and migration rates from genetic samples over space and time. Genetics 163: 429–446.

111. Jorde PE (2012) Allele frequency covariance among cohorts and its use in estimating effective size of age-structured populations. Molecular Ecology Resources 12: 476–480.

112. Jorde PE, Ryman N (2007) Unbiased estimator for genetic drift and effective population size. Genetics 177: 927–935.

113. Pudovkin AI, Zaykin DV, Hedgecock D (1996) On the potential for estimating the effective number of breeders from heterozygote-excess in progeny. Genetics 144: 383–387.

114. Tallmon DA, Koyuk A, Luikart G, Beaumont MA (2008) ONeSAMP: a program to estimate effective population size using approximate Bayesian computation. Molecular Ecology Resources 8: 299–301.

115. Wang JL (2009) A new method for estimating effective population sizes from a single sample of multilocus genotypes. Molecular Ecology 18: 2148–2164.

116. Waples RS (2006) A bias correction for estimates of effective population size based on linkage disequilibrium at unlinked gene loci. Conservation Genetics 7: 167–184.

117. Waples RS, England PR (2011) Estimating Contemporary Effective Population Size on the Basis of Linkage Disequilibrium in the Face of Migration. Genetics 189: 633–644.

118. Hindar K, Fleming IA, McGinnity P, Diserud A (2006) Genetic and ecological effects of salmon farming on wild salmon: modelling from experimental results. Ices Journal of Marine Science 63: 1234–1247.

119. Seeb LW, Antonovich A, Banks AA, Beacham TD, Bellinger AR, et al. (2007) Development of a standardized DNA database for Chinook salmon. Fisheries 32: 540–552.

120. Schwartz MK, Luikart G, Waples RS (2007) Genetic monitoring as a promising tool for conservation and management. Trends in Ecology & Evolution 22: 25–33.

121. Araki H, Schmid C (2010) Is hatchery stocking a help or harm? Evidence, limitations and future directions in ecological and genetic surveys. Aquaculture 308: S2–S11.

122. Glover KA, Ottera H, Olsen RE, Slinde E, Taranger GL, et al. (2009) A comparison of farmed, wild and hybrid Atlantic salmon (*Salmo salar* L.) reared under farming conditions. Aquaculture 286: 203–210.

123. Roberge C, Einum S, Guderley H, Bernatchez L (2006) Rapid parallel evolutionary changes of gene transcription profiles in farmed Atlantic salmon. Molecular Ecology 15: 9–20.

124. Fleming IA, Einum S (1997) Experimental tests of genetic divergence of farmed from wild Atlantic salmon due to domestication. Ices Journal of Marine Science 54: 1051–1063.

125. (1992) Fiske og oppdrett av laks mv. (Fishing and rearing of salmon etc.). Noregs ofisielle statistikk C56 (In Norwegian).

12

Skeletal Anomaly Monitoring in Rainbow Trout (*Oncorhynchus mykiss*, Walbaum 1792) Reared under Different Conditions

Clara Boglione[1], Domitilla Pulcini[1,2]*, Michele Scardi[1], Elisa Palamara[1], Tommaso Russo[1], Stefano Cataudella[1]

1 Laboratory of Experimental Ecology and Aquaculture, Department of Biology, "Tor Vergata" University of Rome, Rome, Italy, 2 Council for Research in Agriculture – Animal Production Centre, Rome, Italy

Abstract

The incidence of skeletal anomalies could be used as an indicator of the "quality" of rearing conditions as these anomalies are thought to result from the inability of homeostatic mechanisms to compensate for environmentally-induced stress and/or altered genetic factors. Identification of rearing conditions that lower the rate of anomalies can be an important step toward profitable aquaculture as malformed market-size fish have to be discarded, thus reducing fish farmers' profits. In this study, the occurrence of skeletal anomalies in adult rainbow trout grown under intensive and organic conditions was monitored. As organic aquaculture animal production is in its early stages, organic broodstock is not available in sufficient quantities. Non-organic juveniles could, therefore, be used for on-growing purposes in organic aquaculture production cycle. Thus, the adult fish analysed in this study experienced intensive conditions during juvenile rearing. Significant differences in the pattern of anomalies were detected between organically and intensively-ongrown specimens, although the occurrence of severe, commercially important anomalies, affecting 2–12.5% of individuals, was comparable in the two systems. Thus, organic aquaculture needs to be improved in order to significantly reduce the incidence of severe anomalies in rainbow trout.

Editor: Athanassios C. Tsikliras, Aristotle University of Thessaloniki, Greece

Funding: This study was funded by the Italian Ministry of the Agricoltural Policy and Forestry. The funders had no role in study design, data collection and analysis, decision to publish, or preparation of the manuscript.

Competing Interests: The authors have declared that no competing interests exist.

* E-mail: domitillapulcini@fastwebnet.it

Introduction

Aquaculture of fish and other aquatic animals has grown rapidly in the last thirty years [1]. Most fish aquaculture production comes from freshwater, with salmonid farming making a significant contribution to global aquaculture production volumes [1,2,3]. Rainbow trout *Oncorhynchus mykiss* (Walbaum 1792) is a dominant farmed salmonids in Europe and North America [1]. Naturally distributed along the Pacific coast of North America and on the Kamchatka Peninsula [4,5], rainbow trout has been extensively introduced for aquaculture practically all over the world since the mid-1800s.

Farmed fish are often affected by skeletal anomalies, with the incidence depending on the species, developmental stage, and rearing methodology. Skeletal anomalies may arise in captivity due to both genetic (such as inbreeding depression due to artificial selection [6–8] and triploidy [9,10]) and environmental causes [11,12]. Rearing conditions different from the species- or developmental stage-specific ones often cause the onset of skeletal anomalies [12–22]. In farmed salmonids, some studies found no relationship between incidence of anomalies and captive conditions [23,24], while others ascribed displacement of vertebral centra, fused and compressed vertebral axis, and decreased bone quality to fast-growing rearing conditions [9,10,25–28]. Among environmental causes, inappropriate rearing densities were reported in previous studies as causative factors of bone malformations [12] in Atlantic salmon (*Salmo salar*, L. 1758) fry and parr [29]. In several species of commercial interest (i.e. *Dicentrarchus labrax*, *Sparus aurata*, *Epinephelus marginatus*, *Dentex dentex*, *Pagrus pagrus*), a reduction in skeletal anomalies (especially commercially relevant ones) has been detected in semi-intensive rearing conditions, characterized by lower densities and larger volumes [11,30–36]. Commercially significant anomalies affect the head and the vertebral axis, thus altering external shape and swimming/feeding performance, with consequent lower growth rate, economic value and welfare status, and higher susceptibility to stress, pathogens and bacteria [13–15,37–41]. Seriously malformed market-size fish have to be discarded or sold at lower than market price due to the consumer's reluctance to buy 'bad-looking' products.

This study tested whether any difference exists in the number (meristic counts) and shape (occurrence of anomalies) of skeletal elements in adult rainbow trout grown under intensive *vs* organic aquaculture.

Materials and Methods

A total of 533 adult rainbow trout (which is not a protected or endangered species) were collected from five European fish farms:

Table 1. Features of the farms where fish were collected (organic ones in grey).

Farm	Pond	Surface	Volume	Water flow	Temperature	Density
INT1 (Italy, Abruzzo)	Rectangular concrete raceways	800×0.7	560	50–100	10–10.5	55
INT2 (Italy, Piedmont)	Rectangular concrete raceways	1000/1300×0.5	500–650	100	12.5	40
ORG1 (Italy, Lazio)	Squared earth ponds	800/1300×0.8	650–1150	50–100	10.5–11	15–30*
ORG2 (Italy, Piedmont)	Rectangular earth ponds	600/1000×0.6	300	300	10–10.5	12
ORG3 (Switzerland)	Rectangular, vegetated earth ponds	720×0.6	430	100	8	10–12

Surface = m^2 · m; Volume = m^3; Water flow = l·s^{-1}; Temperature = °C; Density = kg·m^{-3}. * Individuals were temporarily stocked at high densities (30 kg·m^{-3}).

(1) two intensive ("Az. Agricola Troticoltura Rossi", Abruzzo, Italy; "Az. Agricola Rio Fontane", Veneto, Italy), denoted, respectively, as INT1 and INT2; (2) three organic, ("Az. Agricola Troticoltura Rossi", Lazio, Italy; "Az. Agricola Rio Fontane", Veneto, Italy; "Azienda Agricola Pura" – Switzerland) denoted, respectively, as ORG1, ORG2 and ORG3. The latter followed the standards for organic productions developed by Naturland, ECOCERT and Biosuisse certification bodies, respectively. No specific permissions were required for the activities carried out in the above-mentioned locations as they were not protected areas. The owners of the farms gave permission to collect the samples for this study. The main features (material, shape and size of the rearing ponds, temperature, water flow, density) of the farms are reported in Table 1.

Different strains were collected in the farms: Italian, French, Spanish, Swiss and American. Italian fish were collected from all the farms except ORG3; French fish were collected from INT1, Spanish fish from INT2, American fish from ORG2 and Swiss fish from ORG3.

The number of observed specimens, the total length (TL) range, and the genetic origin of the lots are reported in Table 2.

A detailed description of the historical background of these strains is available in [43].

Commission Regulation (EC) No 710/2009 of 5 August 2009 states that: "Given the early stage of organic aquaculture animal production, organic broodstock is not available in sufficient quantities. Provision should be made for the introduction of non-organic broodstock and juveniles under certain conditions."

For on-growing purposes and when organic aquaculture juvenile animals are not available, non-organic aquaculture juveniles may be brought into a holding. At least the latter two thirds of the duration of the production cycle shall be managed under organic management (*Article 25e*).

Thus, all the specimens collected for this study, in both intensive and organic facilities, shared standardized intensive conditions (water temperature = ~10°C; dissolved oxygen = 12 ppm; density: ~ 13 kg · m^{-3}) until they attained the weight of about 10 g. ORG1 fish originated from the same farm where the INT1 lot was sampled and ORG2 from INT2. The ORG3 lot originated in the hatchery of the same farm.

Samples were euthanatized with a lethal dose of 2-phenoxyethanol (0.5 mg/L), frozen and X-rayed (4 min/5 mA/80 kW) in order to perform meristic counts and skeletal anomalies analysis.

Sampling and killing procedures in this study complied with the Institutional Animal Care and Use Committee (IACUC) guidelines.

The vertebral column was divided into four regions, based on distinct morphological features. Vertebrae were split into cephalic (equipped with epipleural ribs), pre-haemal (with epipleural and pleural ribs and open haemal arch, without haemal spine), haemal (with haemal arch closed by a spine) and caudal (with haemal and neural arches closed by a modified, elongated spine; urostyle was included).

The anatomical terminology is according to [44–46], except for caudal fin structure terminology, which is according to [47].

Table 2. Genetic origin (*Origin*), geographic origin of the source population (*source*), number (*n*) and total length (TL mean ± standard deviation) of observed specimens.

Farm	Origin	Source	Lot	n	TL±S.D. (cm)
INT1	Italy	USA	1	46	28.7±3.2
	France	USA	2	193	30.1±4.4
INT2	Italy	USA	3	16	33.5±4.7
	Spain	France	4	32	12.6±1.1
ORG1	Italy	USA	5	108	31.7±3.4
ORG2	Italy	USA	6	29	24.7±4.4
	USA	USA	7	60	26.8±3.1
ORG3	Switzerland	Germany	8	49	20.2±2.7

Data referring to source populations are from [4] and [42].

Table 3. List of anomalies considered. Bold font indicates commercially severe anomalies.

Region		
	A	Cephalic vertebrae
	B	Pre-hemal vertebrae
	C	Hemal vertebrae
	D	Caudal vertebrae
	E	Pectoral fin
	F	Anal fin
	G	Caudal fin
	H	First dorsal fin
	I	Second dorsal fin
	L	Pelvic fin
Types	**S**	**Scoliosis**
	SB	**Saddle back**
	1	**Lordosis**
	2	**Kyphosis**
	3	Incomplete vertebral fusion
	3*	Complete vertebral fusion
	4	Malformed vertebral body
	5	Malformed neural arch and/or spine
	5*	Extra-ossification in the neural region
	6	Malformed hemal arch and/or spine
	6*	Extra-ossification in the hemal region
	7	Deformed pleural rib
	7*	Extra-ossification of pleural ribs
	8	Malformed pterygophore (deformed, absent, fused, supernumerary)
	9	Malformed hypural (deformed, absent, fused, supernumerary)
	9*	Malformed parahypural (deformed, fused, reduced)
	10	Malformed epural (deformed, absent, fused, supernumerary)
	11	Malformed ray (deformed, absent, fused, supernumerary)
	12	**Swim-bladder anomaly**
	13	Presence of calculi in the terminal tract of the urinary ducts
	14	**Malformed dentale**
	15	**Malformed premaxilla and maxilla**
	16	**Dislocation of glossohyal**
	17sx	**Deformed or reduced left opercular plate**
	17dx	**Deformed or reduced right opercular plate**
	17*sx	Deformed or reduced left branchiostegal ray
	17*dx	Deformed or reduced right branchiostegal ray
	18	Malformed supraneural bones

Vertebrae fusions are considered severe only if affecting at least three consecutive vertebrae.

The following meristic counts were considered: (1) vertebrae; (2) epural and hypurals; (3) anal and dorsal rays; (4) anal and dorsal pterygophores; (5) principal caudal fin rays, divided into upper (UPCR) and lower (LPCR); (6) supraneural bones.

The correlation between meristic counts and total length (TL) was tested by a Spearman rank correlation. The standard descriptive statistics (median and range) for each meristic count were calculated from the raw data. The significance of the differences in the median values of each meristic count was tested by means of the non-parametric Kruskal-Wallis test, with Mann-Whitney pairwise post-hoc comparisons. ANOSIM (Analysis Of SIMilarities) was applied to the overall matrix of meristic counts to compare intensively vsorganically-reared specimens. ANOSIM is a non-parametric test of significant difference between two or more groups, based on any distance measure [48]. In this study, Euclidean distance was selected for meristic counts. Distances were then converted to ranks. The test is based on comparing distances between groups with distances within groups. Let r_b be the mean rank of all distances between groups, and r_w the mean rank of all distances within groups. The test statistic R is then defined as: $R = (r_b - r_w)/[N(N-1)/4]$.

A large positive R (up to 1) signifies dissimilarity between groups. The significance is computed by permutation of group membership (10,000 replicates).

Table 4. Median and ranges of meristic counts.

Lot	Tot	Ceph	Pre-hem	Hem	Caud	Ep	Hyp	UPCR	LPCR	An Pter	An Rays	Supr	Do Pter	Do Rays
1	63	2	37	17.5	7	3	6	10	9	12	15	18	13	16.5
	62-65	2-3	35-39	15-19	5-8	2-3	5-6	10-11	8-9	11-14	14-17	15-20	12-15	15-18
2	63	2	36	18	7	3	6	10	9	13	15	18	13	16
	57-65	1-2	33-38	13-21	6-8	1-3	5-7	9-11	7-10	11-14	14-17	15-20	11-15	13-18
3	62	2	35	18	7	3	6	10	9	12.5	16	18	13	16
	60-64	-	34-36	15-19	6-8	2-3	-	-	9-11	11-14	15-17	16-20	12-14	14-18
4	63	2	36	18	7	3	6	10	9	12	15	18	13	16
	62-64	2-3	35-37	17-19	7-8	-	5-6	9-11	-	11-13	13-16	16-19	12-14	15-17
5	63	2	36	17	7	3	5	10	9	13	16	18	13	16
	59-65	1-3	34-38	16-19	5-9	2-4	5-6	9-11	8-10	11-14	14-17	16-21	11-15	13-18
6	62	2	35	18	7	3	6	10	9	12	15	18	13	16
	61-63	1-3	34-36	16-19	7-9	1-3	5-6	-	9-10	11-13	13-17	16-21	12-14	13-16
7	62	2	35	18	7	3	6	10	9	12	15	18	13	16
	60-64	-	33-36	16-20	6-9	2-3	-	10-11	8-10	11-14	13-17	15-22	11-14	13-17
8	63	2	37	18	7	3	6	10	9	12	15	18	13	16
	60-65	-	35-37	16-20	6-8	2-3	5-6	9-10	8-9	10-14	13-16	16-20	11-14	12-17

Tot: total number of vertebrae; Ceph: cephalic vertebrae; Pre-hem: pre-hemal vertebrae; Hem: hemal vertebrae; Caud: caudal vertebrae; Ep: epurals; Hyp: hypurals; UPCR: upper principal caudal rays; LPCR: lower principal caudal rays; An Pter: anal pterygophores; An Rays: anal pterygophores; Do Pter: dorsal pterygophores; Supr: supraneurals.

Table 5. Mann-Whitney post-hoc pairwise comparisons (Bonferroni corrected).

Vertebrae (H = 85.17; p<0.0001)

	1	2	3	4	5	6	7	8
1		*	*		*	*	*	*
2			*	*		*	*	
3				*	*			*
4					*		*	*
5						*	*	
6								*
7								
8								

Pre-hemal Vertebrae (H = 152.4; p<0.0001)

	1	2	3	4	5	6	7	8
1		*	*	*	*	*	*	
2			*	*	*	*	*	*
3				*	*			*
4						*	*	*
5						*	*	*
6								*
7								*
8								

Hemal Vertebrae (H = 58.09; p<0.0001)

	1	2	3	4	5	6	7	8
1		*		*				
2					*	*	*	*
3					*			
4					*			
5						*	*	*
6								
7								
8								

Caudal Vertebrae (H = 26.55; p<0.0001)

	1	2	3	4	5	6	7	8
1								*
2				*	*	*	*	*
3								*
4								*
5								*
6								*
7								*
8								

Epurals (H = 8.47; p<0.001)

	1	2	3	4	5	6	7	8
1								
2				*	*		*	
3								
4								
5								
6								
7								
8								

UPCR (H = 2.05; p<0.05)

Table 5. Cont.

	1	2	3	4	5	6	7	8
1								
2								
3								
4								
5							*	
6								
7								*
8								

LPCR (H = 122.2; p<0.0001)

	1	2	3	4	5	6	7	8
1			*	*				
2			*	*			*	
3					*	*	*	*
4					*	*	*	*
5								
6								
7								*
8								

Anal Pterygophores (H = 63.09; p<0.0001)

	1	2	3	4	5	6	7	8
1		*		*	*			
2				*		*	*	*
3				*				
4					*			
5						*	*	*
6								
7								
8								

Anal Rays (H = 42.03; p<0.0001)

	1	2	3	4	5	6	7	8
1		*	*	*	*			*
2				*	*	*	*	*
3				*		*	*	*
4					*			
5						*	*	*
6								
7								
8								

Dorsal Pterygophores (H = 63.32; p<0.0001)

	1	2	3	4	5	6	7	8
1		*	*	*		*	*	*
2					*		*	
3							*	
4					*			
5						*	*	*
6								
7								
8								

Dorsal Rays (H = 60.34; p<0.0001)

	1	2	3	4	5	6	7	8

Table 5. Cont.

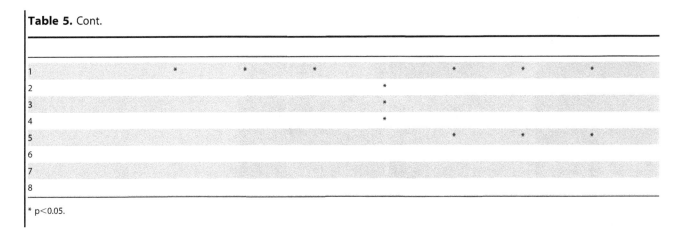

1	*	*	*		*	*	*
2				*			
3				*			
4				*			
5					*	*	*
6							
7							
8							

* p<0.05.

The list of anomalies considered is set out in Table 3. Some anomalies displayed different degrees of alteration (see, for example, C3 and C3* in Table 3) and were indicated as distinctive variables. In this study we chose to distinguish severe anomalies from the biologically severe anomalies as they lead to some commercial (and not only biological) consequences (i.e., unmarketable fish, Table 3): i.e., partial or complete vertebrae fusion is not considered as a commercially severe anomaly if it affects only a few (maximum 3) non-adjacent vertebrae, because this would not influence either growth performance or external shape of the fish. The presence of consecutive fusions involving at least 4 adjacent vertebrae, on the contrary, is likely to stiffen the trunk, so they are considered as commercially and biologically severe anomalies. The presence of deformed vertebrae centra is no longer considered a commercially severe anomaly: the methodology applied actually requires that any axis deviation is considered as an anomaly only if at least one of the vertebrae centra included in the deviation is modified. A deformed centrum leading to axis deviation (kyphosis, lordosis or scoliosis) is scored among commercially severe anomalies, whilst a deformed centrum not involved in axis deviation is considered as a biologically severe anomaly not definitely influencing growth, welfare and health performance.

Paired (pelvic and pectoral) fins were not considered in this study because they were often excessively eroded in the samples examined.

Some assumptions were made in carrying out the analysis: i) non-completely fused bone elements were counted as distinct elements in meristic counts; ii) supernumerary bones with a normal morphology were not considered as an anomaly but as a meristic count variation; conversely, anomalous supernumerary elements were considered anomalies; iii) only the clearly and unquestionably identifiable variations in shape were considered as

Figure 1. Rainbow trout specimens affected by commercially severe anomalies. Specimen with stumpy body due to scoliosis and compressed hemal and caudal vertebrae, and specimen affected by kypho-lordosis in the hemal and caudal vertebrae. Some hemal vertebrae are compressed and fused.

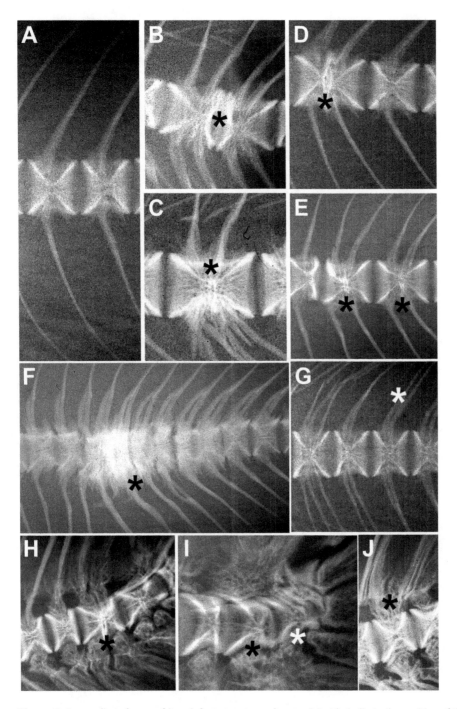

Figure 2. Anomalies observed in rainbow trout specimens. Asterisks indicate the position of the anomalies in the images. A. Normal shaped hemal vertebra; B. one-sided compression of pre-hemal vertebrae (B4), corresponding to type 5 of [58]; C. complete fusion of pre-hemal vertebrae (B3*), corresponding to type 7 of [58]; D. incomplete (C3) fusion of hemal vertebrae; E. complete (C3*) fusion of hemal vertebrae, corresponding to type 7 of [58]; F. compressions and fusions of hemal vertebrae (C3, C3* and C4), corresponding to type 8 of [58]; G. anomalous neural arches (B5); H. incomplete fusion of caudal vertebrae (D3); I. anomalous caudal vertebrae (D4); (J) anomalies of neural spines of caudal vertebrae (D5).

skeletal anomalies: if any doubts arose, then the shape variation was not considered anomalous; iv) misalignments of vertebrae were considered as lordosis and/or kyphosis only if the vertebral bodies involved were deformed.

The data matrix was processed to calculate skeletal anomaly incidence and to perform a descriptive analysis for each anomaly type and lot.

Anomaly data were then converted into binary values (presence or absence of each anomaly type) and frequencies of specimens affected by each anomaly in each lot were calculated. The resulting matrix (32 skeletal typologies x 8 lots) was then subjected to Correspondence Analysis (CA – [49]), in order to visualize the relationships among lots and the role that each anomaly plays in the ordination model.

Table 6. General data on deformed individuals, incidences and typologies of skeletal anomalies in the observed lots.

	1	2	3	4	5	6	7	8
Deformation rate (%)[a]	100	100	100	100	100	100	100	100
Anomaly load[b]	20.3	25.8	26.6	20.7	22.7	24.7	26.7	22.9
No. anomaly types	21	25	20	14	25	22	20	21
Severe anomaly incidence (%)[c]	4.4	0.4	3.8	0.2	2.9	0.1	1.9	0.3
Severe deformation rate (%)[d]	8.7	2.1	12.5	3.1	6.5	3.4	5.0	2.0
Severe anomaly load[e]	2.5	4.5	8.0	1.00	10.3	1.0	10.0	3.0

Organic lots are highlighted with grey background.
[a]Frequency of individuals with at least one anomaly.
[b]Total number of anomalies/number of anomalous individuals.
[c]Number of severe anomalies/total number of anomalies x 100.
[d]Number of individuals with at least one severe anomaly/total number of lot individuals.
[e]Number of severe anomalies/number of individuals with severe anomalies.

ANOSIM was applied to the binary matrix of anomalies to compare intensively and organically-reared specimens, using the Rogers & Tanimoto similarity coefficient [50].

Kruskal-Wallis test, ANOSIM and Correspondence Analysis were performed using PAST (version 2.14 [51]).

Results

Median and ranges of meristic counts are shown in Table 4. No significant correlation was detected between size (TL) and each meristic count, thus excluding any size effect on the observed meristic counts.

The total number of vertebrae varied greatly from 57 (French trout reared in INT1) to 65 (several lots reared both under intensive and organic conditions), but median values ranged from 62 to 63. The number of cephalic and caudal vertebrae were the most canalized, with same medians in all lots (2 and 7, respectively); more variation was observed in the median values of haemal (17–18) and pre-haemal (35–37) vertebrae. Epurals, hypurals, UPCR, LPCR and dorsal pterygophores showed no variation in the median values, while anal pterygophores and rays, supraneurals and dorsal rays showed little variation. Significant differences were detected using the Kruskal-Wallis test in the total number of vertebrae ($H = 85.17$; $p<0.0001$), pre-haemal vertebrae ($H = 152.4$; $p<0.0001$), haemal vertebrae ($H = 58.09$; $p<0.0001$), caudal vertebrae ($H = 26.55$; $p<0.0001$), epurals ($H = 8.47$; $p< 0.001$), UPCR ($H = 2.05$; $p<0.05$), LPCR ($H = 122.2$; $p<0.0001$), anal pterygophores ($H = 63.09$; $p<0.0001$) and rays ($H = 42.03$; $p<0.0001$), dorsal pterygophores ($H = 63.32$; $p<0.0001$) and rays ($H = 60.34$; $p<0.0001$). Mann-Whitney post-hoc pairwise comparisons (Bonferroni corrected) are reported in Table 5. ANOSIM detected highly significant differences in meristic counts between intensive and organic lots ($R = 0.04$; $p<0.0001$).

A total of 32 types of anomaly were observed, some of which are shown in Figs. 1 and 2 (A–J). Intensive lots showed inter-lot variations in the anomaly typologies (14–25) than organic lots (20–25; Table 6). Some severe anomalies, affecting the vertebral column (such as scoliosis or saddle back) and the cephalic region (such as the dislocation of the glossohyal, or anomalous opercular plate), as well as swim bladder anomalies and the presence of calculi in the urinary duct were never observed. Some others (C1: kyphosis in hemal vertebrae; B2: lordosis in pre-hemal vertebrae; D2: lordosis in caudal vertebrae) were extremely rare.

The percentage of individuals with at least one anomaly was 100% in all lots. The anomalies load was very high, ranging from 20.3 to 26.6 anomalies on each deformed individual in intensive lots, and from 22.7 to 26.7 anomalies on each deformed individual in organic ones. The distribution of the number of anomalies per individual was not normal (Shapiro-Wilk's test) in both intensively and organically-reared individuals ($W_{int} = 0.93$, $p_{int}<0.0001$; $W_{org} = 0.83$, $p_{org}<0.0001$), indicating that in both groups the greatest number of individuals was affected by 15–30 anomalies, with fewer affected by a lower (0–10) or higher (35–70) number of anomalies, and rare individuals characterized by a very high number of malformations (≥ 95) (Fig. 1).

Commercially severe anomalies represented 0.2–4.4% and 0.1–2.9% of the total anomalies inspected in intensive and organic lots, respectively. Intensively reared lots showed higher severe deformation rates, ranging from 2.1 to 12.5% of the individuals versus a load ranging from 1 to 10.3% in the semi-intensive individuals. Intensive lot 3 (Italian strain produced from USA eggs and reared in INT2) showed the highest severe deformation rate (12.5%) and severe anomaly load (8.0), while semi-intensive lot 8 (Swiss strain obtained from Germany and reared in ORG3) showed the lowest

Table 7. Frequency (%) of individuals affected by each anomaly in each lot.

	1	2	3	4	5	6	7	8
A5	78.3	1.0	87.5	62.5	64.8	48.3	66.7	71.4
A3	6.5	1.0	6.3		0.9	6.9	5.0	2.0
A4	10.9		18.8			10.3		2.0
B2	2.2							
B3		5.2	6.3	3.1	5.6	10.3	8.3	4.1
B3*		1.6						
B4	4.3	1.6	18.8		2.8		8.3	2.0
B5	100.0	99.5	100.0	100.0	96.3	100.0	98.3	100.0
B7sx	80.4	54.9	62.5	46.9	38.9	41.4	40.0	55.1
C1					0.9			
C3	6.5	1.0			7.4	3.4	1.7	2.0
C4		1.0			3.7	3.4	3.3	
C5	15.2	38.9	18.8	25.0	19.4	31.0	33.3	34.7
C6	39.1	32.1	37.5	46.9	31.5	48.3	48.3	49.0
D2					0.9			
D3	8.7	2.1	12.5		11.1	3.4	1.7	2.0
D3*		0.5						
D4	8.7	4.7	6.3		1.9	3.4	1.7	4.1
D5	58.7	40.4	43.8	46.9	54.6	41.4	38.3	38.8
D5*	13.0	22.3	31.3	15.6	25.9	20.7	26.7	20.4
D6	39.1	35.8	50.0	21.9	54.6	20.7	31.7	30.6
D6*		1.6			0.9			
F8	10.9	10.9	12.5	9.4	12.0	6.9	13.3	16.3
F11		1.0				3.4		
G9	2.2	3.6	6.3		2.8	6.9	3.3	8.2
G10	15.2	23.8	18.8	9.4	24.1	20.7	11.7	20.4
G11								2.0
H8	15.2	9.8	6.3	3.1	26.9	17.2	20.0	6.1
H11		1.6			1.9			
18	56.5	42.0	43.8	34.4	46.3	51.7	56.7	55.1
14				3.1		3.4		
15	2.2				1.9			
Abs^a								

Empty cells indicate 0.0%. Organic lots are highlighted with grey background.
[a]Absence of anomalies.

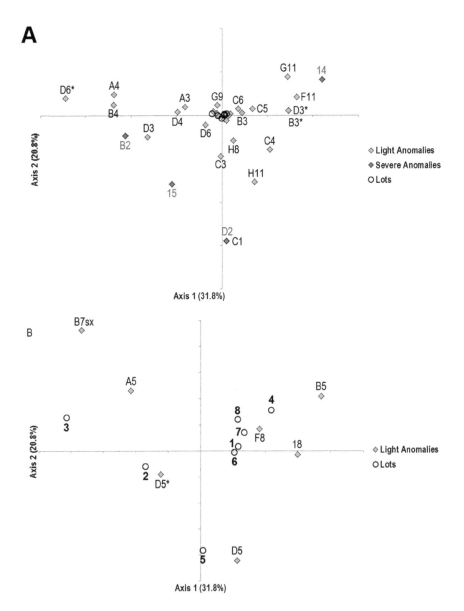

Figure 3. Correspondence Analysis ordination plot. CA (A)ordination of anomalies and lots (axis 1 *vs* 2). Red codes highlight commercially severe anomalies; (B) magnification of (A), in order to visualize lot arrangement more clearly.

severe deformation rate (2.0%). The highest and the lowest severe anomaly loads were both observed in/organic lots, i.e. lots 5 (10.3 severe anomalies/individual; Italian trout reared in ORG1) and 6 (1 severe anomaly/individual; Italian strain from USA, reared in ORG2), respectively.

Some types of anomaly were observed only in a few intensively reared lots (Table 7): lordosis in pre-haemal vertebrae (B2), complete fusion between the bodies of the same vertebrae (B3*), or of the caudal vertebrae (D3*). However, kyphosis in haemal vertebrae (C1), lordosis in the caudal vertebrae (D2), and deformed caudal rays (G11) were observed only in organic lots. All these anomalies were very rare and not evenly distributed among the lots.

The most frequent anomaly was B5 (deformation of neural arches and spines in pre-hemal vertebrae – Fig. 2G) in all the observed lots, followed by A5 (deformation of neural arches and spines in cephalic vertebrae) and B7 (deformed pleural ribs).

Neural arches and spines of all vertebrae were often anomalous in all lots.

The pre-haemal region of the vertebral column was the most affected by anomalies as it was the only region affected by severe anomalies in all lots, except for lots 4, 6 and 8 (Table 8). Also commonly affected were the cephalic and caudal regions, with no clear pattern of linkage with rearing methodology or strain. Fin anomalies and head malformations were evenly distributed in organic and intensive lots. Head malformations were quite rare (1.9–3.4% of individuals affected).

The CA ordination plot of lots and descriptors (anomalies) on the first two correspondence axes is shown in Figure 3A and 3B. The first two axes accounted for 31.8% and 20.8% of the overall variance, respectively. As the lot centroids were much closer to the axis origin than most of the descriptor points, the ordination of lots was also plotted on a separate enlarged figure (Fig. 3B), in order to visualize lot arrangements more satisfactorily. Intensive lots were more scattered in the space described by the first two axes

Table 8. Frequency of individuals (%) affected by anomalies/severe anomalies (in bold) in the different body regions. Empty cells indicate 0.0%. Organic lots are highlighted with grey background. VT = vertebrae.

	1	2	3	4	5	6	7	8
Cephalic VT	80.4	73.1	100.0	62.5	65.7	62.1	70.0	71.4
Pre-Hem. VT	100.0 **2.2**	100.0 **1.0**	100.0 **18.7**	100.0	99.1 **3.7**	100.0	100.0 **1.7**	100.0
Hemal VT	50.0	53.9 **0.5**	50.0	56.3	39.8 **5.6**	58.6	65.0 **5.0**	65.3
Caudal VT	78.3 **4.3**	70.5 **0.5**	75.0	59.4	77.8 **3.7**	65.5	66.7	59.2 **2.0**
Anal Fin	10.9	10.9	12.5	9.4	12.0	10.3	13.3	16.3
Caudal Fin	15.2	25.9	25.0	9.4	25.9	24.1	13.3	30.6
Dorsal Fin	15.2	11.4	6.3	3.1	26.9	17.2	20.0	6.1
Head	2.2			3.1	1.9	3.4		

compared with the organic ones, which are all located in the negative portion of the first axis. Lots coming from ORG2 (6 and 7) and ORG3 (8) were very close to each other and located in the second quadrant, while the lot sampled in ORG1 (lot 5) was positioned in the third quadrant.

No farm-related patterns (lots coming from the same farm, such as 1 and 2 or 3 and 4, were very far from each other) or related to the genetic origin (e.g., Italian lots were not closer to each other than to the other lots) were clearly detectable.

Anomalies clustered in four main groups (Fig. 3A):

(1) anomalies of the vertebral bodies (A4, B4– Fig. 2B – and D4– Fig. 2I) and the presence of extra-ossifications in the haemal arches of the caudal vertebrae (D6*) in the first quadrant, fusions of the cranial (A3) and caudal vertebrae (D3– Fig. 2H), kyphosis in the pre-haemal region (B2) and malformed premaxilla and maxilla (15) in the negative region of CA1;

(2) anomalies of the caudal and anal fin rays (G11, F11), complete fusion of pre-haemal vertebrae (B3*–Fig. 2C) and malformed dentale (14) in the second quadrant of the ordination;

(3) anomalies of the haemal vertebrae (C1, C3–Fig. 2D–and C4– Fig. 2F), those affecting the second dorsal fin (H8 and H11) and kyphosis in caudal vertebrae (D2) located in the third quadrant;

(4) descriptors closer to the axis origin, common to all the observed lots.

ANOSIM detected the mean inter-group distances for lots reared in intensive and organic farms as significant (R = 0.02, p< 0.0001). The anomalies thus seemed to be related to the rearing conditions. However, no significant differences between intensive and organic lots were detected (R = 0.0005, p = 0.12) when only severe anomalies were considered.

Discussion

This study represents one of the first attempts [52] to characterize and compare the skeletal quality of rainbow trout reared under intensive and organic aquaculture. In salmonids, vertebral axis deviations appear dramatically only after smoltification, and are only rarely observed in early juveniles [12]. One exception is that the displacement of vertebral bodies has been reported in under yearling smolts of fast growing intensively-reared salmon [28]. Fin anomalies other than fin erosion are rarely reported in salmonids [12]. Because anomalies are the subject of significant economic [24] and animal welfare concern [53–55], it is important to identify their potential causes and find appropriate rearing conditions for ensuring correct skeletal development.

Recent effort to rear rainbow trout under organic aquaculture is an opportunity to analyse if this methodology can produce trout of higher morphological quality than the intensive rearing technology.

In this investigation, several lots of adult rainbow trout from intensive and organic farms were inspected for the presence of skeletal anomalies. Unlike the majority of available studies on rainbow trout [6,8,26,27,29,56–62], the presence of anomalies affecting the vertebral axis, the unpaired fins, and the *splanchno-cranium* were scored. The frequencies of each kind of anomaly in each body region were described, and a detailed computation made of the meristic characters.

The rainbow trout spine normally consists of 59–63 vertebrae [63], even if some previously analysed hatchery lots [64] and wild populations [65,66] showed wider ranges of variation (the

Table 9. Summary of meristic ranges in previously analysed reared [64] and wild [65,66] rainbow trout. MX: Mexico; BC: Bogota Columbia; AK: Alaska; AJ: S. Africa Jonkershoek; AP: S. Africa Pirie; SP: Spain; PG: Poland; NS: Normandale Spring; ID: Idaho; NF: Normandale fall.

Origin	Strain	Vertebrae	Dorsal Rays	Caudal Rays	Anal Rays	Source
Native		60–66	12–16	18–20	12–16	[65,66]
Farmed	MX	58–64	14–17	19	13–16	[64]
	BC	60–65	13–17	19–20	13–16	
	AK	60–63	13–15	17–19	12–15	
	AJ	60–64	13–17	19–20	13–16	
	AP	60–64	14–17	18–20	13–16	
	SP	60–63	14–18	18–19	13–16	
	PG	61–64	14–17	19–20	12–16	
	NS	61–65	15–17	18–19	14–16	
	ID	63–67	14–16	18–19	13–15	
	NF	60–64				

minimum and maximum values reported are 58 and 67, respectively – Table 9). In this study, the number of vertebrae varied from 57 to 65 in intensive lots, and from 59 to 65 in organic ones. Although organically-reared lots show a narrower range of variation, the interquartile distribution of the number of vertebrae was nearly the same for the two groups (61–64 vs. 61–63, respectively). The wider range in the intensive lots was therefore due to the presence of a few outliers. The pre-haemal region was the most variable portion of the vertebral column, with 33–39 and 33–38 elements in intensive and organic lots, respectively. The cephalic and caudal regions were very conservative. Rainbow trout is a subcarangiform generalist swimmer, propelled by the undulatory motion of the body with the caudal peduncle acting as a single unit (BCF) [67]: the whole body is involved in the undulatory propulsion, but wave amplitude is maximum near the tail or in the posterior third of the body. Joined to a relatively deep caudal peduncle, there is a caudal fin characterized by a low aspect ratio [68]. The involvement of the caudal peduncle in the swimming propulsion, i.e. one of the most important and adaptive functions of fish, is probably the reason leading to the high degree of canalization of the number of vertebrae in this region. Conversely, the number of thoracic vertebrae (above all the pre-haemal ones) is probably less strictly controlled, as this region of the body is not directly involved in generating thrust.

Previous meristic counts in wild and reared rainbow trout revealed range values of dorsal (12–18), caudal (17–20) and anal rays (12–16) that substantially overlapped those recorded in the lots analysed in this study (Table 9– [64–66]). Moreover, in the hatchery lots observed by MacGregor & MacCrimmon [64], some of the meristic characters analysed (i.e. vertebrae, anal and dorsal rays) showed significant different mean values as they are useful characters for stock discrimination. In this study, beyond vertebrae and anal and dorsal fin rays, anal pterygophores and supraneurals showed median value differences among lots, corroborating previous investigations.

All individuals displayed at least one anomaly in all lots. Such high rates of anomalous individuals in reared lots of rainbow trout, never previously described in literature, could be explained by applying the methodology applied in this study, which has now been amply standardized and already applied to other farmed, mostly marine, fish [11–13,15,16,33–36]. The detailed and mass

monitoring of all anomalies affecting the *splanchnocranium*, vertebral axis and fins was never applied to salmonids, often scored only for vertebrae centra anomalies, or inspected only for externally detectable anomalies (see Table 10 for a brief review of some studies on salmonids anomalies), of furnishing lower deformation rates. For instance, some authors found that up to 55% of normally shaped rainbow trout (i.e. showing no external anomalies) of market size were found to be affected by vertebral anomalies on French farms [59]. Others reported that a certain number of Atlantic salmon were affected to a different degree by a variable number of compressed vertebrae that were not externally visible [28].

No differences in the occurrence of deformed individuals were detected between intensive and organic conditions. However, ANOSIM found significant mean inter-group distances for lots reared on intensive and semi-intensive organic farms. This is due to differences in the anomaly pattern; intensive lots showed higher inter-lot differences in the anomaly typologies (14–25 types) than the organic ones (20–25 types – Table 6), as highlighted by their scattered distribution in the CA ordination plot with respect to the organic lots (Fig. 3B). The most frequent anomalies were B5 (deformed neural arches and spines in pre-haemal vertebrae) and the presence of extra-ossifications of pleural ribs (B7*). Some anomalies were detected only in organic lots, i.e. anomalies of caudal (G11) and anal (F11) fins rays and axis deviations (scoliosis and kyphosis) of the haemal and caudal region (C1 and D2). All these anomalies were detected in three different semi-intensive lots, so they cannot be ascribed to specific sub-lots of the organic group. In particular, the only individuals affected by C1 and D2 typologies were both detected in the same lot- the Italian lot reared in ORG1 (Lot 5), which is the organic farm characterized by the highest rearing densities and the lowest water renewal. The peculiar pattern of anomalies in the individuals reared on this farm was also emphasized by its isolated position in the CA ordination plot with respect to the ORG2 and ORG3 lots, which were closer to each other.

No clear patterns of skeletal anomalies distinguishing between the different lots on a genetic basis were found. The observed differences in the anomaly typologies and frequencies in the intensive and organic lots were statistically significant (ANOSIM), thus indicating the presence of an effect of rearing methodology on

Table 10. Summary of some previous studies on salmonid skeletal anomalies. Occurrence refers to the percentage of affected individuals (mean±S.D., range or maximum).

Species	Developmental stage	Types of anomalies considered	Inspection methodology	Occurrence (%)	Source
O. mykiss	Juvenile	Vertebral axis	External visual inspection	3–10	[6]
O. mykiss	Juvenile	Splanchnocranium, vertebral axis and fins	In toto staining	62.8±26.9	[69]
S. trutta	Adult	Vertebral axis	External visual inspection	8.9	[8]
O. mykiss	Sub-adult	Vertebrae centra	X-rays	9.8±3.1	[70]
S. salar	Juvenile and adult	Vertebrae centra	X-rays	0–100*	[71]
S. salar	Pre- and post-smolt	Splanchnocranium	External visual inspection	20–65	[9]
O. mykiss	Sub-adult	Vertebrae centra	X-rays	50.6	[56]
S. salar	Adult	Vertebral axis	External visual inspection	2.3–21.5	[24]
S. salar	Embryo	Vertebral axis	Not specified	14	[72]
S. salar	Sub-adult	Vertebral axis	X-rays	27–34	[73]
S. salar	Adult	Vertebral axis (short-tail phenotype)	X-rays	35	[26]
S. salar	Juvenile	Vertebral axis	X-rays	45–60	[74]
S. salar	Pre- and post-smolt	Vertebrae centra	X-rays	12	[56]
S. salar	Juvenile and smolt	Splanchnocranium and vertebral axis	X-rays	7.0–12.4	[75]
O. mykiss	Adult	Splanchnocranium and vertebral axis	External visual inspection	7.1±9.5	[76]
O. mykiss	Adult	Vertebrae centra	X-rays	21.1±16.1	[59]
O. mykiss	Sub-adult	Vertebrae centra	X-rays	60.0	[27]
S. salar	Juvenile	Vertebrae centra	X-rays	33.7**	[60]
O. mykiss	Juvenile	Vertebral axis	External visual inspection	10–45	[77]
S. salar	Juvenile	Vertebral axis	X-rays	8.9–13.9	[29]
O. mykiss	Adult	Rib and vertebrae centra	X-rays	39.3	[78]
S. salar	Post-smolt	Vertebrae centra	X-rays	37	[79]
S. salar	Juvenile	Vertebrae centra	X-rays	25–92**	[80]
S. salar	Post-smolt	Vertebrae centra	X-rays	2.5–16.4	[81]
S. salar	Juvenile	Vertebral axis	In toto staining	29.6	[82]
S. salar	Juvenile	Splanchnocranium and vertebral axis	External visual inspection	<2.5%	[83]

*Percentage of columnal length with changes in *centra*.
**Range/Maximum percentage of anomalous vertebrae, not individuals.

skeletal anomalies, even if a clear pattern characteristic of intensive or organic lots has not been identified.

A non-significant higher average percentage of individuals affected by severe anomalies was detected in intensive lots (6.6% vs. 4.2%). Fused and anomalous cephalic vertebrae (A3 and A4) were absent (A4) or quite rare (A3) in all lots, except for Italian strains, both intensively and organically-reared. Anomalies affecting fin rays were rarer than those involving pterygophores. No clear relationship between the degree of anomaly and the rearing conditions or genetic origin was evidenced.

Dentale, pre-maxilla and maxilla anomalies were found in a few individuals and lots, and were not related to specific rearing conditions. These data suggest that, in rainbow trout, unlike marine reared fish [11,12,84–88], anomalies affecting skeletal elements other than vertebrae and the vertebral axis are quite rare.

In this study, vertebrae arches and centra were the most commonly affected elements, varying from only a single abnormal vertebra to various compressed and/or fused vertebrae. This reveals a wide range of plastic responses of the salmonid axial skeleton to environmental factors [26,53,54,57,68,71,74,89]. Previous studies [27,56,59] reported caudal vertebrae as being the most likely to be affected by severe anomalies. Also in this study it was common for caudal vertebrae to be anomalous,

especially in intensively-reared lots. This is probably due to the sub-carangiform swimming of this species, in which the muscles located in this region ensure propulsion [90–93], but also exert strong mechanical forces, which could determine intervertebral joint failures and then vertebrae compression and fusion [26,57]. Mechanical forces exerted by extra-activity of muscles on the column under intensive rearing conditions may lead to bone and cartilage remodelling, thus generating spinal anomalies. Moreover, stressful handling procedures (e.g., vaccination) in intensive farming conditions could induce inflammation [94], which has been hypothesized to induce bone and/or cartilage remodelling [62] leading to vertebrae compression.

As organic production is based on non-organic aquaculture juveniles, it is necessary to compare adult stages of the same origin in order to analyse whether rearing conditions affect skeletal anomaly pattern and/or occurrence. In this study, only Italian strains had adults both in semi-intensive organic and intensive conditions. This suggests that organic adults showed a larger number of anomaly typologies and a lower ratio of severe anomalies and a lower occurrence of severely deformed individuals compared with the same lot reared in intensive conditions. On close examination (Table 7), it appears that the observed differences are very small, refer to 1–2 individuals, and are a probable consequence of sampling.

The lack of significant differences in the incidence of severe anomalies in intensive and semi-intensive lots, in contrast to what had previously been observed in some reared marine fish (i.e., *Pagrus pagrus* [31], *Sparus aurata* [36]), suggests that factors other than stocking density and water volume influence the skeletogenetic processes in rainbow trout.

The lack of significant differences between rainbow trout adults on-grown under traditional intensive and organic aquaculture could be explained by a variety of hypotheses.

Common conditions shared during embryonic, larval and early juvenile developmental stages could be the most likely cause of such a lack of significant differences in the occurrence of anomalies and in the pattern of severe anomalies. It has been emphasized that spinal anomalies can develop at all life stages of Atlantic salmon [57]. Indeed, several critical stages for the development of bone anomalies have been identified, such as egg incubation, the period between yolk sac alevins and first feeding juveniles, first feeding period to smoltification and later, the seawater period [95]. These results would suggest the need for the establishment of protocols for the organic rearing of larvae and juveniles and for organic broodstocks in order to produce high quality fish. The possibility of introducing non-organic juveniles in organic farms for on-growing will be banned in the next two years (EC 710/2009) thus making it essential for fish farmers to make an effort in this direction.

Another hypothesis that should be considered and tested in the future is the loss of adaptive potential of fully domesticated strains of rainbow trout and the consequent reduced ability to phenotypically react to new environmental cues due to both decreased genetic variability and phenotypic plasticity [96–98]. Genetic variability in captive populations is generally subject to intense reduction due both to non-directional (i.e., inbreeding and genetic drift) and directional mechanisms (i.e., artificial selection, reduction of natural selection) [99–100]. Loss of genetic variation in hatchery stocks maintained in captivity for a long time has harmful effects on a variety of important traits related to fitness (e.g.,

survival of eggs and larvae, growth rate, feed conversion efficiency, risk-taking behavior and swimming performance) [101–104], thus impairing the ability to adapt to changes in environmental conditions. Adaptive response to changes in environmental conditions may also depend on phenotypic plasticity: the genotype, through interactions with the environment, generates different phenotypes, depending on the external conditions [105]. Historically, environmentally affected phenotypes were scarcely considered because of their apparent lack of a genetic basis. The modern view rejects this notion and, in many circumstances, phenotypic plasticity is considered adaptive. This view can be summarized in the statement that "phenotypic plasticity evolves to maximize fitness in variable environments" [106]. On the basis of this assumption, it could be hypothesized that the constant biotic and abiotic conditions experienced in captive environments make the high maintenance costs of phenotypic plasticity pointless, thus impairing genotype skill to generate different phenotypes under the thrust of changing external cues (a phenomenon denoted as *environmental robustness* [97,107-112], that is to say the insensitivity of the phenotypic outcome to environment). This could be considered as a new kind of homeorhetic trajectory [113], where the fluctuation of physiological variables is stabilized. Very little is yet known, however, about how developmental systems generate robustness when exposed to variation in ecologically relevant conditions [114].

Acknowledgments

The Authors wish to thank Dr.Grispan, Dr.Fuselli and Dr.Gattoni, who provided us with rainbow trout samples. The manuscript was edited by Prof. Thorgaard of the Washington State University (Pullman) and by Prof. MacGilvray.

Author Contributions

Conceived and designed the experiments: CB DP SC. Performed the experiments: DP EP TR. Analyzed the data: CB DP MS TR. Contributed reagents/materials/analysis tools: CB SC. Wrote the paper: CB DP TR.

References

1. FAO (2012) The State of World Fisheries and Aquaculture. Food and Agriculture Organization of the United Nations, Rome, ISSN 1020–5489. 209 pp.

2. Naylor R, Goldburg R, Primavera J, Kautsky N, Beveridge M, et al. (2000) Effect of aquaculture on world fish supplies. Nature 405: 1017–1024.

3. Naylor R, Burke M (2005) Aquaculture and ocean resources: raising tigers of the sea. Annu Rev Env Resour 30: 185–218.

4. MacCrimmon HR (1971) World distribution of rainbow trout (*Salmo gairdneri*). J Fish Res Board Can 28: 663–704.

5. Behnke RJ (1992) Native Trout of Western North America. Bethesda, Maryland: American Fisheries Society.

6. Aulstad D, Kittelsen A (1971) Abnormal body curvatures of rainbow trout *Salmo gairdneri* inbred fry. J Fish Res Board Can 28: 1918–1920.

7. Kinkaid HL (1976) Inbreeding in rainbow trout (*Salmo gairdneri*). J Fish Res Board Can 33: 2420–2426.

8. Poynton SL (1987) Vertebral column abnormalities in brown trout, *Salmo trutta*, L. J Fish Dis 10: 53–57.

9. Sadler J, Pankhurst PM, King HR (2001) High prevalence of skeletal deformity and reduced gill surface area in triploid Atlantic salmo (*Salmo salar* L.). Aquaculture 198(3–4): 369–386.

10. Fjelldal PG, Hansen T (2010) Vertebral deformities in triploid Atlantic salmon (*Salmo salar* L.) under yearling smolts. Aquaculture 309(1–4): 131–136.

11. Boglione C, Costa C (2011) Skeletal deformities and juvenile quality. In: Pavlidis M, Mylonas C, editors. Biology and aquaculture of gilthead sea bream and other species. Oxford, United Kingdom:Wiley-Blackwell. pp. 233–294.

12. Boglione C, Gisbert E, Gavaia P, Witten PE, Moren M, et al. (2013) Skeletal anomalies in reared European fish larvae and juveniles. Part 2: main typologies, occurrences and causative factors. Rev Aquacult 5 (Suppl. 1): S121–S167.

13. Boglione C, Gagliardi F, Scardi M, Cataudella S (2001) Skeletal descriptors and quality assessment in larvae and post-larvae of wild-caught and hatchery-reared gilthead sea bream (*Sparus aurata* L. 1758). Aquaculture 192: 1–22.

14. Boglione C, Giganti M, Selmo C, Cataudella S (2003a) Morphoecology in larval finfish: a new candidate species for aquaculture, *Diplodus puntazzo* (Sparidae). Aquacult Int 11: 17–41.

15. Boglione C, Costa C, Di Dato P, Ferzini G, Scardi M, et al. (2003b) Skeletal quality assessment of reared and wild sharpsnout sea bream and pandora juveniles. Aquaculture 227(1–4): 373–394.

16. Boglione C, Costa C, Giganti M, Cecchetti M, Di Dato P, et al. (2005) Biological monitoring of wild thicklip grey mullet (*Chelon labrosus*), golden grey mullet (*Liza aurata*), thinlip mullet (*Liza ramada*) and flathead mullet (*Mugil cephalus*) (Pisces: Mugilidae) from different Adriatic sites: meristic counts and skeletal anomalies. Ecol Indic 6: 712–732.

17. Abdel I, Abellán E, López-Albors O, Valdés P, Nortes MJ, et al. (2005) Abnormalities in the juvenile stage of sea bass (*Dicentrarchus labrax* L.) reared at different temperatures: types, prevalence and effect on growth. Aquacult Int 12: 523–538.

18. Jezierska B, Lugowska K, Witeska M (2009) The effects of heavy metals on embryonic development of fish (a review). Fish Physiol Biochem 35: 625–640.

19. Sfakianakis DG, Georgakopoulou E, Papadakis IE, Divanach P, Kentouri M, et al. (2006) Environmental determinants of haemal lordosis in European sea bass, *Dicentrarchus labrax* (Linnaeus, 1758). Aquaculture 254: 54–64.

20. Georgakopoulou E, Katharios P, Divanach P, Koumoundouros G (2010) Effect of temperature on the development of skeletal deformities in gilthead seabream (*Sparus aurata* Linnaeus, 1758). Aquaculture 308: 13–19.

21. Grini A, Hansen T, Berg A, Wargelius A, Fjelldal PG (2011) The effect of water temperature on vertebral deformities and vaccine-induced abdominal lesions in Atlantic salmon, *Salmo salar* L. J Fish Dis 34: 531–546.

22. Hamre K, Yúfera M, Rønnestad I, Boglione C, Conceição LEC, et al. (2013) Fish larval nutrition and feed formulation: knowledge gaps and bottlenecks for advances in larval rearing. Rev Aquacult 5: S26–S58.

23. MacCrimmon HR, Bidgood B (1965) Abnormal vertebrae in the rainbow trout with particular reference to electrofishing. T Am Fish Soc 94: 84–88.

24. Gjerde B, Pante MJR, Baeverfjord G (2005) Genetic variation for a vertebral deformity in Atlantic salmon (*Salmo salar*). Aquaculture 244: 77–87.

25. Gil-Martens L, Obach A, Ritchie G, Witten PE (2005) Analysis of a short tail type in farmed Atlantic salmon (*Salmo salar*). Fish Vet Journal 8: 71–79.

26. Witten PE, Gil-Martens L, Hall BK, Huysseune A, Obach A (2005) Compressed vertebrae in Atlantic salmon (*Salmo salar*): Evidence for metaplastic chondrogenesis as a skeletogenic response late in ontogeny. DAO64: 237–246.

27. Deschamps MH, Labbe L, Baloche S, Fouchereau-Peron M, Dufour S, et al. (2009) Sustained exercise improves vertebral histomorphometry and modulates hormonal levels in rainbow trout. Aquaculture296: 337–346.

28. Gil-Martens L (2012) Is inflammation a risk factor for spinal deformities in farmed Atlantic salmon? PhD Dissertation, University of Bergen, Norway.

29. Gil-Martens L, Lock EJ, Fjelldal PG, Wargelius A, Araujo P, et al. (2010) Dietary fatty acids and inflammation in the vertebral column of Atlantic salmon, *Salmo salar* L., smolts: a possible link to spinal deformities. J Fish Dis 33: 957–972.

30. Koumoundouros G, Divanach P, Kentouri M (2001) The effect of rearing conditions on development of saddleback syndrome and caudal fin deformities in *Dentex dentex* (L.). Aquaculture 200: 285–304.

31. Izquierdo MS, Socorro J, Roo J (2010) Studies on the appearance of skeletal anomalies in red porgy: effect of culture intensiveness, feeding habits and nutritional quality of live preys. J Appl Ichthyol 26(2): 320–326.

32. Roo J, Socorro J, Izquierdo MS (2010) Effect of rearing techniques on skeletal deformities and osteological development in red porgy *Pagrus pagrus* (Linnaeus, 1758) larvae. J Appl Ichthyol 26: 372–376.

33. Russo T, Prestinicola L, Scardi M, Palamara E, Cataudella S, et al. (2010a) Progress in modeling quality in aquaculture: an application of the Self – Organizing Map to the study of skeletal anomalies and meristic counts in gilthead seabream (*Sparus aurata*, L. 1758). J Appl Ichthyol 26: 360–365.

34. Russo T, Scardi M, Boglione C, Cataudella S (2010b) Rearing methodologies and morphological quality in aquaculture: an application of the Self – Organizing Map to the study of skeletal anomalies in dusky grouper (*Epinephelus marginatus* Lowe, 1834) juveniles reared under different methodologies. Aquaculture 315(1–2): 69–77.

35. Prestinicola L, Boglione C, Cataudella S (2013a) Environmental effects on the skeleton in reared gilthead seabream (*Sparus aurata*). Invited speaker at LARVI '13– FISH & SHELLFISH LARVICULTURE SYMPOSIUM, 6th fish & shellfish larviculture symposium Gent, Belgium, September 2–5, 2013. Book of abstract: 382–283.

36. Prestinicola L, Boglione C, Makridis P, Spanò A, Rimatori V, et al. (2013b) Environmental conditioning of skeletal anomalies typology and frequency in gilthead seabream (*Sparus aurata* L., 1758) juveniles. PLoS ONE8(2): e55736. doi:10.1371/journal.pone.0055736.

37. Balbelona MC, Morinigo MA, Andrades JA, Santamaria JA, Becerra J, et al. (1993) Microbiological study of gilthead sea bream (*S. aurata* L.) affected by lordosis (a skeletal deformity). Bull Eur Ass Fish Pathol 13: 33.

38. Andrades JA, Becerra J, Fernández-Llébrez P (1996) Skeletal deformities in larval, juvenile and adult stages of cultured gilthead sea bream (*Sparus aurata* L.). Aquaculture 141: 1–11.

39. Hilomen-Garcia GV (1997) Morphological abnormalities in hatchery-bred milkfish (*Chanos chanos* Forsskal) fry and juveniles. Aquaculture 152: 155–166.

40. Koumoundouros G, Gagliardi F, Divanach P, Boglione C, Cataudella S, et al. (1997) Normal and abnormal osteological development of caudal fin in *Sparus aurata* L fry. Aquaculture 149: 215–226.

41. Matsuoka M (2003) Comparison of meristic variations and bone abnormalities between wild and laboratory-reared red sea bream. JARQ 37(1): 21–30.

42. Crawford SS, Muir AM (2008) Global introductions of salmon and trout in the genus Oncorhynchus: 1870–2007. Rev Fish Biol Fisheries 18: 313–344.

43. Pulcini D, Russo T, Reale P, Massa-Gallucci A, Brennan G, et al. (2014) Rainbow trout (*Oncorhynchus mykiss*, Walbaum 1792) develop a more robust body shape under organic rearing. Aquac Res 45: 397–409.

44. Harder W (1975) *Anatomy of Fishes.Part I. Text.*E. Stuttgart: Schweizerbart'sche Verlagsbuchhandlung (Nägele u. Obermiller).

45. Matsuoka M (1987) Development of skeletal tissue and skeletal muscle in the Red sea bream, *Pagrus major*. Jap J Ichthyol 65: 1–114.

46. Mabee PM (1988) Supraneural and predorsal bones in fishes: development and homologies. Copeia 4: 827–838.

47. Schultze HP, Arrantia G (1989) The composition of the caudal skeleton of Teleosts (*Actinopterygii. Osteichthyes*). Zool J Linn Soc 97: 189–231.

48. Clarke KR (1993) Non-parametric multivariate analysis of changes in community structure. Aust J Ecol 18: 117–143.

49. Benzécri JP (1973) *L'Analyse des Données*. 2 vols. Paris, France: Dunod.

50. Rogers DJ, Tanimoto TT (1960) A computer program for classifying plants. Science 132: 1115–1118.

51. Hammer Ø, Harper DAT, Ryan PD (2001) PAST: Paleontological Statistics Software Package for Education and Data Analysis. Palaeontol Electron 4: 9. Available: http://folk.uio.no/ohammer/past/download.html.

52. Pulcini D, Boglione C, Palamara E, Cataudella S (2010) Use of meristic counts and skeletal anomalies to assess developmental plasticity in farmed rainbow trout (*Oncorhynchus mykiss*, Walbaum 1792). J Appl Ichthyol 26: 298–302.

53. Sullivan M, Hammond G, Roberts RJ, Manchester NJ (2007a) Spinal deformation in commercially cultured Atlantic salmon, *Salmo salar* L. A clinical and radiological study. J Fish Dis 30: 745–752.

54. Sullivan M, Reid SWJ, Ternent H, Manchester NJ, Roberts RJ, et al. (2007b) The aetiology of spinal deformity in Atlantic salmon, *Salmo salar* L.: influence of different commercial diets on the incidence and severity of the preclinical condition in salmon parr under two contrasting husbandry regimes. J Anim Physiol An N 30: 759–767.

55. Hansen T, Fjelldal PG, Yurtseva A, Berg A (2009) The effect of vertebral deformities on growth in Atlantic salmon (*Salmo salar* L.). Abstract presented at the Workshop Interdisciplinary Approaches in Fish Skeletal Biology. 27–29 April 2009. Tavira, Portugal.

56. Kacem A, Meunier FJ, Aubin J, Haffray P (2004) Caractérisationhisto-morphologique des malformations du squelette vertébral chez la truite arc-en-ciel (*Oncorhynchus mykiss*) après different traitements de triploïdisation. Cybium 28: 15–23.

57. Witten PE, Obach A, Huysseune A, Baeverfjord G (2006) Vertebrae fusion in Atlantic salmon (*Salmo salar*): Development, aggravation and pathways of containment. Aquaculture 258: 164–172.

58. Witten PE, Gil-Martens L, Huysseune A, Takle H, Hjelde K (2009) Towards a classification and an understanding of developmental relationships of vertebral body malformations in Atlantic salmon (*Salmo salar* L.). Aquaculture 295: 6–14.

59. Deschamps MH, Kacem A, Ventura R, Gourty G, Haffray P, et al. (2008) Assessment of "discreet" vertebral abnormalities, bone mineralization and bone compactness in farmed rainbow trout. Aquaculture 279: 11–17.

60. Fjelldal PG, Hansen T, Breck O, Sandvik R, Waagbø R, et al. (2009) Supplementation of dietary minerals during the early seawater phase increase vertebral strength and reduce the prevalence of vertebral deformities in fast growing under-yearling Atlantic salmon (*Salmo salar* L.) smolt. Aquacult Nutr 15: 366–378.

61. Deschamps MH, Sire JY (2010) Histomorphometrical studies of vertebral bone condition in farmed rainbow trout, *Oncorhynchus mykiss*. J Appl Ichthyol 26: 377–380.

62. Gil-Martens L (2010) Inflammation as a potential risk factor for spinal deformities in farmed Atlantic salmon (*Salmo salar* L.). J Appl Ichthyol 26(2): 350–354.

63. Spillman CJ (1961) *Faune de France: Poissonsd'eau douce*. Tome 65. Paris: Fédération Française des Sociétés Naturelles.

64. MacGregor RB, MacCrimmon HR (1977) Meristic variation among world hatchery stocks of rainbow trout, *Salmo gairdneri* Richardson. Environ Biol Fish 1: 127–143.

65. Bidgood BF(1965) A comparative taxonomic study of rainbow trout *Salmo gairdneri* Richardson in selected watersheds in Ontario. M.Sc. thesis, University of Guelph, Guelph, Ontario.pp. 107.

66. MacGregor RB (1975) A taxonomic study of rainbow trout, *Salmo gairdneri* Richardson, stressing meristics. M.Sc. thesis, Univ. Of Guelph, Guelph, Ontario. pp. 200.

67. Webb PW (1977) Effects of median-fin amputation on fast-start performance of rainbow trout (*Salmo gairdneri*). J Exp Biol 68: 123–135.

68. Lindsey CC (1978) Form, function and locomotory habits in fish. In: Hoar WS, Randall DJ, editors. Ecomorphology of fishes. Kluwer Academic Publishers. pp. 199–211.

69. Hose JE, Hannah JB, Puffer HW, Landol ML (1984) Histologic and skeletal abnormalities in benzo(a)pyrene-treated rainbow trout alevins. Arch Environ Contam Toxicol 13: 675–684.

70. Madsen L, Arnberg J, Dalsgaard I (2001) Radiological examination of the spinal column in farmed rainbow trout *Oncorhynchus mykiss* (Walbaum). Experiments with *Flavobacterium psychrophilum* and oxytetracycline. Aquacult Res 32: 235–241.

71. Kvellestad A, Høie S, Thorud K, Tørud B, Lyngøy A (2000) Platyspondyly and shortness of vertebral column in farmed Atlantic salmon *Salmo salar* in Norway - Description and interpretation of pathological changes. Dis Aquat Organ 39: 97–108.

72. Takle H, Baeverfjord G, Lunde M, Kolstad K, Andersen Ø (2005) The effect of heat and cold exposure on HSP70 expression and development of deformities during embryogenesis of Atlantic salmon (*Salmo salar*). Aquaculture 249: 515–524.

73. Wargelius A, Fjelldal PG, Hansen T (2005) Heat shock during early somitogenesis induces caudal vertebral column defects in Atlantic salmon (*Salmo salar*). Dev Genes Evol 215: 350–357.

74. Helland S, Refstie S, Espmark Å, Hjelde K, Baeverfjord G (2005) Mineral balance and bone formation in fast-growing Atlantic salmon parr (*Salmo salar*) in response to dissolved metabolic carbon dioxide and restricted dietary phosphorus supply. Aquaculture 250: 364–376.

75. Fjelldal PG, Hansen TJ, Berg AE (2007) A radiological study on the development of vertebral deformities in cultured Atlantic salmon (*Salmo salar* L.). Aquaculture 273: 721–728.

76. Kause A, Ritola O, Paananen T (2007) Changes in the expression of genetic characteristics across cohorts in skeletal deformations of farmed salmonids. Genet Sel Evol 39: 529–543.

77. Fontagné S, Silva N, Bazin D, Ramos A, Aguirre P, et al. (2009) Effects of dietary phosphorus and calcium level on growth and skeletal development in rainbow trout (*Oncorhynchus mykiss*) fry. Aquaculture 297: 141–150.

78. Gislason H, Karstensen H, Christiansen D, Hjelde K, Helland S, et al. (2010) Rib and vertebral deformities in rainbow trout (*Oncorhynchus mykiss*) explained by a dominant-mutation mechanism. Aquaculture 309: 86–95.

79. Hansen T, Fjelldal PG, Yurtseva A, Berg A (2010) A possible relation between growth and number of deformed vertebrae in Atlantic salmon (*Salmo salar* L.). J Appl Ichthyol 26: 355–359.

80. Wargelius A, Fjelldal PG, Grini A, Gil-Martens L, Kvamme PO, et al. (2010) MMP-13 (matrix metalloproteinase 13) expression might be and indicator of increased ECM remodeling and early signs of vertebral compression in farmed Atlantic salmon (*Salmo salar* L.). J Appl Ichthyol 26: 366–371.

81. Gil-Martens L, Fjelldal PG, Lock EJ, Wargelius A, Wergeland H, et al. (2012) Dietary phosphorus does not reduce the risk for spinal deformities in a model of adjuvant-induced inflammation in Atlantic salmon (*Salmo salar*) post-smolts. Aquac Nut 18: 12–20.

82. Sánchez RC, Obregón EB, Rauco MR (2011) Hypoxia is like an ethiological factor in vertebral column deformity of salmon (*Salmo salar*). Aquaculture 316: 13–19.

83. Taylor JF, Preston AC, Guy D, Migaud H (2011) Ploidy effects on hatchery survival, deformities, and performance in Atlantic salmon (*Salmo salar*). Aquaculture 315: 61–68.

84. Gavaia PJ, Dinis MT, Cancela ML (2002) Osteological development and abnormalities of the vertebral column and caudal skeleton in larval and juvenile stages of hatchery-reared Senegal sole (*Solea senegalensis*). Aquaculture 211: 305–323.

85. Engrola S, Figueira L, Conceição LEC, Gavaia P, Dinis MT (2009) Co-feeding in Senegalese sole larvae with inert diet from mouth opening promotes growth at weaning. Aquaculture 288: 264–272.

86. Fernández I, Pimentel MS, Ortiz-Delgado JB, Hontoria F, Sarasquete C, et al. (2009) Effect of dietary vitamin A on Senegalese sole (*Solea senegalensis*) skeletogenesis and larval quality. Aquaculture 295: 250–265.

87. Fernández I, Gisbert E (2010) Senegalese sole bone tissue originated from chondral ossification is more sensitive than dermal bone to high vitamin A content in enriched *Artemia*. J Appl Ichthyol 26: 344–349.

88. Koumoundouros G (2010) Morpho-anatomical abnormalities in Mediterranean marine aquaculture. In: Koumoundouros G, editor. Recent Advances in Aquaculture Research. Kerala, India: Transworld Research Network. pp. 125–148.

89. Helland S, Denstadli V, Witten PE, Hjelde K, Storebakken T, et al. (2006) Occurrence of hyper dense vertebrae in Atlantic salmon (*Salmo salar* L.) fed diets with graded levels of phytic acid. Aquaculture 261: 603–614.

90. Ramzu M, Meunier FJ (1999) Descripteurs morphologiques de la zonation de la colonne vertébrale chez la truite arc-en-ciel *Oncorhynchus mykiss* (Walbaum, 1792) (Teleostei, Salmoniformes). Ann Sci Nat 3: 87–97.

91. Westneat MW, Wainwright SA (2001) Mechanical design for swimming: muscle, tendon, and bone. In: Block B, Stevens E, editors. Fish Physiology. San Diego: Academic Press. pp. 271–311.

92. Coughlin DJ, Spiecker A, Schiavi JM (2004) Red muscle recruitment during steady swimming correlates with rostral-caudal patterns of power production in trout. Comp Biochem Physiol 137: 151–160.

93. Meunier FJ, Ramzu MY (2006) La regionalisation morphofonctionnelle de l'axe vertébral chez les Téléostéens en relation avec le mode de nage. ComptesRendulPalevol5: 499–507.

94. Ostland VE, McGrogan DG, Ferguson HW (1997) Cephalic osteochondritis and necrotic scleritis in intensively reared salmonids associated with *Flexibacter psychrophilus*. J Fish Dis 20: 443–451.

95. Waagbø R, Kryvi H, Breck O, Ørnsrud R (2005) Final report of the workshop Bone Disorders in Intensive Aquaculture of Salmon and Cod. NIFES, Bergen, Norway, pp. 1–42.

96. Dudley S, Schmitt J (1996) Testing the adaptive plasticity hypothesis: density-dependent selection on manipulated stem light in *Impatiens capensis*. Am Nat 147: 445–465.

97. Debat V, David P (2001) Mapping phenotypes: canalization, plasticity and developmental stability. Trends Ecol Evol 16(10): 555–561.

98. Van Buskirk J (2002) A comparative test of the adaptive plasticity hypothesis: relationships between habitat and phenotype in anuran larvae. Am Nat 160: 87–102.

99. Ollivier L (1981) Eléments de génétique quantitative. Massons, Paris.

100. Price EO (1998) Behavior genetics and the process of animal domestication. In: Grandin T (ed.), Genetica and the behavior of domestic animals. Academic Press, New York, pp. 31–35.

101. Berejikian BA (1995) The effects of hatchery and wild ancestry and experience on the ability of steelhead trout fry (*Oncorhynchus mykiss*) to avoid a benthic predator. Can J Fish Aq Sci 52: 2476–2482.

102. Alvarez D, Nicieza AG (2005) Is metabolic rate a reliable predictor of growth and survival of brown trout (*Salmotrutta*) in the wild? Can J Fish Aq Sci 62: 643–649.

103. Araki H, Cooper B, Bluoin MS (2007) Genetic effects of captive breeding cause arapid, cumulative fitness decline in the wild. Science 318: 100–103.

104. Reinbold D, Thorgaard GH, Carter PA (2009) Reduced swimming performance and increased growth in domesticated rainbow trout, *Oncorhynchus mykiss*. Can J Fish Aq Sci 66: 1025–1032.

105. Stearns SC (1989) The evolutionary significance of phenotypic plasticity. BioScience 39: 436–445.

106. Via S, Gomulkiewicz R, De Jong G, Scheiner SM, Schlichting CD, et al. (1995) Adaptive phenotypic plasticity: consensus and controversy. Trends Ecol Evol 10(5): 212–217.

107. Waddington CH (1942) Canalization of development and the inheritance of acquired characters. Nature (London) 150: 563–565.

108. Nijhout HF (2002) The nature of robustness in development. Bioessays 24: 553–563.

109. Kitano H (2004) Biological robustness. Nat Rev Genet 5: 826–837.

110. Flatt T (2005) The evolutionary genetics of canalization. Q Rev Biol 80: 287–316.

111. de Visser JA, Hermisson J, Wagner GP, Ancel Meyers L, Bagheri-Chaichian H, et al. (2003) Perspective: Evolution and detection of genetic robustness. Evolution 57: 1959–1972.

112. Wagner A (2005) Robustness and evolvability in living systems. Princeton, Oxford: Princeton University Press.

113. Waddington CH (1975) The evolution of an evolutionist. Edinburgh: Edinburgh University Press.

114. Braendle C, Félix MA (2009) The other side of phenotypic plasticity: a developmental system that generates an invariant phenotype despite environmental variation. J Biosci 34(4): 543–551.

13

Selection of Reliable Biomarkers from PCR Array Analyses Using Relative Distance Computational Model: Methodology and Proof-of-Concept Study

Chunsheng Liu*[◊], Hongyan Xu[◊], Siew Hong Lam, Zhiyuan Gong*

Department of Biological Sciences, National University of Singapore, Singapore, Singapore

Abstract

It is increasingly evident about the difficulty to monitor chemical exposure through biomarkers as almost all the biomarkers so far proposed are not specific for any individual chemical. In this proof-of-concept study, adult male zebrafish (*Danio rerio*) were exposed to 5 or 25 µg/L 17β-estradiol (E2), 100 µg/L lindane, 5 nM 2,3,7,8-tetrachlorodibenzo-*p*-dioxin (TCDD) or 15 mg/L arsenic for 96 h, and the expression profiles of 59 genes involved in 7 pathways plus 2 well characterized biomarker genes, *vtg1* (*vitellogenin1*) and *cyp1a1* (*cytochrome P450 1A1*), were examined. Relative distance (RD) computational model was developed to screen favorable genes and generate appropriate gene sets for the differentiation of chemicals/concentrations selected. Our results demonstrated that the known biomarker genes were not always good candidates for the differentiation of pair of chemicals/concentrations, and other genes had higher potentials in some cases. Furthermore, the differentiation of 5 chemicals/concentrations examined were attainable using expression data of various gene sets, and the best combination was the set consisting of 50 genes; however, as few as two genes (e.g. *vtg1* and *hspa5* [*heat shock protein 5*]) were sufficient to differentiate the five chemical/concentration groups in the present test. These observations suggest that multi-parameter arrays should be more reliable for biomonitoring of chemical exposure than traditional biomarkers, and the RD computational model provides an effective tool for the selection of parameters and generation of parameter sets.

Editor: Raya Khanin, Memorial Sloan Kettering Cancer Center, United States of America

Funding: This work was supported by the Singapore National Research Foundation under its Environmental & Water Technologies Strategic Research Programme and administered by the Environment & Water Industry Programme Office (EWI) of the PUB, grant number R-154-000-328-272. The funders had no role in study design, data collection and analysis, decision to publish, or preparation of the manuscript.

Competing Interests: The authors have declared that no competing interests exist.

* E-mail: liuchunshengidid@126.com (CL); dbsgzy@nus.edu.sg (ZG)

◊ These authors contributed equally to this work.

Introduction

Increasing attention has been drawn to the wide occurrence of natural and man-made chemicals in the aquatic environment. Many chemicals can be bioaccumulated in the aquatic organisms and magnified in the food chains, thus threatening human health. The Minamata disease is a typical case, where methylmercury (MeHg) poisoning occurred in human due to the ingestion of fish and shellfish contaminated by MeHg [1]. Such scenarios have promoted researchers to develop early-warning methods for monitoring contaminants in the aquatic system through both chemical monitoring and biomonitoring.

As new pollutants in the environment are emerging rapidly, it becomes increasingly unfeasible to monitor all contaminants in the environment. Since the presence of a foreign chemical in a segment of the environment does not always indicate adverse biological effects [2], it is important to combine chemical monitoring with the biomonitoring for a reliable environmental risk assessment. An ideal approach is to examine biological responses that can reflect the contaminants in the exposed organisms [2]. Under this concept, various biomarkers from fish have been proposed and used for biomonitoring aquatic contaminants. However, most of biomarkers proposed were not specific

for individual chemicals. For example, biomarker for estrogen, *vtg1* mRNA could be induced not only by the native female hormone, 17β-estradiol (E2), but also by many other compounds that can interact with estrogen receptors, including many xenobiotics, such as lindane [3]. The expression of *cyp1a1* was up-regulated by 2,3,7,8-tetrachlorodibenzo-*p*-dioxin (TCDD) as well as by other chemicals such as arsenic in mice [4].

It has been demonstrated that exposure to single chemicals generated unique gene expression signature in experimental animals [5–9]. Therefore, a multi-parameter quantitative real-time PCR (qRT-PCR) array could be developed as a useful tool to differentiate a complicated set of chemical groups. However, in previous studies, the parameters (genes) were selected only based on responsive difference of gene expression among chemicals after exposure [10–11] and did not represent the best parameter (gene) set for the discrimination of chemicals. Therefore, a proof-of-concept study was designed and conducted in the present study, with the objective of finding the best parameter (gene) set for the discrimination of chemicals tested. Especially, a relative distance (RD) computational model was developed to select gene sets from 61 gene examined for chemical discrimination. Therefore, it is feasible to integrate qRT-PCR arrays and RD computational model to develop a reliable biomonitoring tool for chemical exposure.

Materials and Methods

Chemicals and reagents

E2, lindane, TCDD and arsenic ($Na_2HAsO_4 \cdot 7H_2O$) were purchased from Sigma (St. Louis, MO, USA). Arsenic was dissolved in deionized water directly and the other three chemicals were dissolved in dimethyl sulfoxide (DMSO) as stock solutions. The TRizol reagent and LightCycle FastStart DNA Master SYBR Green I were obtained from Invitrogen (New Jersey, NJ, USA) and Roche Applied Science (Mannheim, Germany), respectively.

Fish and chemical exposure

In this study, experimental procedures were carried out following the approved protocol by Institutional Animal Care and Use Committee of National University of Singapore (Protocol 079/07). Adult male zebrafish (*Danio rerio*, 5-month old) were purchased from a local aquarium farm (Mainland Tropical Fish Farm, Singapore), and acclimated for at least two weeks in our aquarium before chemical treatment. After acclimation, fish were exposed to 5 nM TCDD, 5 μg/L E2, 50 μg/L E2, 100 μg/L lindane or 15 mg/L arsenic for 96 h in a static condition. Each tank (5 L size) included 3 L exposure solution and 3 fish, and each concentration included 3 replicated tanks. During the exposure period, fish were fed once a day with commercial frozen bloodworms (Hikari) as described before [12]. The concentrations of these chemicals were chosen based on previous studies of ours and others [12–16], where biological effects of these concentrations have been confirmed by significant changes of some mRNAs examined. For E2, two concentrations were used to test the feasibility to develop a gene expression based model to differentiate exposure concentrations besides different chemicals. Fresh chemical solutions were daily replaced during the exposure experiment. For E2, lindane and TCDD exposure experiments, treatment and control groups received 0.01% DMSO, and for arsenic exposure experiments, treatment and control groups received 0.01% deionized water in this study. After 96-h exposure, the fish were anesthetized with MS-222 (1 mM) and livers were collected and preserved in TRizol reagent at –80°C until RNA isolation.

Selection of target genes for PCR array

A PCR array of sixty-one zebrafish genes was designed as follows. First, seven well characterized pathways commonly affected by chemicals were selected: oxidative and metabolic stress [17–18], apoptosis signaling [19–20], proliferation and carcinogenesis [21–22], DNA damage and repair [23–24], growth arrest and senescence [25–26], heat shock [27–28], and inflammation pathways [29–30]. Representative genes from these pathways were selected by referring Molecular Toxicology PathwayFinder PCR array from SABioscience Gene Network Central (http://www.sabiosciences.com/rt_pcr_product/HTML/PAHS-3401Z.html). Second, annotated zebrafish orthologues of human genes were searched from Ensemble website and confirmed using online synteny tool [31]; unannotated zebrafish orthologues were manually determined first by amino acid sequence comparison with human candidate sequences through UCSC website (http://genome.ucsc.edu/) and then confirmed by comparison of genomic organization, chromosomal locations and chromosomal synteny analysis as conducted in a previously study [32]. Finally the zebrafish orthologues of 59 human genes were obtained for designing of PCR primers. In addition, two well-established biomarker genes, *vtg1* and *cyp1a1*, were also included in order to compare the potentials of biomonitoring between traditional biomarkers and genes/gene sets developed in this study, as inducers of *vtg1* and *cyp1a1* such as E2 and TCDD were also used in the present exposure experiments. The complete list of genes in PCR array and their PCR primeer sequences are presented in Table S1. The number of genes in each pathway was 14, 10, 10, 6, 4, 13 and 2 for oxidative and metabolic stress, apoptosis signaling, DNA damage and repair, proliferation and carcinogenesis, growth arrest and senescence, heat shock and inflammation pathways, respectively.

Quantitative real-time PCR (qRT-PCR)

Total RNA was isolated from zebrafish livers with TRizol reagent and used for cDNA synthesis. Real time qPCR was performed using the LightCycler system (Roche Applied Science, Mannheim, Germany) with LightCycler FastStart DNA Master SYBR Green I following manufacturer's instruction. The primer sequences were designed using Primer 3 software (http://frodo.wi.mit.edu/as). The amplicon efficiencies of primers were >90%. Three housekeeping genes, *β-actin* (*beta-actin*), *β-2m* (*beta-2-microglobulin*) and *rpl13a* (*ribosomal protein L13a*), were used as internal control and the geometric means the expression of the three housekeeping genes were used as the normalized factor by $2^{-\Delta\Delta Ct}$ method. Each group included three biological replicates and each replicate included a pool of three fish.

Statistical analysis

Gene expression values were logarithmically transformed (log2) before statistical analysis. The homogeneity and normality of data were examined using the Kolmogorov-Smirnov and Levene's test, respectively. Statically significant differences between treatment and corresponding control groups were evaluated by ANOVA based on a p-value <0.05. Average linkage ($p < 0.05$) was used to examine the cluster relationships of different treatment groups based on mRNA expression profiles. The statistical analyses were performed using Kyplot Demo 3.0 software (Tokyo, Japan).

Relative distance (RD) computational model

The differentiation of two chemical/concentration groups not only depends on Euclidean distance between the two groups but also depends on the distance among individual replicates within each group. In this study, the RD computational model was developed to quantitatively describe the potential that three biological replicates from group A can be differentiated from the three replicates in group B based on mRNA expression profiles (fold change), and RD between one replicate from group A treatment and three replicates from group B (rd_{a1b})

$$rd_{a1b} = md_{a1b} - md_{aa} - 1/2 \times SD_{a1b} - 1/2 \times SD_{aa} \qquad (1)$$

$$md_{a1b} = (d_{a1b1} + d_{a1b2} + d_{a1b3})/3 \qquad (2)$$

$$md_{aa} = (d_{a1a1} + d_{a1a3})/2 \qquad (3)$$

$$d_{a1b1} = \sqrt{\sum_{j=1}^{j} (a_{1j} - b_{1j})^2}$$

$$d_{a1b2} = \sqrt{\sum_{j=1}^{j} (a_{1j} - b_{2j})^2} \qquad (4)$$

$$d_{a1b3} = \sqrt{\sum_{j=1}^{j} (a_{1j} - b_{3j})^2}$$

$$d_{a1a3} = \sqrt{\sum_{j=1}^{j} (a_{1j} - a_{3j})^2} \qquad (5)$$

$$SD_{a1b} = \sqrt{((d_{a1b1} - md_{a1b})^2 + (d_{a1b2} - md_{a1b})^2 + (d_{a1b3} - md_{a1b})^2)/(3-1)} \qquad (6)$$

$$SD_{aa} = \sqrt{((d_{a1a2} - md_{aa})^2 + (d_{a1a3} - md_{aa})^2)/(2-1)} \qquad (7)$$

where j is the total number of genes examined; a and b are gene expression values in the treatment groups A and B, respectively; md_{a1b} is the mean Euclidean distance between one biological replicate from treatment group A (a_1) and three replicates from treatment group B (b_1, b_2, b_3); md_{aa} is the mean Euclidean distance between one biological replicate from treatment group A (a_1) and other two biological replicates from the same group (a_2, a_3); SD_{a1b} is the standard deviation of Euclidean distance between one biological replicate from treatment group A (a_1) and three replicates from treatment group B (b_1, b_2, b_3); SD_{aa} is the standard deviation of Euclidean distance between one biological replicate from treatment group A treatment and other two biological replicates from the same group; d_{a1b1}, d_{a1b2} and d_{a1b3} are the Euclidean distance between one biological replicate from treatment group A (a_1) and three replicates from treatment group B (b_1, b_2, b_3); d_{a1a2} and d_{a1a3} are the Euclidean distance of biological responses between one biological replicate from treatment group A (a_1) and other two biological replicates from the same group (a_2, a_3).

In this study, first, we calculated all the RD values between two chemical treatment groups using expression data of individual genes. When all six RD values were >0 for each pair of chemicals, it was considered that the gene could be used to differentiate the two chemicals/concentrations. The cluster analyses (average linkage) were performed using commercial software (Kyplot Demo 3.0, Tokyo, Japan) (p-value <0.05) to confirm the feasibility of RD model in screening genes for the differentiation of chemical/concentration treatments. Second, the mean RD values were calculated to quantitatively compare the potentials of individual genes in differentiating two chemicals/concentrations. Finally, a C-language computational program (see Program S1) was edited for selecting genes and generating gene sets that could be used to differentiate all of five chemical/concentration treatments simultaneously using the RD model developed in this study, and maximum mean RD of each gene sets with the same amount of genes and the corresponding components of genes were outputted.

Results

Broad changes of gene expression patterns in the seven selected pathways in response to chemical insults

Adult male zebrafish were treated with 5 nM TCDD, 5 µg/L E2, 50 µg/L E2, 100 µg/L lindane or 15 mg/L arsenic for 96 hours and no mortalities were observed throughout the exposure experiment. As shown in Figure 1 and Table S2, exposure to different chemicals led to different gene expression profiles. TCDD exposure significantly down-regulated the expression of most selected genes involved in the oxidative and metabolic stress, apoptosis signaling, DNA damage and repair, proliferation and carcinogenesis, growth arrest and senescence, heat shock and inflammation pathways, while the expression of *cyp1a1*, *hspa5* and *hsp70* (*heat shock protein 70-kDa*) was among the highest up-regulated. Treatment with arsenic significantly altered the expression of most selected genes in the seven pathways, such as up-regulation of expression of *ptgs1* (*prostaglandin-endoperoxide synthase 1*), *cyp1a1* and *hsp90aa1* (*heat shock protein 90, alpha, class A member 1, tandem duplicate 1*), and down-regulation of *b1p1* (*Bcl-XL-like protein 1*), *tnfr* (*tumor necrosis factor receptor*) and *vtg1*. A significant up-regulation in the expression of *vtg1* was observed upon exposure to 5 or 50 µg/L E2, clearly showing estrogenic activity. Similar to TCDD, exposure to E2 (5 or 50 µg/L) significantly down-regulated the expression of most selected genes included in the seven pathways investigated. In contrast, exposure to lindane up-regulated the expression of most selected genes in the seven pathways; with exception of only few down-regulated genes, notably *cdkn1a* (*cyclin-dependent kinase inhibitor 1A, transcript variant 1*) in the growth arrest and senescence pathway and *fmo5* (*flavin containing monooxygenase 5*) in the oxidative and metabolic stress pathway.

Correlation of RD and potential differentiation of chemical treatment pairs

Using an RD computational model, we calculated all of RD values between two chemical/concentration treatment groups based on expression fold change of individual genes and the results are presented in Figure 2 (see details in Table S3) for all of the 10 possible chemical/concentration pairs. The ability of each of the 61 genes to discriminate the chemical/concentration pairs was tested by the software Kyplot Demo 3.0 program and the findings are presented in Figure 2. There was a good correlation of the RD and the ability to discriminate pair of chemicals/concentrations. All the genes with top and high RD values were found to be able to discriminate pair of chemicals/concentrations. For example, the two best known biomarker genes, *vtg1* and *cyp1a1*, were able to discriminate eight of the ten pairs: TCDD/arsenic, TCDD/E2_high, E2_high/lindane, E2_high/arsenic, TCDD/E2_low, TCDD/lindane, lindane/E2_low, and arsenic/E2_low. However, for the lindane/arsenic pair, *cyp1a1* could not be used to discriminate them, while for the E2_low/E2_high concentration pair, both *vtg1* and *cyp1a1* failed to discriminate them. Interestingly, *vtg1* and *cyp1a1* were not always among the top of the list based on the calculated RD. There were also many other genes (even with better RD) that could be also used to differentiate the corresponding pair of chemicals.

Selection of discriminating gene sets based on RD computational model

While it is relatively easy to discriminate a pair of chemical treatment groups based on expression data from one or few genes, it is more challenging to discriminate multiple treatment groups.

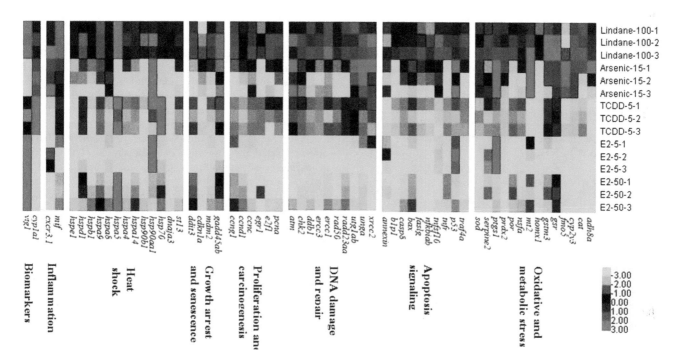

Figure 1. Gene expression profiles included in seven selected pathways in male zebrafish livers after exposure to 100 μg/L lindane, 5 nM 2,3,7,8-tetrachlorodibenzo-*p*-dioxin (TCDD), 5 μg/L 17β-estradiol (E2), 25 μg/L E2, or 15 mg/L arsenic for 96 h. There were 3 biological replicates, and each replicate were pooled from 3 fish. Gene expressions were expressed as fold change relative to the corresponding control. The full names of genes can be found in Tables S1 or S2.

In the current dataset, no single gene can be used to discriminate all of the five chemical/concentration groups. Thus, it was necessary to select a gene set for discriminating the chemical/concentration groups. Here, we further explored the RD model to select best gene sets for differentiating all of the five chemical/concentration groups. RDs were computed for all possible gene combinations from one to 61 genes and the highest mean distances for gene sets from 1 to 65 genes are presented in Figure 3. For example, the 2-gene set of the highest mean RD was *hspa5* and *vtg1* with a value of 10.57 (Figure 3 and Table S4) and the two genes can be used to discriminate the five chemical/concentration groups perfectly (Figure 4A). In comparison, the gene pair of best

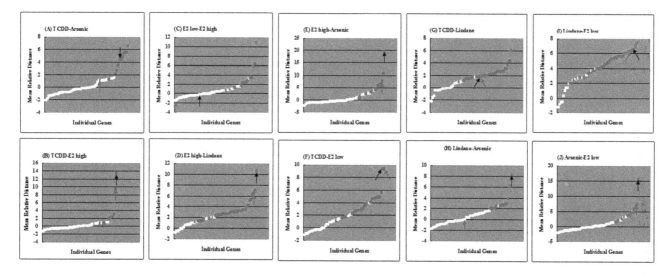

Figure 2. Mean Relative Distances (RDs) between two chemicals/concentration groups. (A) TCDD vs. Arsenic; (B) TCDD vs. E2_high; (C) E2_low vs. E2_high; (D) E2_high vs. Lindane; (E) E2_high vs. Arsenic; (F) TCDD vs. E2_low; (G) TCDD vs. Lindane; (H) Lindane vs. Arsenic; (I) Lindane vs. E2_low; (J) Arsenic vs. E2_low. Black arrows indicate the positions of *vtg1*, and red arrows indicate the positions of *cyp1a1*; White boxes indicate the positions of genes that did not pass the model test and could not be used to discriminate the corresponding two chemicals/concentrations; Pink boxes indicate the positions of genes that passed the model test and could be used to discriminate the corresponding two chemicals/concentrations. TCDD: 5 nM 2,3,7,8-tetrachlorodibenzo-*p*-dioxin; lindane: 100 μg/L lindane; arsenic: 15 mg/L arsenic; E2_low: 5 μg/L 17β estradiol; E2_high: 50 μg/L 17β-estradiol. The information of RDs and the corresponding genes can be found in Table S3.

known biomarkers, *vtg1* and *cyp1a1*, has a value of 10.33 and they could not correctly discriminate all of the five groups, particularly the two concentration groups of E2 treatment (Figure 4B). All other gene sets (3 or more genes) of the highest mean RD were also capable of differentiating all the five chemical/concentration groups correctly (Figure 3). In general, there was an increase of mean RD with the number of genes in gene sets and the maximal mean RD (19.153) was observed in the set with 50 genes, where chemicals were also completely differentiated, including different concentrations (Figure 4C).

Discussion

The environment is continuously loaded with natural and man-made chemicals, and the effects of contaminant exposure to human health have been extensively documented [33–37]. In general, adverse effects of contaminants at population levels in wildlife and human tend to be delayed; when the effects finally become clear, the destructive processes may have been beyond the point where it can be reversed by available remedial actions [2]. Therefore, various biomonitoring methods have been developed in the past few decades for the purpose of early warning. However, most of these methods focused on one or several biological parameters (e.g., biomarkers vitellogenins and cytochrome P450 enzymes 1A1) [38–43]. To search for more biomarker genes to predict chemical contamination, it is common to use high throughput and large scale analyses such as DNA microarray and more recently RNA-seq platform [8,12,44]. However, the methodology for selecting biomarkers from thousands of genes could be a great challenge. Here we performed a proof-of-concept study by selecting a handful of biomarker genes to develop a practical assay with the aid of RD computational model.

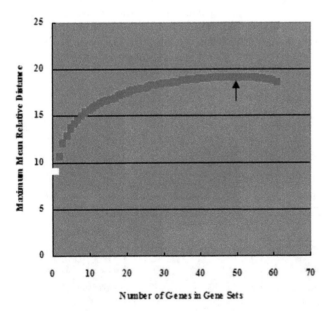

Figure 3. Maximum mean RD of gene sets with different numbers of genes among 5 chemicals/concentrations. Black arrow indicates the position of gene set (50 genes), where maximum RD was achieved. White box indicates the position of gene set (1 gene) that did not pass the model test and could not be used to differentiate the corresponding five chemicals/concentrations; Pink boxes indicate the positions of gene sets that passed the model test and could be used to differentiate the corresponding five chemicals/concentrations. The information about maximum mean RDs and the corresponding components of genes can be found in Table S4.

Here four chemicals including E2, lindane, TCDD and arsenic were tested. Both E2 and lindane exposures caused up-regulation of hepatic *vtg1* expression; similarly, treatment with TCDD or arsenic showed up-regulation of *cyp1a1* expression. These observations are consistent with previous studies [3–4,45], suggesting the effectiveness of these chemical exposure experiments. In general, exposure to different chemicals resulted in different gene expression profiles in the seven biological pathways examined. For example, both of E2 and lindane induced *vtg1* expression, but E2 down-regulated the expression of essentially all of the selected genes in the seven pathways while lindane up-regulated the expression of most of these genes. Similarly, TCDD down-regulated the expression of most of genes and arsenic up-regulated many of the genes, especially in two pathways, oxidative_and_metabolic_stress and DNA_damage_and_repair, suggesting a molecular basis for their discrimination.

In the current data set, we found that none of the 61 genes could be used to correctly discriminate all of the five chemical/concentration groups; thus, it has to rely on multiple gene sets for successful discrimination, which should be the direction for future development of multiple gene signatures for discrimination of a multiple chemical groups, as previously proposed [8,12]. To systematically select the best discriminator genes, here we developed a computational model using RD to determine the prediction power of each gene or in combination with others. First, we demonstrated that there was a positive correlation between the RD values and the discrimination of different treatments groups (Fig. 2). In our data set, a minimum of two genes (e.g. *vtg1* and *hspa5*) could be used to successfully discriminate all of the five chemical/concentration groups. There is a general increase of mean RD values with the number of genes added to the gene set, which indicate the power of using more genes for discriminating more complicated data set. In our dataset, we also found that the 50-gene set had the highest mean RD values, indicating that there is an optimal gene number used for the discrimination. From a practical viewpoint, the used of minimal number of genes will minimize workload and ease downstream data analysis. However, using more genes, especially those representing different molecular pathways, provides additional important biological information in molecular-marker based biomonitoring.

In summary, the data of this study demonstrated chemicals that induced similar responses in biomarker (e.g., TCDD and arsenic, E2 and lindane) could cause different biological responses depending on the parameters examined, and the use of parameter sets consisting of different biological responses for biomonitoring should be more appropriate. Furthermore, the computational model based on RD may be useful to select appropriate gene sets to develop efficient biomarker-based biomonitoring. Considering the rapid, sensitive, convenient and high-throughput properties of PCR, a PCR array including multiple gene parameters should be a feasible tool to develop for biomonitoring of chemical exposure.

Supporting Information

Table S1   Sequences of primers for selected genes.

Table S2   mRNA expression profiles in the livers of zebrafish after chemical exposure.

Table S3   Mean relative distances (MRDs) of individual genes between chemicals.

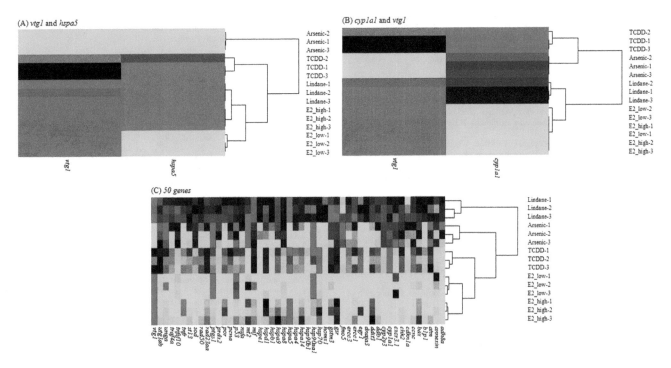

Figure 4. Clustering relationships among chemicals/concentrations using mRNA expression data of (A) *cyp1a1* **and** *vtg1*, **(B)** *vtg1* **and** *hspa5* **and (C) 50 genes with the marximum RD.** TCDD: 5 nM 2,3,7,8-tetrachlorodibenzo-*p*-dioxin; lindane: 100 μg/L lindane; arsenic: 15 mg/L arsenic; E2_low: 5 μg/L 17β-estradiol; E2_high: 50 μg/L 17β-estradiol. The full names of genes can be found in Table S1 or S2.

Table S4　Maximum mean relative distances (MMRDs) of gene sets with different amounts of genes among 5 chemicals/concentrations and the corresponding components of genes.

Author Contributions

Conceived and designed the experiments: CL HX SHL ZG. Performed the experiments: HX. Analyzed the data: CL HX SHL ZG. Contributed reagents/materials/analysis tools: CL HX SHL ZG. Wrote the paper: CL HX SHL ZG.

References

1. Harada M (1995) Minamata disease: Methylmercury poisoning in Japan caused by environmental pollution. Crit Rev Toxicol 25: 1–24.
2. van der Oost R, Beyer J, Vermeulen NPE (2003) Fish bioaccumulation and biomarkers in environmental risk assessment: a review. Environ Toxicol Phar 13: 57–149.
3. Flouriot G, Pakdel F, Ducouret B, Valotaire Y (1995) Influence of xenobiotics on rainbow trout liver estrogen receptor and vitellogenin gene expression. J Mol Endocrino 15: 143–151.
4. Wu JP, Chang LW, Yao HT, Chang H, Tsai HT, et al. (2009) Involvement of oxidative stress and activation of aryl hydrocarbon receptor in elevation of CYP1A1 expression and activity in lung cells and tissues by arsenic: an *in vitro* and *in vivo* study. Toxicol Sci 107: 385–393.
5. Amin RP, Hamadeh HK, Bushel PR, Bennett L, Afshari CA, et al. (2002) Genomic interrogation of mechanism(s) underlying cellular responses to toxicants. Toxicology 181–182: 555–563.
6. Bartosiewicz M, Penn S, Buckpitt A (2001) Applications of gene arrays in environmental toxicology: fingerprints of gene regulation associated with cadmium chloride, benzo(a)pyrene, and trichloroethylene. Environ Health Perspect 109: 71–74.
7. Hamadeh HK, Bushel PR, Jayadev S, Martin K, DiSorbo O, et al. (2002) Gene expression analysis reveals chemical-specific profiles. Toxicol Sci 67: 219–231.
8. Hook SE, Skillman AD, Small JA, Schultz IR (2006) Gene expression patterns in rainbow trout, *Oncorhynchus mykiss*, exposed to a suite of model toxicants. Aquat Toxicol 77: 372–385.
9. Lam SH, Mathavan S, Tong Y, Li H, Karuturi RKM, et al. (2008) Zebrafish whole-adult-organism chemogenomics for large-scale predictive and discovery chemical biology. PLoS Genetics 4: e1000121.
10. Garcia-Reyero N, Poynton HC, Kennedy AJ, Guan X, Escalon BL, et al. (2009) Biomarker discovery and transcriptomic responses in *Daphnia magna* exposed to munitions constituents. Environ Sci Technol 43: 4188–4193.
11. Osborn HL, Hook SE (2013) Using transcriptomic profiles in the diatom *Phaeodactylum tricornutum* to identify and prioritize stressors. Aquat Toxicol 138–139: 12–25.
12. Lam SH, Winata CL, Tong Y, Korzh S, Lim WS, et al. (2006) Transcriptome kinetics of arsenic–induced adaptive response in zebrafish liver. Physiol Genomics 27: 351–361.
13. Cuesta A, meseguer J, Esteban MÁ (2008) Effects of the organochlorines *p*,*p*?-DDE and lindane on gilthead seabream leucocyte immune parameters and gene expression. Fish Shellfish Immun. 25: 682–688.
14. Mattingly C, Toscano WA (2011) Posttranscriptional silencing of cytochrome P4501A1 (CYP1A1) during zebrafish (*Danio rerio*) development. Dev Dynam 222: 645–654.
15. Yamaguchi A, Ishibashi H, Kohra S, Arizono K, Tominaga N (2005) Short-term effects of endocrine-disrupting chemicals on the expression of estrogen-responsive genes in male medaka (*Oryzias latipes*). Aquat Toxicol 72: 239–249.
16. Xu H, Lam S.H, Shen Y, Gong Z (2013) Genome-wide identification of molecular pathways and biomarkers in response to arsenic exposure in zebrafish liver. PLoS ONE 8: e68737.
17. Di Giulio R.T, Washburn P.C, Wenning R.J, Winston G.W, Jewell C.S (1989) Biochemical responses in aquatic animals: A review of determinants of oxidative stress. Environ Toxicol Chem 8: 1103–1123.
18. Lackner R (1998) "Oxidative stress" in fish by environmental pollutants. Fish Ecotoxicol 86: 203–224.
19. Franco R, Sánchez-Olea R, Reyes-Reyes E.M, Panayiotidis M.I (2009) Environmental toxicity, oxidative stress and apoptosis: Ménage à trois. Mutat Res-Rev Mutat 674: 3–22.
20. Roberts R.A, Nebert D.W, Hickman J.A, Richburg J.H, Goldsworthy T.L (1997) Perturbation of the mitosis/apoptosis balance: A fundamental mechanism in toxicology. Toxicol Sci 38: 107–115.
21. Murata M, Midorkawa K, Koh M, Umezawa K, Kawanishi S (2004) Genistein and daidzein induce cell proliferation and their metabolites cause oxidative

DNA damage in relation to isoflavone-induced cancer of estrogen-sensitive organs. Biochemistry 43: 2569–2577.

22. Soto A.M, Sonnenschein C (2010) Environmental causes of cancer: endocrine disruptors as carcinogens. Nat Rev Endocrinol 6: 363–370.

23. Simic M.G (1991) DNA damage, environmental toxicants, and rate of aging. J Environ Sci Heal C 9: 113–153.

24. Hartwig A, Schwerdtle T (2002) Interactions by carcinogenic metal compounds with DNA repair processes: toxicological implications. Toxicol Lett 127: 47–54.

25. Caino M.C, Oliva J.L, Jiang H, Penning T.M, Kazanietz M.G (2007) Benzo[a]pyrene-7,8-dihydrodiol promotes checkpoint activation and G_2/M arrest in human bronchoalveolar carcinoma H358 cells. Mol Pharmacol 71: 744–750.

26. Pomati F, Castiglioni S, Zuccato E, Fanelli R, Vigetti D, et al. (2006) Effects of a complex mixture of therapeutic drugs at environmental levels on human embryonic cells. Environ Sci Technol 40: 2442–2447.

27. Gupta S.C, Sharma A, Mishra M, Mishra R.K, Chowdhuri D.K (2010) Heat shock proteins in toxicology: How close and how far? Life Sci 86: 377–384.

28. Lee S, Lee S, Park C, Choi J (2006) Expression of heat shock protein and hemoglobin genes in *Chironomus tentans* (Diptera, chironomidae) larvae exposed to various environmental pollutants: a potential biomarker of freshwater monitoring. Chemosphere 65: 1074–1081.

29. Handzel Z.T (2000) Effects of environmental pollutants on airways, allergic inflammation, and the immune response. Rev Environ Health 15: 325–336.

30. Khalaf H, Salste L, Karlsson P, Ivarsson P, Jass J, et al. (2009) In vitro analysis of inflammatory responses following environmental exposure to pharmaceuticals and inland waters. Sci Total Environ 407: 1452–1460.

31. Catchen J.M, Conery J.S, Postlethwait J.H (2009) Automated identification of conserved synteny after whole-genome duplication. Genome Res 19: 1497–1505.

32. Xu H, Li Z, Li M, Wang L, Hong Y (2009) Boule is present in fish and bisexually expressed in adult and embryonic germ cells of medaka. PLoS ONE 4: e6097.

33. Harley K.G, Marks AR, Chevrier J, Bradman A, Sjödin A, et al. (2010) PBDE concentrations in women's serum and fecundability. Environ Health Perspect 118: 699–704.

34. Meeker JD, Stapleton HM (2013) House dust concentrations of organophosphate flame retardants in relation to hormone levels and semen quality parameters. Environ Health Perspect 118: 318–323.

35. Nelson JW, Hatch EE, Webster TF (2009) Exposure to polyfluoroalkyl chemicals and cholesterol, body weight, and insulin resistance in the general U.S. population. Environ Health Perspect 118: 197–202.

36. Stapleton HM, Eagle S, Anthopolos R, Wolkin A, Miranda ML (2011) Associations between polybrominated diphenyl ether (PBDE) flame retardants, phenolic metabolites, and thyroid hormones during pregnancy. Environ Health Perspect 119: 1454–1459.

37. Vested A, Ramlar-Hansen CH, Olsen SF, Bonde JP, Kristensen SL, et al. (2013) Associations of in utero exposure to perfluorinated alkyl acids with human semen quality and reproductive hormones in adult men. Environ Health Perspect 121: 453–458.

38. Ariese F, Kok SJ, Verkaik M, Gooijer C, Velthorst NH, et al. (1993) Synchronous fluorescence spectrometry of fish bile: A rapid screening method for the biomonitoring of PAH exposure. Aquat Toxicol 26: 273–286.

39. Lucarelli F, Authier L, Bagni G, Marrazza G, Baussant T, et al. (2003) DNA biosensor investigations in fish bile for use as a biomonitoring tool. Anal Lett 36: 1887–1901.

40. Schnurstein A, Braunbeck T (2001) Tail moment *versus* tail length–Application of an *in vitro* version of the comet assay in biomonitoring for genotoxicity in native surface waters using primary hepatocytes and gill cells from zebrafish (*Denio rerio*). Ecotox Environ Sate 49: 187–196.

41. Thomas M, Florion A, Chrétien D, Terver D (1996) Real-time biomonitoring of water contamination by cyanide based on analysis of the continuous electric signal emitted by a tropical fish: *Apteronotus albifrons*. Water Res 30: 3083–3091.

42. Vindimian E, Namour P, Migeon B, Garric J (1991) In situ pollution induced cytochrome *P*450 activity of freshwater fish: barbell (*Barbus barbus*), chub (*Leuciscus cephalus*) and nase (*Chondrostoma nasus*). Aquat Toxicol 21: 255–266.

43. Zeng Z, Shan T, Tong Y, Lam SH, Gong Z (2006) Development of estrogen-responsive transgenic medaka for environmental monitoring of endocrine disrupters. Environ Sci Technol 39: 9001–9008.

44. Zheng W, Xu H, Lam SH, Luo H, Karuturi RK, et al. (2013) Transcriptomic analyses of sexual dimorphism of the zebrafish liver and the effect of sex hormones. PLoS One 8: e53562.

45. Ankley GT, Miller DH, Jensen KM, Villeneuve DL, Martinović D (2008) Relationship of plasma sex steroid concentrations in female fathead minnows to reproductive success and population status. Aquat Toxicol 88: 67–74.

14

RNA-Sequencing Analysis of TCDD-Induced Responses in Zebrafish Liver Reveals High Relatedness to *In Vivo* Mammalian Models and Conserved Biological Pathways

Zhi-Hua Li[1,2], Hongyan Xu[1], Weiling Zheng[1], Siew Hong Lam[1], Zhiyuan Gong[1]*

1 Department of Biological Sciences, National University of Singapore, Singapore, **2** Yangtze River Fisheries Research Institute, Chinese Academy of Fishery Sciences, Wuhan, China

Abstract

TCDD is one of the most persistent environmental toxicants in biological systems and its effect through aryl hydrocarbon receptor (AhR) has been well characterized. However, the information on TCDD-induced toxicity in other molecular pathways is rather limited. To fully understand molecular toxicity of TCDD in an in vivo animal model, adult zebrafish were exposed to TCDD at 10 nM for 96 h and the livers were sampled for RNA-sequencing based transcriptomic profiling. A total of 1,058 differently expressed genes were identified based on fold-change>2 and TPM (transcripts per million) >10. Among the top 20 up-regulated genes, 10 novel responsive genes were identified and verified by RT-qPCR analysis on independent samples. Transcriptomic analysis indicated several deregulated pathways associated with cell cycle, endocrine disruptors, signal transduction and immune systems. Comparative analyses of TCDD-induced transcriptomic changes between fish and mammalian models revealed that proteomic pathway is consistently up-regulated while calcium signaling pathway and several immune-related pathways are generally down-regulated. Finally, our study also suggested that zebrafish model showed greater similarity to *in vivo* mammalian models than *in vitro* models. Our study indicated that the zebrafish is a valuable in vivo model in toxicogenomic analyses for understanding molecular toxicity of environmental toxicants relevant to human health. The expression profiles associated with TCDD could be useful for monitoring environmental dioxin and dioxin-like contamination.

Editor: Zhanjiang Liu, Auburn University, United States of America

Funding: This work was supported by the Singapore National Research Foundation under its Environmental & Water Technologies Strategic Research Programme and administered by the Environment & Water Industry Programme Office (EWI) of the PUB (grant number R-154-000-328-272). The funders had no role in study design, data collection and analysis, decision to publish, or reparation of the manuscript.

* E-mail: dbsgzy@nus.edu.sg

Introduction

Dioxin-like compounds are major environmental contaminants that could pose serious threats to public health and the ecosystem [1]. Since 1980s, it has been increasingly documented that dioxin-like compounds cause various biological effects in laboratory animals and human [2,3]. Among them, 2,3,7,8-tetrachlorodibenzo-*p*-dioxin (TCDD) is the most potent toxicant and it is produced from both natural and anthropogenic processes including incineration of chlorine-containing substances, bleaching of paper, manufacturing of specific organochlorine chemicals, volcanoes, and forest fires [4]. As an aromatic hydrocarbon, TCDD has a long biological half-life and is heavily accumulated in the food chain, which causes adverse effects on human health at environmental levels [4,5]. Till now, many animal models, including fishes, have been used in TCDD studies [6,7]. However, most of these studies have focused on the physiological-biochemical parameters and typical molecular markers, such as the gene *cyp1a* and the Aryl hydrocarbon receptor (AhR) signaling pathway [7] and detailed molecular toxicity of TCDD remains to be elucidated.

Genomic approaches have been increasingly used in toxicological research in the past decade. Previous toxicogenomic studies mostly used DNA microarray technology to capture the global gene expression data and to evaluate the effects of toxicant exposure [8,9]. However, there are several drawbacks in microarray analysis, e.g. limited sensitivity, probe cross-hybridization, incomplete genome coverage and a prerequisite for sequence information in order to include new probes [10]. Recently, the advent of next-generation sequencing (NGS) technologies has significantly accelerated genomic research and provided a better alternative for transcriptomic analysis [11]. By high-throughput RNA sequencing, it is feasible to measure transcript abundance and transcriptomic profiles with a broad dynamic range, therefore providing a powerful tool to determine the potential adverse effects of environmental contaminants on public health [8].

Comparative studies across different taxonomic groups are important not only in understanding of organism diversity but also for inferring important biological responses due to evolutionary conservation [4]. In toxicology, animal models are widely used to infer human responses to chemical exposure. Particularly, it is valuable to use multiple animal models for comparative analyses in order to determine conserved adverse responses or molecular events. As lower vertebrates in evolution, fishes are particularly important models for such comparative studies [12,13]. Now the zebrafish (*Danio rerio*) has become an increasingly popular model in human disease studies because of its many advantageous properties in laboratory experiments, such as its small size and easy availability in a large number, short generation time for genetic manipulation, and cost effectiveness for high-throughput studies. More importantly, many molecular and developmental studies have shown that zebrafish and human share many common genes in conserved developmental pathways in organogenesis and related physiological processes as well as in carcinogenesis [14,15].

As there is so far no NGS based transcriptomic analysis for molecular response to TCDD treatment, in the present study we carried out RNA sequencing analyses of TCDD-treated in order to carry out a genome-wide identification of novel TCDD responsive genes and pathways, which should provide a comprehensive

understanding of the molecular mechanism of TCDD-induced toxicity in mammals and human. We further compared the zebrafish response with those from mammalian systems and thus identified common molecular pathways deregulated in both fish and mammals. We found that the zebrafish model is more similar to mammalian in vivo models than in vitro models, thus indicating a validity of the zebrafish as an emerging in vivo model in comparative toxicological research.

Materials and Methods

Zebrafish

Experimental procedures were carried out following the approved protocol by Institutional Animal Care and Use Committee of

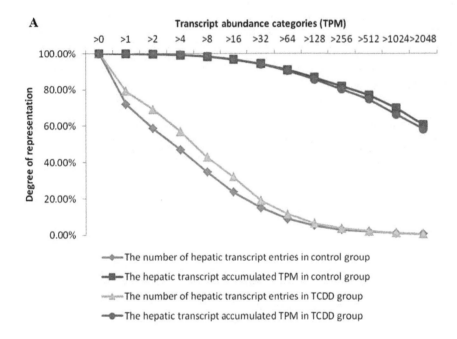

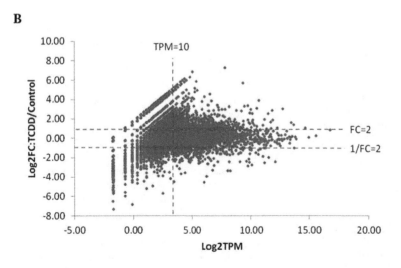

Figure 1. Comparison of transcriptomic profiles between TCDD and control groups. (A) Distribution of transcript entries and total transcript counts over different tag abundance categories in liver of zebrafish. The percentages of total transcript counts and number of different transcript entries per category are plotted on a log scale (base10). (B) Relationship between the hepatic transcriptome changing range and its expression level in zebrafish after TCDD treatment. The base of log value is 2.

National University of Singapore (Protocol 079/07). Adult zebrafish (3-month old), which are Singapore wild type zebrafish, were purchased from a local aquarium farm (Mainland Tropical Fish Farm, Singapore) and acclimated for at least one week in our aquarium before chemical exposure experiment. Fish were maintained at ambient temperature of around 28°C with a 14-h light and 10-h dark cycle in a flow-through water system.

Chemical exposure

TCDD was purchased from Sigma-Aldrich (USA) and dissolved in dimethyl sulfoxide (DMSO). Male adult fish were used in the exposure experiments by immersing in the TCDD water (10 nM) for 96 h at ambient temperature (28°C) in a static condition. Control fish were kept in water with 0.01% DMSO (vehicle) under the same condition. Water was changed daily throughout the treatment. After 96 hours of chemical exposure, treated and control fish were sacrificed and liver samples were dissected from each fish.

RNA sample preparation and SAGE library sequencing

Total RNA was extracted from livers (excluding gall bladders) of individual fish using TRIzol® Reagent (Invitrogen) and treated with DNase I (Invitrogen) to remove genomic DNA contamination. For RNA sequencing, RNA was pooled equally from 9 fish for each group (TCDD and control). Poly A+ RNA was purified using Dynabeads® Oligo (dT) EcoP (Invitrogen) and subjected to cDNA synthesis. Synthesized cDNA was digested by NlaIII and sequencing adapters were added to the cDNA fragments. SAGE (serial analysis of gene expression) sequencing (tag length = 27 nucleotides) was performed by Mission Biotech (Taiwan) with ABI SOLiD™ System 2.0 (Applied Biosystems). The RNA-sep data

reported in the present study was submitted to Gene Expression Omnibu with an access number GSE49915.

Gene annotation and selection of differentially expressed genes

All SAGE tags were mapped to the zebrafish Reference Sequence database (http://www.ncbi.nlm.nih.gov/RefSeq) with maximum 2 nucleotide mismatch. Uniquely mapped tag counts for each transcript were normalized to TPM (transcripts/tags per million). For biological implication analyses, genes with only marginal expression, as defined by TPM<10 in both control and TCDD groups, were excluded. As it has been previously reported that the actual measured quantity of differential expression (fold change or ratio) is more consistent and reproducible in identifying differentially expressed genes than the statistical significance (p-value) [16,17], in the present study, we selected differentially expressed genes based on fold change>2 and TPM>10.

Gene Ontology enrichment analysis

Gene ontology enrichment analysis was performed using DAVID (The Database for Annotation, Visualization and Integrated Discovery) with the total zebrafish genome information as the background and p-values based on a modified Fisher's exact t-test. Gene Ontology Fat categories were used for this analysis and the cut-off p-value is 0.05.

Real-time PCR

Real-time PCR was performed using the LightCycler system (Roche Applied Science) with LightCycler FastStart DNA Master SYBR Green I (Roche Applied Science) according to the manufacturer's instruction. For comparison between real-time

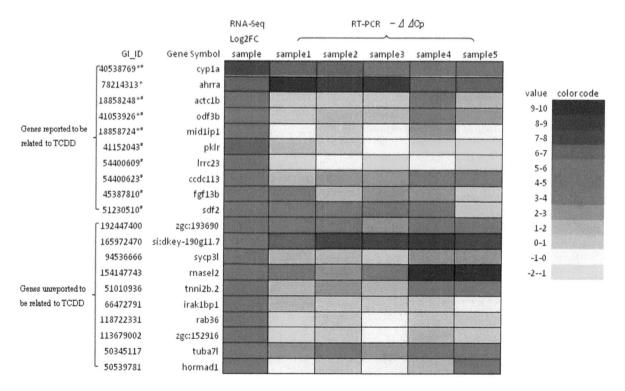

Figure 2. RT-qPCR validation of top 20 up-regulated TCDD-induced genes. RT-qPCR was performed from five individual liver samples collected from five fish treated by TCDD from a new experiment. The relative expression level of the genes was shown in color code as indicated on the fight and the value is in log2 fold change as compared with a housekeeping gene, *β-actin1*. *-only reported in fish model; #-only reported in mammalian model.

Table 1. Top up-regulated genes by TCDD in zebrafish liver.

GI_ID	Gene Symbol	Gene Name	TPM Value		Log2FC
			Control	TCDD	
40538769	cyp1a	cytochrome P450, family 1, subfamily A	1.49	225.60	7.24
41152043	pklr	pyruvate kinase, liver and RBC	0.37	28.99	6.28
78214313	ahrra	aryl-hydrocarbon receptor repressor a	0.30	20.48	6.09
118722331	rab36	RAB36, member RAS oncogene family	0.37	23.32	5.97
94536666	sycp3l	synaptonemal complex protein 3 like	0.30	18.27	5.93
113679002	zgc:152916	zgc:152916	0.37	20.80	5.80
154147743	rnasel2	ribonuclease like 2	12.65	641.82	5.67
192447400	zgc:193690	zgc:193690	0.30	13.55	5.50
54400609	lrrc23	leucine rich repeat containing 23	0.30	12.29	5.36
66472791	irak1bp1	interleukin-1 receptor-associated kinase 1 binding protein 1	0.30	11.97	5.32
51010936	tnni2b.2	troponin I, skeletal, fast 2b.2	0.30	11.34	5.24
54400623	ccdc113	coiled-coil domain containing 113	0.37	13.86	5.22
18858248	actc1b	actin, alpha, cardiac muscle 1b	1.12	39.07	5.13
41053926	odf3b	outer dense fiber of sperm tails 3B	0.37	11.66	4.97
165972470	si:dkey-190g11.7	si:dkey-190g11.7	0.74	23.00	4.95
51230510	sdf2	stromal cell-derived factor 2	1.49	45.06	4.92
45387810	fgf13b	fibroblast growth factor 13b	0.37	10.40	4.80
50539781	hormad1	HORMA domain containing 1	1.49	39.39	4.73
50345117	tuba7l	tubulin, alpha 7 like	0.74	19.53	4.71
18858724	mid1ip1	MID1 interacting protein 1	0.74	18.27	4.62

PCR and RNA-seq results, Cp (crossing point value) and TPM values for each gene were normalized against Cp and TPM of β-actin1 (GI_ID:18858334).

Analysis of the hepatic enriched gene list by GSEA and IPA analysis

GSEA (Gene Set Enrichment Analysis) pre-ranked option was used to analyze the entire set of differentially expressed genes (651 up- and 407 down-regulated genes). Briefly, the gene symbols of human homologs of the enriched zebrafish Unigene clusters were ranked using logarithm transformed fold change (base 2). The number of permutation used was 1000. Pathways with false discovery rate (FDR) <0.25 were considered statistically significant. For IPA (Ingenuity Pathway Analysis), the same set of differentially expressed genes was uploaded to online Ingenuity Pathways Knowledge Base for functional implication analyses.

Cross-species comparison

Six sets of transcriptomic data for *in vivo* and *in vitro* mammalian models treated by TCDD (GSE10082, GSE10083, GSE10769, GSE10770, GSE14555 and GSE34251) were retrieved from GEO (Gene Expression Omnibus). GSEA was used to establish the relatedness between zebrafish and mammalian models. The zebrafish hepatic transcriptome lists were converted into human and mouse homolog Unigene clusters. The statistical significance of the enrichment score was estimated by using an empirical phenotype-based permutation test. An FDR value was provided by introducing adjustment of multiple hypothesis testing.

Results and Discussion

General features of SAGE tags in TCDD-treated and control groups

There were 11.7 million and 17.9 million SAGE sequence tags generated from the DMSO vehicle control and TCDD-treated groups, respectively. About 2.7 million tags in the control group and 3.2 million tags in the TCDD group could be mapped with the known transcripts. The mapped tags were normalized to transcripts per million (TPM) and the expression level of genes ranged from 0.3 to 110844.2 TPM in the two groups, indicating a dynamic range of more than six orders of magnitude in transcript abundance. As shown in Figure 1A, transcript abundance profiles in both groups were very similar. The transcriptomes consist of a small number of high abundant transcripts and a large number of low abundant transcripts, similar to those reported in many previous RNA-seq studies [18,19]. The percentages of the transcripts with relatively high expression level (TPM>256) were 3.8% in the TCDD group and 3.1% in the control group, but contributed to 80.4% and 83.0% of the total transcripts respectively. In contrast, the percentages of the transcripts with the low expression level (TPM<16) were 67.1% and 76.3% in the TCDD group and control group, contributing only 3.1% and 3.2% of the total transcripts.

Differentially expressed genes in response to TCDD and their gene ontology profile

Figure 1B shows the relationship between the dynamic transcript fold-change of hepatic transcriptome and its expression level in zebrafish treated by TCDD. Differentially expressed genes

Table 2. Enriched GO terms in response to TCDD treatment in zebrafish liver (p<0.01).

Category	Term	Count	%	P-Value	Fold Enrichment
Biological	protein folding	19	2.98	1.05E-06	3.94
process	cellular macromolecule catabolic process	24	3.77	4.61E-05	2.59
	macromolecule catabolic process	26	4.08	5.61E-05	2.44
	proteolysis involved in cellular protein catabolic process	22	3.45	7.17E-05	2.65
	cellular protein catabolic process	22	3.45	7.17E-05	2.65
	modification-dependent protein catabolic process	20	3.14	1.65E-04	2.66
	modification-dependent macromolecule catabolic process	20	3.14	1.65E-04	2.66
	protein catabolic process	22	3.45	2.53E-04	2.43
	cell cycle	19	2.98	4.11E-04	2.56
	protein maturation by peptide bond cleavage	4	0.63	8.39E-04	17.76
	proteolysis	41	6.44	1.40E-03	1.66
	protein transport	26	4.08	1.59E-03	1.96
	establishment of protein localization	26	4.08	1.59E-03	1.96
	protein localization	27	4.24	1.80E-03	1.91
	response to bacterium	8	1.26	2.08E-03	4.33
	intracellular transport	19	2.98	2.14E-03	2.22
	cellular protein localization	15	2.35	4.82E-03	2.33
	cellular macromolecule localization	15	2.35	5.13E-03	2.31
Cellular	endoplasmic reticulum	42	6.59	1.19E-13	3.67
component	proteasome complex	16	2.51	2.67E-09	6.93
	cytosol	21	3.30	6.39E-07	3.65
	endoplasmic reticulum part	14	2.20	6.83E-06	4.57
	endoplasmic reticulum membrane	11	1.73	1.89E-04	4.25
	nuclear envelope-endoplasmic reticulum network	11	1.73	3.44E-04	3.96
	proteasome core complex	6	0.94	4.41E-03	5.31
Molecular	unfolded protein binding	12	1.88	2.85E-05	4.83
function	oligosaccharyl transferase activity	4	0.63	5.94E-04	19.98
	acid-amino acid ligase activity	12	1.88	1.74E-03	3.06
	threonine-type peptidase activity	6	0.94	2.23E-03	6.24
	threonine-type endopeptidase activity	6	0.94	2.23E-03	6.24
	dolichyl-diphosphooligosaccharide-protein lycotransferase activity	3	0.47	4.64E-03	24.98
	ligase activity, forming carbon-nitrogen bonds	13	2.04	5.65E-03	2.50
	translation initiation factor activity	8	1.26	5.92E-03	3.63
	heme binding	12	1.88	8.31E-03	2.50
	RNA binding	20	3.14	9.61E-03	1.89

after TCDD treatment were first determined by comparison of the two sets of mapped SAGE tags with fold-change>2 and TPM >10. In total, 1,058 genes were identified, including 651 up-regulated genes and 407 down-regulated genes. The top 20 most up-regulated transcripts based on fold-change are listed in the Table 1. After compared with TCDD responsive genes in CTD (The Comparative Toxicogenomics Database, http://ctdbase. org/), ten gens (*cyp1a, ahrra, actc1b, odf3b, mid1ip1, pklr, lrrc23, ccdc113, sdf2,* and *fgf13*) have been reported to be related in fish and/or mammalian models after TCDD exposure among the top 20 genes. Not surprisingly, *cyp1a*, the best known molecular marker for TCDD exposure [20], was the most up-regulated gene (151.4 fold) in the list, indicating that our TCDD treatment in the

experiment was effective. It is interesting to note that *adrra* (aryl hydrocarbon receptor repressor a) is the second highest up-regulated gene, while the two aryl hydrocarbon receptor genes (*ahr1a* and *ahr2*) were not significantly up-regulated in our RNA-seq data with a modest 50% increase for *adr2* and slightly decrease for *adr1a* (data not shown), 10 novel TCDD-responsive genes, eight annotated (*rab36, sycp3l, rasel2, tnni2b.2, hormad1, tuba7l, irak1bp1* and zgc:152916,) and two unannotated (zgc:193690 and si:dkey-190g11.7), have been identified from the top 20 most up-regulated genes. To confirm their inducibility of the 20 genes by TCDD, real-time PCR was carried out with individual fish liver samples from an independent set of TCDD treatment experiment. As shown in Figure 2, 17 out of 20 genes showed up-regulation in

Table 3. Significantly regulated pathways in zebrafish after TCDD treatment, by GSEA analysis with cutoff FDR<0.25.

Pathways	NES	NOM p-value	FDR q-value
G1 to S cell cycle reactome	1.94	<0.001	0.024
HSA04110 cell cycle	1.9	<0.001	0.025
Cell cycle KEGG	1.86	0.003	0.031
Proteasome pathway	1.81	0.007	0.053
HSA03050 Proteasome	1.74	0.007	0.115
HSA03010 Ribosome	−2.31	<0.001	<0.001
Ribosomal proteins	−2.21	<0.001	0.001
HSA04020 Calcium signaling pathway	−1.85	0.005	0.126
Prostaglandin synthesis regulation	−1.81	0.005	0.167
ST FAS signaling pathway	−1.81	0.007	0.139
ST T cell signal transduction	−1.79	<0.001	0.135
HSA04650 Natural killer cell mediated cytotoxicity	−1.76	0.007	0.160
HSA04660 T cell receptor signaling pathway	−1.75	0.008	0.152

Note: NES, normalized enrichment scores; NOM p-value, nominal p-value for enrichment; FDR, false discovery rate.

at least three individual samples; among them, 13 genes had up-regulation in all of the five individual samples. Thus, our RT-qPCR data indicated a strong agreement with the RNA-sequencing data and, more importantly, the validation was from five independent biological samples. Interestingly, all of the 10 novel TCDD responsive genes found in the present study were validated by RT-qPCR (Figure 2). Thus, our RNA-seq data provided additional biomarker genes for TCDD exposure. Among these validated biomarker genes, several of them apparently encodes secreted protein (e.g. *fgf13b* and *masel2*) and their protein products are also likely present in the circulating blood and may offer convenient non-invasive assays for detection of TCDD exposure. A notable exception in the validation experimental data was *lrrc23* that was constantly down-regulated in all five individuals, indicating that *lrrc23* is not a reproducibly up-regulated gene by TCDD. In future, these novel molecular markers for TCDD exposure can be further tested for their time- and dose-responsiveness.

Besides *cyp1a*, some other genes with high abundance were also up-regulated by TCDD exposure in zebrafish (Table S1). The up-regulation of *pklr* (pyruvate kinase, liver and red blood cell) in our experiment was consistent with several previous reports that TCDD affects pyruvate utilization for energy and thus glycolysis and gluconeogenesis [21,22]. Gene expression of *alas1*, the first and rate-limiting enzyme involved in heme biosynthesis [23], was up-regulated in our study, suggesting that the heme biosynthesis was deregulated in hepatic cells of fish exposed to TCDD. Hspa5 is a member of the heat shock protein 70 (Hsp70) family and it has been used as an endoplasmic reticulum (ER) stress sensor [24]. In our study, the expression of *hspa5* was up-regulated with 10.2 fold, implying TCDD led to the ER stress and this event was a highly conserved adaptive response induced by environmental toxicants. Moreover, the gene *pck1* (phosphoenolpyruvate carboxykinase 1), as a key factor in gluconeogenesis, was up-regulated 6.8 fold, indicating that TCDD exposure also affected the gluconeogenesis by altering the expression of genes encoding key gluconeogenic enzymes [25].

Through gene ontology analysis, the differentially expressed genes were classified into different categories based on GO database using DAVID software (Table 2). Among the up-regulated GO

categories, the most significant ones included Endoplasmic reticulum (p-value = 1.19E-13) and Proteasome complex (p-value = 2.67E-09). Since most P450s (e.g. CYP1A enzymes) are involved in xenobiotic metabolism and primarily located in endoplasmic reticulum, it is not surprising that the TCDD-induced physiological stress occurred mainly in these cellular components, which was also supported by the up-regulated 19 genes of CYP family in our study (Table S2). Moreover, other histological studies also found TCDD led to hypertrophy of hepatocytes, glycogen depletion and lipidosis of liver in zebrafish and other fish [26], which were related to the endoplasmic reticulum (ER) stress via some cellular receptor or/and protein [27,28]. In subsequent analysis, our results also showed that the proteasome related pathways (Proteasome pathway and HSA03050 Proteasome, see Table 3) were significantly up-regulated in zebrafish after TCDD treatment. Consistent with this, several GO categories involved in Proteasome deregulation, such as Protein folding and Cellular protein catabolic process in the Biological Process category, as well as Unfolded protein binding in the Molecular Function category, further indicating that the proteasome related bio-functions were deregulated.

Change of transcription factor networks by TCDD exposure

By IPA analysis, six significant transcription factors were found to be related to TCDD-induced hepatotoxicity (Table 4), including Xbp1, Nfe2l2, Nr5a2, Ptf1a, Tp53 and Mycn. In particular, the two up-regulated Xbp1 and Nfe2l2 networks were highly enriched with P-values of 1.47E-16 and 4.73E-11 respectively, In the Xbp1 network, 38 target genes were found from the differentially expressed gene list and almost all of them (36) were up-regulated (Figure 3A), indicating that Xbp1 plays a central role in regulating a battery of genes responsible for protein trafficking and secretion [29]. In the Nfe2l2 network, 36 up- and 14 down-regulated genes were induced in the differentially expressed gene list (Figure 3B). This is in consistence with a previous study [30] that most of the regulated genes (such as *mgst1*, *usp14*, *herpud1*, *dnajc3*, *actg1*, *hmox1*, *dnajb11*, etc.) by Nfe2l2 are related to oxidative stress and mediate transcriptional events that facilitate protective responses in animal models exposed to xenobiotic.

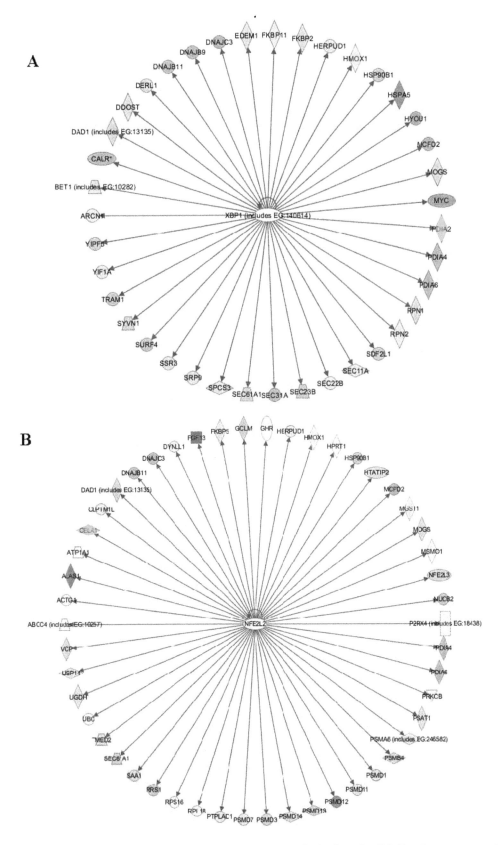

Figure 3. Two most significant transcription factor networks in the zebrafish liver in response to TCDD exposure. (A) Xbp1 network. (B) Nfe2l2 network, Target genes in the differentially expressed gene list (Table S1) are shown with red color indicating up-regulation and green color down-regulation. The intensity of the color corresponds to the relatively levels of up- and down-regulation.

Table 4. Most significant transcription factors in zebrafish liver after TCDD treatment.

GI_ID	Gene Symbol	Gene Name	Regulation Z-Score	P-value of Overlap
18859572	xbp1	X-box binding protein 1	4.07	1.47E-16
33504556	nfe2l2	nuclear factor (erythroid-derived 2)-like 2	2.45	4.73E-11
24158438	nr5a2	nuclear receptor subfamily 5, group A, member 2	−2.59	1.05E-03
124249093	ptf1a	pancreas specific transcription factor, 1a	−2.79	7.64E-06
18859502	tp53	tumor protein p53	−3.035	2.73E-02
47271377	mycn	v-myc myelocytomatosis viral related oncogene, neuroblastoma derived (avian)	−3.55	4.47E-04

Deregulated pathways in zebrafish livers treated by TCDD

To investigate the change of molecular pathways induced by TCDD treatment, GSEA was performed using the set of differentially expressed genes (651 up and 407 down) and 163 up- and 123 down-regulated pathways were identified (Table S3). Among all pathways, five up- and eight down-regulated pathways had significant FDR values less than 0.25 and are shown in Table 4. The first three pathways are associated with cell cycle progression, which have also been reported in mammalian models under TCDD stress [31,32]. However in the present study, cell cycle and related pathways were significantly enhanced while some mammalian studies showed a decrease of these activities, suggesting different regulatory mechanism in various models as well as by different treatment regimes.

The mechanisms responsible for TCDD toxicity are always associated with its ability to disrupt endocrine functions. In our data, ubiquitin-proteasome pathway was significantly up-regulated in fish liver after TCDD treatment. Consistent with another study in the mammalian system [33], the up-regulated proteasome and related pathways in our study further indicated that the TCDD-induced ubiquitin-proteasomal degradation of AhR influenced the nucleus transcription by controlling the level of ligand-activated AhR. As ubiquitin-proteasome pathway is significantly induced in zebrafish and all mammalian models analyzed in our study, it is an apparently conserved mechanism of TCDD-induced toxicity between fish and mammals. Prostaglandin synthesis regulation pathway was suppressed in TCDD-treated zebrafish, consistent with a previous report that prostanoid synthesis pathway could be regulated by COX2-TBXS-TP (cyclooxygenase2-thromboxane A synthase1-thromboxane receptor) in zebrafish after TCDD treatment [34].

In the present study, several pathways related to signal transduction were significantly inhibited in zebrafish after TCDD treatment, including Ca^{2+} regulation pathways, Fas signaling pathway, T-cell signal transduction and T-cell receptor signaling pathway, ribosomal protein and its related pathways. TCDD has been reported to significantly increase intracellular free Ca^{2+} in many cell culture systems [35] and to trigger Ca^{2+}-mediated endonuclease activity leading to apoptosis [36]. Our study, consistent with several previous reports [37–40], indicated that Ca^{2+} regulation pathways could be commonly involved in TCDD action. Furthermore, Fas signaling pathway, T-cell signal transduction and T-cell receptor signaling pathway were also suppressed in our study. Fas, a member of the tumor necrosis factor receptor (TNFR) family, contains a death domain which is essential for the delivery of the death signal [41]. Thus, Fas signaling regulation could be a mechanism of impaired T-cell

related pathway induced by TCDD, which are consistent with another previous report [41]. Moreover, ribosomal proteins, in conjunction with rRNA, are involved in the cellular process of translation [42]. Interestingly, the ribosomal protein and its related pathway were down-regulated in zebrafish liver after TCDD exposure, but were induced in other mammalian models [43].

Furthermore, three immune-related pathways were significantly repressed in the zebrafish liver after TCDD exposure (Table 4), including natural killer cell mediated cytotoxicity, T-cell signal transduction and T-cell receptor signaling pathway. Moreover, several B-cell related pathways were also found to be inhibited with FDR>0.25 (Table S3). Collectively, our data indicate that hepatic immune-related functions were impaired in fish exposed to TCDD.

Toxicogenomic comparison between zebrafish and mammalian models

In order to gain insight into the common molecular toxicity of TCDD between fish and mammals as well as the validity of the zebrafish model to predict chemical toxicity for risk assessment for human health, our transcriptomic data were compared with available transcriptomic data from both in vivo and in vitro mammalian studies with TCDD. Among 26 related series in the GEO database using human cell lines, rat and mouse tissues and cells based on microarray studies, we found six series (GSE10082, GSE10083, GSE10769, GSE10770, GSE14555 and GSE34251) with effective comparability (Table 5) while other series were not included in the comparative analysis due to incompatibility in data uploading, platform, experimental strategy, etc. The list of 650 up-regulated genes from the current zebrafish study was used to represent the zebrafish transcriptome in GSEA. As shown in Figure 4A, based on normalized enrichment scores (NES) and false discovery rate (FDR), the zebrafish hepatic genes expression showed more resemblance to the in vivo models (average NES is 1.54) than in vitro models (average NES is 0.93). The comparison with all of the four in vivo data, FDR was smaller than 0.25, while none of the in vitro data showed such significance. These observations indicate that the zebrafish data is more similar to the in vivo mammalian data than in vitro data, further enforcing the validity of the zebrafish model as a potentially high-throughput and economic in vivo experimental models for studies relevant to human health.

To further analyze the correlation of zebrafish and other mammalian models after TCDD treatment, comparison of pathways were made by GSEA pre-ranked function. One up-regulated (Proteasome pathway) and three down-regulated pathways (HSA04650 Natural killer cell mediated cytotoxicity, HSA04660 T cell receptor signaling pathway, HSA04020

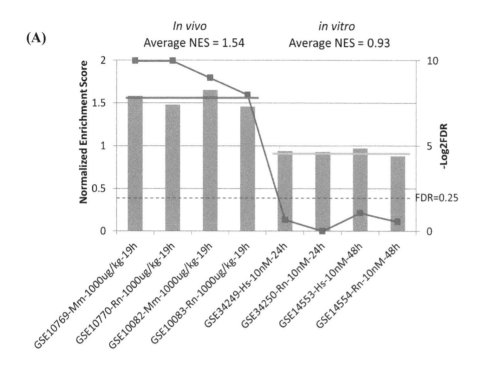

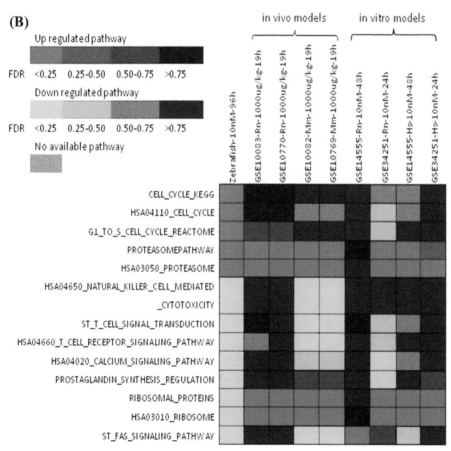

Figure 4. Comparative analyses of zebrafish and mammalian transcriptomic data from TCDD treatments. (A) Correlation of hepatic transcriptome changes in the zebrafish with the mammalian in vivo and in vitro models by GSEA analysis. –Log2FDR = 2, that means FDR = 0.25. (B) Comparison of the pathways in zebrafish and the mammalian models by GSEA analysis. The heat map includes the significant pathways in zebrafish and mammalian models treated by TCDD, the criteria of zebrafish pathways is FDR<0.25.

Table 5. Detail information of GSE series from GEO database used in this study.

GSE series no.	Animal models	Treated concentrations	Treated time	Strategy
GSE10083	Rat, *Rattus norvegicus* (liver)	1000ug/kg	19h	in vivo
GSE10770	Rat, *Rattus norvegicus* (liver)	1000ug/kg	19h	in vivo
GSE10082	Mouse, *Mus musculus* (liver)	1000ug/kg	19h	in vivo
GSE10769	Mouse, *Mus musculus* (liver)	1000ug/kg	19h	in vivo
GSE14555	Rat, *Rattus norvegicus* (Primary Rat Hepatocytes)	10nM	48h	in vitro
GSE34251	Rat, *Rattus norvegicus* (Primary Rat Hepatocytes)	10nM	24h	in vitro
GSE14555	Human, *Homo sapiens* (Primary Human Hepatocytes)	10nM	48h	in vitro
GSE34251	Human, *Homo sapiens* (Primary Human Hepatocytes)	10nM	24h	in vitro

Calcium signaling pathway) in zebrafish showed the same regulation direction as those in all other mammalian models, indicating that the four pathways were well conserved in vertebrates. Meanwhile, HSA03010 Ribosome pathway displayed completely opposite regulated trends between zebrafish and all mammalian models we compared (Figure 4B), indicating the species-specific patterns between fish and mammals after TCDD treatment, which need further analysis in future. Moreover, after comparing the conserved pathways (the same regulation direction pathways) among zebrafish model, *in vivo* and *in vitro* mammalian models, we noted that the degree of regulation is more consistent between the zebrafish and in vivo mammalian models than between the zebrafish and in vitro mammalian models, as indicated by the color codes in Figure 4B.

In summary, by RNA-sequencing based transcriptomic analyses, we have carried out detailed analyses of TCDD-induced molecular changes in zebrafish. Other than the well characterized AhR pathway and *cyp1a1* biomarker gene induced by TCDD, we also found some new biomarker genes that have been validated from independent experiments. Interestingly, some of the new biomarker genes encode secreted proteins such as Fgf13b, Sdf2 and RNasel2, which may by analyzed from serum samples by a non-invasive approach. Further and comparative transcriptomic analyses of our zebrafish data and available mammalian data from TCDD treatment experiments indicate several well conserved TCDD-responsive pathways, including up-regulated proteomic pathway and several down regulated pathways such as calcium

signaling pathway, natural killer cell mediated cytotoxicity, T-cell receptor signaling pathway etc. Furthermore, GSEA analyses indicate that the TCDD-induced zebrafish transcriptomic data is more similar to in vivo mammalian data than in vitro data, thus indicating the validity of the zebrafish model as a valuable in vivo model to infer molecular toxicity relevant to human health.

Supporting Information

Table S1 Complete list of differentially expressed genes in zebrafish liver exposed to TCDD with a cut off of fold change>2 and TPM >10.

Table S2 Up-regulated CYP family genes in zebrafish treated by TCDD.

Table S3 Complete list of pathways in zebrafish exposed to TCDD as revealed by GSEA analysis.

Author Contributions

Conceived and designed the experiments: HX SHL ZG. Performed the experiments: HX. Analyzed the data: ZHL WZ ZG. Contributed reagents/materials/analysis tools: ZHL HX WZ SHL. Wrote the paper: ZHL ZG.

References

1. Ovando BJ, Ellison CA, Vezina CM, Olson JR (2010) Toxicogenomic analysis of exposure to TCDD, PCB126 and PCB153: identification of genomic biomarkers of exposure to AhR ligands. BMC Genomics 11: 583.
2. Mandal PK (2005) Dioxin: a review of its environmental effects and its aryl hydrocarbon receptor biology. J Comp Physiol B 175: 221–230.
3. Consonni D, Sindaco R, Bertazzi PA (2012) Blood levels of dioxins, furans, dioxin-like PCBs, and TEQs in general populations: A review, 1989–2010. Environ Int 44: 151–162.
4. Black MB, Budinsky RA, Dombkowski A, Cukovic D, LeCluyse EL, et al. (2012) Cross-species comparisons of transcriptomic alterations in human and rat primary hepatocytes exposed to 2,3,7,8-tetrachlorodibenzo-*p*-dioxin. Toxicol Sci 127: 199–215.
5. Ryu YM, Kim KN, Kim YR, Sohn SH, Seo SH, et al. (2005) Gene expression profiles related with TCDD-induced hepatotoxicity. Mol Cell Toxicol 1: 164–171.
6. Cao ZJ, Tanguay RL, McKenzie D, Peterson RE, Aiken JM (2003) Identification of a putative calcium-binding protein as a dioxin-responsive gene in zebrafish and rainbow trout. Aquat Toxicol 63: 271–282.
7. Volz DC, Bencic DC, Hinton DE, Law JM, Kullman SW (2005) 2,3,7,8-Tetrachlorodibenzo-p-dioxin (TCDD) induces organ-specific differential gene expression in male Japanese medaka (*Oryzias latipes*). Toxicol Sci 85: 572–584.
8. Su ZQ, Li ZG, Chen T, Li QZ, Fang H, et al. (2011) Comparing next-generation sequencing and microarray technologies in a toxicological study of the effects of aristolochic acid on rat kidneys. Chem Res Toxicol 24: 1486–1493.
9. Davey JW, Hohenlohe PA, Etter PD, Boone JQ, Catchen JM, et al. (2011) Genome-wide genetic marker discovery and genotyping using next-generation sequencing. Nat Rev Genet 12: 499–510.
10. Marioni JC, Mason CE, Mane SM, Stephens M, Gilad Y (2008) RNA-seq: An assessment of technical reproducibility and comparison with gene expression arrays. Genome Res 18: 1509–1517.
11. Metzker ML (2010) Sequencing technologies – the next generation. Nat Rev Genet 11: 31–46.
12. Kasahara M, Naruse K, Sasaki S, Nakatani Y, Qu W, et al. (2007) The medaka draft genome and insights into vertebrate genome evolution. Nature 447: 714–719.
13. Venkatesh B (2003) Evolution and diversity of fish genomes. Curr Opin Genet Dev 13: 588–592.

14. Lam SH, Wu YL, Vega VB, Miller LD, Spitsbergen J, et al. (2006) Conservation of gene expression signatures between zebrafish and human liver tumors and tumor progression. Nat Biotechnol 24: 73–75.

15. Pichler FB, Laurenson S, Williams LC, Dodd A, Copp BR, et al. (2003) Chemical discovery and global gene expression analysis in zebrafish. Nat Biotechnol 21: 879–883.

16. Guo L, Lobenhofer EK, Wang C, Shippy R, Harris SC, et al. (2006) Rat toxicogenomic study reveals analytical consistency across microarray platforms. Nat Biotechnol 24: 1162–1169.

17. Shi LM, Reid LH, Jones WD, Shippy R, Warrington JA, et al. (2006) The microarray quality control (MAQC) project shows inter- and intraplatform reproducibility of gene expression measurements. Nat Biotechnol 24: 1151–1161.

18. Hegedus Z, Zakrzewska A, Agoston VC, Ordas A, Racz P, et al. (2009) Deep sequencing of the zebrafish transcriptome response to mycobacterium infection. Molecular Immunology 46: 2918–2930.

19. Zheng W, Wang Z, Collins JE, Andrews RM, Stemple D, et al. (2011) Comparative transcriptome analyses indicate molecular homology of zebrafish swimbladder and mammalian lung. PLoS One 6: e24019.

20. Kim S, Dere E, Burgoon LD, Chang CC, Zacharewski TR (2009) Comparative analysis of AhR-mediated TCDD-elicited gene expression in human liver adult stem cells. Toxicol Sci 112: 229–244.

21. Ruiz-Aracama A, Peijnenburg A, Kleinjans J, Jennen D, van Delft J, et al. (2011) An untargeted multi-technique metabolomics approach to studying intracellular metabolites of HepG2 cells exposed to 2,3,7,8-tetrachlorodibenzo-p-dioxin. BMC Genomics 12: 251.

22. Forgacs AL, Kent MN, Makley MK, Mets B, DelRaso N, et al. (2012) Comparative metabolomic and genomic analyses of TCDD-elicited metabolic disruption in mouse and rat liver. Toxicol Sci 125: 41–55.

23. May BK, Dogra SC, Sadlon TJ, Bhasker CR, Cox TC, et al. (1995) Molecular regulation of heme-biosynthesis in higher vertebrates. Prog Nucl Res Molec Biol 51: 1–51.

24. Falahatpisheh H, Nanez A, Montoya-Durango D, Qian YC, Tiffany-Castiglioni E, et al. (2007) Activation profiles of HSPA5 during the glomerular mesangial cell stress response to chemical injury. Cell Stress Chaperones 12: 209–218.

25. Boverhof DR, Burgoon LD, Tashiro C, Chittim B, Harkema JR, et al. (2005) Temporal and dose-dependent hepatic gene expression patterns in mice provide new insights into TCDD-mediated hepatotoxicity. Toxicol Sci 85: 1048–1063.

26. Zodrow JM, Stegeman JJ, Tanguay RL (2004) Histological analysis of acute toxicity of 2,3,7,8-tetrachlorodibenzo-p-dioxin (TCDD) in zebrafish. Aquat Toxicol 66: 25–38.

27. Jo H, Choe SS, Shin KC, Jang H, Lee JH, et al. (2012) ER stress induces hepatic steatosis via increased expression of the hepatic VLDL receptor. Hepatology DOI: 10.1002/hep.26126.

28. Gentile CL, Frye M, Pagliassotti MJ (2011) Endoplasmic reticulum stress and the unfolded protein response in nonalcoholic fatty liver disease. Antioxid Redox Signal 15: 505–521.

29. Shaffer AL, Shapiro-Shelef M, Iwakoshi NN, Lee AH, Qian SB, et al. (2004) XBP1, downstream of Blimp-1, expands the secretory apparatus and other organelles, and increases protein synthesis in plasma cell differentiation. Immunity 21: 81–93.

30. Xu SH, Weerachayaphorn J, Cai SY, Soroka CJ, Boyer JL (2010) Aryl hydrocarbon receptor and NF-E2-related factor 2 are key regulators of human MRP4 expression. Am J Physiol-Gastroint Liver Physiol 299: G126–G135.

31. Dere E, Boverhof DR, D Burgoon L, Zacharewski TR (2006) In vivo – in vitro toxicogenomic comparison of TCDD-elicited gene expression in HepaIcIc7 mouse hepatoma cells and C57BL/6 hepatic tissue. BMC Genomics 7.

32. Marlowe JL, Knudsen ES, Schwemberger S, Puga A (2004) The aryl hydrocarbon receptor displaces p300 from E2F-dependent promoters and represses S phase-specific gene expression. J Biol Chem 279: 29013–29022.

33. Ma Q, Baldwin KT (2000) 2,3,7,8-tetrachlorodibenzo-p-dioxin-induced degradation of aryl hydrocarbon receptor (AhR) by the ubiquitin-proteasome pathway – Role of the transcription activaton and DNA binding of AhR. J Biol Chem 275: 8432–8438.

34. Teraoka H, Kubota A, Kawai Y, Hiraga T (2008) Prostanoid signaling mediates circulation failure caused by TCDD in developing zebrafish. Biological Responses to Chemical Pollutants, in: Murakami, H, Nakayama, K, Kitamura, SI, Iwata, H and Tanabe, S (Eds), Interdisciplinary Studies on Environmental Chemistry Terra Pub, Tokyo, 61–80.

35. Puga A, Hoffer A, Zhou SY, Bohm JM, Leikauf GD, et al. (1997) Sustained increase in intracellular free calcium and activation of cyclooxygenase-2 expression in mouse hepatoma cells treated with dioxin. Biochem Pharmacol 54: 1287–1296.

36. Mcconkey DJ, Hartzell P, Duddy SK, Hakansson H, Orrenius S (1988) 2,3,7,8-Tetrachlorodibenzo-p-dioxin kills immature thymocytes by Ca^{2+}-mediated endonuclease activation. Science 242: 256–259.

37. Cantrell SM, Lutz LH, Tillitt DE, Hannink M (1996) Embryotoxicity of 2,3,7,8-tetrachlorodibenzo-p-dioxin (TCDD): The embryonic vasculature is a physiological target for TCDD-induced DNA damage and apoptotic cell death in medaka (Orizias latipes). Toxicol Appl Pharmacol 141: 23–34.

38. Toomey BH, Bello S, Hahn ME, Cantrell S, Wright P, et al. (2001) 2,3,7,8-Tetrachlorodibenzo-p-dioxin induces apoptotic cell death and cytochrome P4501A expression in developing Fundulus heteroclitus embryos. Aquat Toxicol 53: 127–138.

39. Dong W, Teraoka H, Yamazaki K, Tsukiyama S, Imani S, et al. (2002) 2,3,7,8-Tetrachlorodibenzo-p-dioxin toxicity in the zebrafish embryo: Local circulation failure in the dorsal midbrain is associated with increased apoptosis. Toxicol Sci 69: 191–201.

40. Chen YH, Tukey RH (1996) Protein kinase C modulates regulation of the CYP1A1 gene by the aryl hydrocarbon receptor. J Biol Chem 271: 26261–26266.

41. Dearstyne EA, Kerkvliet NI (2002) Mechanism of 2,3,7,8-tetrachlorodibenzo-p-dioxin (TCDD)-induced decrease in anti-CD3-activated CD4(+) T cells: the roles of apoptosis, Fas, and TNF. Toxicology 170: 139–151.

42. Komili S, Farny NG, Roth FP, Silver PA (2007) Functional specificity among ribosomal proteins regulates gene expression. Cell 131: 557–571.

43. Handley-Goldstone HM, Grow MW, Stegeman JJ (2005) Cardiovascular gene expression profiles of dioxin exposure in zebrafish embryos. Toxicol Sci 85: 683–693.

15

Phospholipid-Derived Fatty Acids and Quinones as Markers for Bacterial Biomass and Community Structure in Marine Sediments

Tadao Kunihiro[1,2]*, Bart Veuger[1], Diana Vasquez-Cardenas[1,2], Lara Pozzato[1], Marie Le Guitton[1], Kazuyoshi Moriya[3], Michinobu Kuwae[4], Koji Omori[4], Henricus T. S. Boschker[2], Dick van Oevelen[1]

1 Department of Ecosystem Studies, Royal Netherlands Institute of Sea Research (NIOZ), Yerseke, The Netherlands, 2 Department of Marine Microbiology, Royal Netherlands Institute of Sea Research (NIOZ), Yerseke, The Netherlands, 3 School of Natural Science & Technology, Kanazawa University, Kakuma-machi, Kanazawa, Japan, 4 Center for Marine Environmental Studies (CMES), Ehime University, Matsuyama, Ehime, Japan

Abstract

Phospholipid-derived fatty acids (PLFA) and respiratory quinones (RQ) are microbial compounds that have been utilized as biomarkers to quantify bacterial biomass and to characterize microbial community structure in sediments, waters, and soils. While PLFAs have been widely used as quantitative bacterial biomarkers in marine sediments, applications of quinone analysis in marine sediments are very limited. In this study, we investigated the relation between both groups of bacterial biomarkers in a broad range of marine sediments from the intertidal zone to the deep sea. We found a good log-log correlation between concentrations of bacterial PLFA and RQ over several orders of magnitude. This relationship is probably due to metabolic variation in quinone concentrations in bacterial cells in different environments, whereas PLFA concentrations are relatively stable under different conditions. We also found a good agreement in the community structure classifications based on the bacterial PLFAs and RQs. These results strengthen the application of both compounds as quantitative bacterial biomarkers. Moreover, the bacterial PLFA- and RQ profiles revealed a comparable dissimilarity pattern of the sampled sediments, but with a higher level of dissimilarity for the RQs. This means that the quinone method has a higher resolution for resolving differences in bacterial community composition. Combining PLFA and quinone analysis as a complementary method is a good strategy to yield higher resolving power in bacterial community structure.

Editor: David W. Pond, Scottish Association for Marine Science, United Kingdom

Funding: This work was partly supported by Grant-in-Aid for Young Scientists (B: 21710081 and 23710010) of MEXT of Japan, a Sasakawa Scientific Research Grant (21-748M) from the Japan Science Society, the Global COE Program of the Ministry of Education, Culture, Sports, Science and Technology (MEXT) of Japan, the joint research program of the EcoTopia Science Institute, Nagoya University (Project No.2-5), and the oversea research program from the Institute for Fermentation, Osaka (IFO) to T.K. The funders had no role in study design, data collection and analysis, decision to publish, or preparation of the manuscript.

Competing Interests: The authors have declared that no competing interests exist.

* E-mail: tadao.kunihiro@nioz.nl

Introduction

Microbial biomass in marine sediments accounts for 0.18–3.6% of Earth's total biomass (4.1 petagram carbon), and their community composition is highly diverse due to variation in oxygen concentrations in the overlying water, sediment carbon content, and sediment depth [1,2]. Sediment bacteria fulfill an important role in organic matter remineralization [3,4] and nutrient cycling [5], and are an integral component of food-webs, particularly those that are detritus-based [6,7]. Sediment bacterial communities are more diverse than planktonic communities, and respond actively to environmental conditions of their habitat [8–10]. Studies on the role of bacteria in sediment biogeochemistry particularly require a quantitative assessment of both bacterial biomass and community composition.

Nevertheless, studies on estimates of bacterial biomass, community composition, and diversity are constrained by the methodological limitation that over 99% of the total bacterial population cannot be cultivated by traditional culture techniques [11–13]. In the past few decades, the rise of culture-independent techniques (molecular approach and chemical analysis-based method) has allowed us to further reveal sediment microbial ecology. Molecular approaches that are highly suited for high resolution description of bacteria communities in marine sediment are rDNA clone libraries, denaturing gradient gel electrophoresis (DGGE), and terminal restriction fragment length polymorphism (T-RFLP). In recent years, the advent of high-throughput sequencing technologies (e.g. pyrosequencing and Illumina) has greatly enhance the knowledge on bacterial community structure [14,15]. As most powerful quantitative molecular approaches, the Q-PCR approach has been widely applied to quantify gene copy number as a proxy of bacterial abundance [16–18], and the FISH technique has been used for visualizing and quantifying bacterial cells in sediments [19,20]. Both quantitative approaches are distinctly suitable for targeting specific phylogenetic groups but less suitable for analysis of the full bacterial community, because quantitative application for analysis of all bacterial groups requires the use of many target-specific primers and probes and also need to optimize its protocol for each target group. Moreover, the PCR-based approaches cannot eliminate methodological biases, and nucleic acid extraction from sediment samples has inherent biases, for instance extraction efficiency from sample and bacterial species [21].

Despite the superiority of molecular approaches for the analysis of bacterial community structure, also phospholipid-derived fatty acid (PLFA) analysis [22,23] and the quinone profiling method [24,25] have been successfully used as a chemotaxonomic analytical-based method to quantify bacterial biomass and to profile the bacterial community composition in marine sediments. PLFAs are essential membrane lipids of microbial cells and therefore proxies for bacterial biomass. Microorganisms contain numerous PLFAs with some being "general" and unspecific, while others are more specific and found in higher abundance in some microbial groups [23,26]. Respiratory quinones (RQ, including ubiquinone [UQ] and menaquinone [MK]), and photosynthetic quinones (including phylloquinone [K1] and plastquinone [PQ]) are lipid coenzymes used for electron transfer in microbial cell membranes. A bacterial phylum has generally only one dominant molecular species of respiratory quinone (e.g. [27]). The main advantage of both chemotaxonomic methods is that there are established and standardized quantitative extraction protocols available [28,29] that allow rapid and quantitative extraction from various types of sediment samples. Therefore, the lipid analysis is easily applicable to quantify bacterial biomass for a wide range of marine sediments without optimizing the extraction method. In fact, concentrations of bacteria-specific PLFAs have been used as a proxy for total bacterial biomass in various marine sediments (e.g. [30]), and the total RQ concentration has been found to correlate with bacterial biomass in soil [31], with the total bacterial cell count in various environments [32], and with the bacterial cell volume in lake water [33].

Moreover, analysis of PLFA- and quinone profiles has been widely utilized as a valuable tool for showing differences between samples and also community shift during experimental/monitoring periods (e.g. [34,35]). Cluster analysis for characterizing bacterial community structure based on dissimilarity- or similarity value matrix of PLFA- and quinone profile for marine sediments showed a similar clustering pattern with that of the molecular techniques [34,35]. One major disadvantage of both lipid analyses for studies of microbial ecology is the lower phylogenetic resolution for identifying bacterial groups than molecular approach. Thus, the lipid analysis has been combined with molecular approach as a means of overcoming the limitation of low phylogenetic resolution [24,35,36]. A major advantage of the PLFA technique is that it can be combined with carbon stable isotope analysis to identify active bacterial groups and to trace carbon flows in both benthic and pelagic food webs via bacteria and other microbial groups, such as microalgae, to higher trophic levels (e.g. [23]). In addition, the quinone profiling method is also possible to identify active bacterial groups by combination with carbon radioisotope labelling [37].

For marine sediments, PLFAs have been widely used as quantitative bacterial biomarkers [30], but applications of quinones as biomarkers for sedimentary bacteria are very few [24,25]. In addition to their potential as a proxy for bacterial biomass, the ratio between quinones and PLFAs may also provide a proxy for the level of activity of the bacterial community [38], because PLFAs are structural biomass components while quinone concentrations are related to biomass and respiratory activity as they are part of electron transport chains (e.g. [39]). The ratio between total RQ and total PLFA was firstly applied in estuarine and deep sea sediments by Hedrick and White [38]. Until now, very few studies have applied this potential proxy in other aquatic systems [40,41].

In this study we compared and evaluated PLFAs and quinones as quantitative bacterial biomarkers for bacterial biomass and community structure in marine sediments. We also explored whether the ratio between these two bacterial biomarkers could be used as a potential proxy for bacterial activity, by examining the concentrations of PLFAs and quinones and quantity and quality of organic matter in a wide range of marine sediments from the intertidal zone to the deep sea.

Materials and Methods

Study Areas and Sampling Procedure

Samples were collected from a wide variety of sediments, ranging from intertidal to coastal, shelf and deep-sea sediments (see Table 1 for the sites and sampling depth details and Table S1). Samples selected for this study came from previous published and unpublished studies were either PLFA or quinone analysis had already been performed and we completed the data set by additional analyses. No specific permissions for all sampling were required for these locations. Intertidal sediments were collected from 3 locations (Oude Bietenhaven, Zandkreek, and Rattekaai) in the Oosterschelde, a marine embayment in the SW of the Netherlands, and one location (Kapellebank) in the nearby Scheldt Estuary. Sediment was sampled manually at low tide using cores (30 cm height and 6 cm in diameter) and cores were sliced. Another tidal flat location was sampled for long-term incubations in the laboratory. In short, lab incubation sediments were collected from the surface (0~2 cm) of a tidal flat (Biezelingse Ham) in the Scheldt Estuary, homogenized, and incubated for up to 261 days in vitro with regular sampling in a similar manner of the experiment that is described in [42]. North Sea sediments were collected from three stations in November 2010. Stations NS-1 and -3, located close to the Dutch coast and on the Dogger Bank respectively, are non-depositional areas, while station NS-2, situated on the Oyster Ground, is a semi-depositional area. Sediment was sampled with cores by multi-corer (Octopus type).

Japanese natural coastal sediments were collected from nine different bays and embayments in the Seto Inland Sea using a Smith-McIntyre Grab sampler or an Ekman-Berge grab sampler, and then subsampled by collecting surface sediment samples from late September to early October 2008 and from early May to early June 2009. Japanese fish farm sediments were collected from 14 stations located in and around fish farming areas in the north part of Sukumo Bay, located in Sikoku, Japan in the same manner as the coastal sediments collected from the Seto Inland Sea.

Deep sea sediments were collected from the Arabian Sea and the Atlantic Ocean. Samples from the Arabian Sea were obtained from two stations, with one station (AS-1) being situated within the oxygen minimum zone (OMZ) (i.e. <9 μM O_2 in the overlying water) and the other station (AS-2) below the OMZ (i.e. oxic bottom water) in January 2010 [43]. Sediment from the Atlantic Ocean was sampled at the Galicia Bank off the coast of Spain in September–October 2008 [44].

All sediment samples were either directly stored frozen (−20°C) or freeze-dried and subsequently stored frozen (−20°C) until extraction and analysis of PLFAs, quinones, and organic carbon.

Chemotaxonomic Markers of PLFAs and RQs in Different Groups of Bacteria

Important chemotaxonomic PLFA and RQ markers for bacteria are listed in Table 2. In this study, we defined the sum of saturated fatty acid (SFAs, C_{13}–C_{18}), branched fatty acids (BFAs), and mono-unsaturated fatty acids (MUFAs, ≤C_{19}) as total bacterial PLFA. In addition, there are various other bacteria-

Table 1. Sample codes and characteristics.

Site	Code	Water depth (m)	Sediment depth* (cm)	n**	Analysis OC***	DI****
Dutch intertidal (DI):						
Oude bietenhaven	DI-N-OB	-	0–2	2	n	n
Zandkreek	DI-N-Z	-	0–5	2	n	n
Rattekaai	DI-N-R	-	0–1.5	2	n	n
Kapellebank	DI-N-K	-	0–2	2	n	n
Lab incubations	DI-L	-	0–2	1–12	y	n
North Sea (NS):						
Station 1	NS-1	12	0–1	1	y	n
Station 2	NS-2	45	0–9	1–6	y	n
Station 3	NS-3	27	0–9	1–6	y	n
Japanese coast (JC):						
Natural	JC-N	6–83	0–1 or 0–2	1–9	y	y
Fish farm	JC-FF	30–75	0–2	1–14	y	y
Arabian Sea (AS):						
Station 1	AS-1	989	0–2	1	y	n
Station 2	AS-2	1700	0–2	1	y	n
Galicia Bank	GB	1900	0–1	1	y	n

*Total sampled depth range;
**n, sample number;
***OC: organic carbon content,
****DI: degradation index. Additional information is shown in Table S1.

specific PLFAs, for instance i17:1ω7 is for a marker for the genus *Desulfovibrio* [45], but these compounds are typically present in low concentrations, which precluded analysis of these compounds in most sample sets in the present study.

PLFA Extraction and Analysis

PLFAs were extracted from freeze-dried sediment (~4 g) and analyzed as described in [28]. In short, total lipids were extracted from the sample in chloroform–methanol–water (1:2:0.8, v/v) using a modified Bligh and Dyer method and fractionated on silicic acid into different polarity classes. The methanol fraction, containing phospholipids, was derivatized using mild alkaline methanolysis to yield fatty acid methyl esters (FAMEs), which were recovered by hexane extraction. FAME concentrations were determined by gas-chromatography-combustion-isotope ratio mass spectrometry (GC-c-IRMS) for all samples except for Japanese samples that were analysed by gas chromatography-flame ionization detection (GC-FID). The concentrations obtained by both methods are comparable from our previous experience ($r^2 = 0.99$, unpubished data). Identification of individual FAME was based on comparison of retention times with known reference standards.

Quinone Extraction and Analysis

Quinones were extracted from freeze-dried or frozen sediment (~6 g) as described previously [25,29]. The types and concentrations of each quinone were determined using a HPLC equipped with an ODS column (Eclipse Plus C18, 3.0 (I.D.) ×150 mm, pore size 3.5 μm, Agilent technologies) and a photodiode array detector (SPD-M20A, Shimadzu: for the Japanese samples, and Waters 996 for the samples from the Dutch intertidal zone, North Sea, and the

deep sea). A mixture of 18% isopropyl ether in methanol was used as the mobile phase at a flow rate of 0.5 mL min^{-1}. The quinone molecular species were identified by the linear relationship between the logarithm of the retention times of quinones and the number of their isoprene units, using the identification-supporting sheet, which is available upon request from T. K, based on the equivalent number of isoprene units (ENIU) of quinone components as described by [46]. Details on the analytical conditions have been described by [33].

Organic Carbon Content

For determination of the organic carbon (OC) content, sediment samples were first freeze-dried or dried at 60°C in an oven overnight, acidified to remove carbonate, and further vacuum-dried. The OC content of the sediment was determined with an elemental analyzer (NA-1500n, Fisons, Rodano-Milan, Italy: for the Japanese sediments and FlashEA 1112, Thermo Electron, Bremen, Germany: for the sediments of other samples).

HAA Extractions, Analysis and Calculation of Degradation Index

For the Japanese sediments, concentrations of hydrolysable amino acids (HAAs) were analyzed as described in [47] and used to calculate the degradation index (DI), a proxy for the quality, or "freshness", of the organic matter in the sediment. Briefly, samples (~1 g) of freeze-dried sediment were washed with 2 M HCl and Milli-Q water and then hydrolyzed in 6 M HCl at 110°C for 24 h. After neutralization by 1 M NaOH, amino acids were derivatized with *o*-phthaldialdehyde (OPA) [48] prior to injection to reverse-phase high-pressure liquid chromatography (HPLC). Amino acid concentrations were measured by HPLC and further details on the

Table 2. Major fingerprints of PLFA and quinone as a marker for different bacterial groups in this study.

Biomarker	Proteobacteria					Bacteroidetes	Actinobacteria
	Alpha-	Beta-	Gamma-	Delta-	Epsilon-		
PLFA[a]							
SFA (C_{12}–C_{19})[c]	G	G	G	G	G	G	G
i14:0							
i15:0				++M		+++	++
i16:0						+	+++
i17:0				++M		+	+
a15:0				+M		+M	+
a17:0				+M			++
10Me16:0				++M			+M
10Me17:0							+M
10Me18:0							++M
cy17:0	+[d]	+	+	+++M		+	
cy19:0	+	+	+	+M			
16:1ω7c	G	G	G	G	G	G	G
18:1ω9c	G	G	G	G	G	G	G
18:1ω7c	+++	+	+++	++M	+++		
Ref. no.	[67–69]	[70–72]	[73,74]	[45,75]	[76,77]	[70,78,79]	[80]
Quinone[b]							
UQ-8	+++*	+++M	++++M				
UQ-9	+++*		++++M				
UQ-10	+++M						
MK-n (n≤8)		+++*	++++*	++++M	++++M (MK-6)	++++M	++++*
MK-n (n≥9)						++++*	++++M
MK-n(Hx)				++++*			++++M
Ref. no.	[67–69]	[72,81]	[74,82,83]	[84,85]	[86]	[70,79,87]	[80,88,89]

In this study, we refer to the different quinones with the following abbreviations: ubiquinone - UQ-n; and menaquinone - MK-n. The number (n) indicates the number of the isoprene unit in the side chain of the quinone. Partially hydrogenated MKs were expressed as MK-n(Hx), where x indicates the number of hydrogen atoms saturating the side chain.
[a]PLFA data were modified mainly from [23,26,90,91].
[b]Quinone data were modified mainly from [56,92–94].
[c]Saturated fatty acids.
[d]+, 1–5%; ++, 5–15%; +++, 15–40%; ++++, >40% of total PLFA pool or total quinone pool; *, present in few species; G, a maker found in a broad range of bacteria and algae, and M, a marker can be used specifically as an indicator for specific bacterial group with the phylum.

analytical conditions have been described by [47]. The DI was calculated following [47]:

$$DI = \sum_i \left[\frac{var_i - AVGvar_i}{STDvar_i} \right] \times fac.coef_i$$

where var_i, $AVGvar_i$, $STDvar_i$, and $fac.coef_i$ are the mol%, mean, standard deviation and factor coefficient of amino acid i, respectively. The factor coefficient was described in [49].

Cluster Analysis of the Pattern of Differences Among Samples in Individual PLFAs and RQs

We conducted a cluster analysis to identify groups of similar bacterial PLFA and RQ patterns. We first normalized the mole fraction of bacterial PLFA and RQ ($Z_{j,i}$), because this analysis depends on the absolute values of the data, using the following normalization equations [50]:

$$Z_{j,i} = \frac{P_{j,i} - \overline{P}_j}{S_j}$$

With:

$$\overline{P}_j = \frac{\sum_{i=1}^N P_{j,i}}{N}$$

$$S_j = \left[\frac{\sum_{i=0}^N \left(P_{j,i} - \overline{P}_j \right)}{N-1} \right]^{1/2}$$

where $P_{j,i}$ is the mole fraction of bacterial PLFA or RQ component j and sample i, N is the number of samples, and $\overline{P}_j$ and S_j are the average value and the standard deviation of the mole fraction of bacterial PLFA or RQ among samples, respectively. After normalization, the average value $\overline{P}_k$ and the standard deviation S_k are shown respectively as 0 and 1, where k is the normalized component of bacterial PLFA or RQ. We used both >1 mol% of component to the bacterial PLFA (without general bacterial compounds (SFAs ($C_{12}-C_{19}$), 16:1ω7c, and 18:1ω9c), MUFAs ($\geq C_{20}$), and PUFAs) or RQ profile, >30% of coefficient of variance of compound among all samples for this data analysis, and reconstructed profiles, because general and minor components interfere with the result of this analysis. As results, the cluster analysis was conducted based on the mole fraction of 12 bacterial PLFAs and 16 RQ molecular species among all samples (see "Cluster analyses of bacterial PLFA and RQ profiles"). The normalized values were used to produce a cluster dendrogram based on the Euclidean distance matrix, and the dendrogram was constructed using Ward's method with the graphing program KyPlot version 5.0 (KyensLab Inc., Tokyo, Japan).

Cluster Analysis Based on the Full Profiles of the Bacterial PLFAs and RQ

We conducted another cluster analysis to compare sample discrimination and its resolution based on bacterial PLFA or RQ profiles. A dissimilarity index (D) of profile was calculated using the following equation [51].

$$D = \frac{1}{2} \sum_{k=1}^n |f_{ki} - f_{kj}|$$

where n is the number of PLFA or RQ component. In the PLFA profiles, f_{ki} and f_{kj} are the mole fractions of the k PLFA component for the i and j samples, respectively. In RQ profiles, f_{ki} and f_{kj} are the mole fractions of the k RQ component for the i and j samples, respectively (f_{ki}, f_{kj}>1 mol%; $\Sigma f_{ki} = \Sigma f_{kj} = 100$ - mol%). Cluster analysis was performed with the program KyPlot version 5.0 based on the D distance matrix and a dendrogram was constructed using the between-groups linkage method. Values ≤0.1 of D of RQs are not recognized as different RQ profiles according to the analytical precision based on the duplicate analytical results including extraction and measurement process (97% statistical reliability) [52]. For the PLFA analysis, we determined the threshold value, 0.13, in the same manner as the value of the quinone profiling method (see [52]) using 12 duplicate results of the incubation sediment samples (Fig. S1).

Statistical Analysis

Spearman's rank correlations (r_s) were used to show the relationships among bacterial PLFA concentration, RQ concentration and organic carbon content and the relationships between OC content and DI. Pearson's correlation coefficients (r) were used to show the relationships between OC content and RQ/bacterial PLFA ratio and between DI and RQ/bacterial PLFA ratio. Analysis with Spearman's rank correlation and Pearson's correlation coefficient was performed using the statistical program PASW Statistics for Windows version 18J (IBM Japan, Tokyo, Japan). Mantel tests were used to test the significance of the correlation between dissimilarity matrices based on bacterial PLFA or RQ profiles, using the R package [53].

Results

PLFA and Quinone Concentrations

Total bacterial PLFA concentrations (i.e. the sum of SFAs [$C_{13}-C_{18}$], BFAs, and MUFAs [$\leq C_{19}$]) in the sediment varied over three orders of magnitude (range 1.2–834 nmol gdw^{-1}) with lowest values for Japanese natural coastal sediment (JC-N-9) and highest values for Dutch intertidal natural sediment (Fig. 1 and Table 3). Total RQ concentrations in the sediment ranged from 0.01 to 28 nmol gdw^{-1} with lowest values for the Galicia bank (GB) and highest values for Japanese fish farm sediment, and were one to two orders of magnitude lower than the bacterial PLFA concentrations (Fig. 1 and Table 3). RQ concentration showed a positive log-log correlation with the bacterial PLFA concentration for the full dataset as well as within the individual sample sets (Fig. 1 and Table 4). However, there were clear differences in slopes of the fits for the individual sample sets with the highest slope for the Japanese fish farm sediments (1.487) (i.e. relatively rich in RQs) and the lowest slope for the Dutch intertidal incubation sediments (0.716) (i.e. relatively rich in PLFAs) (Table 4). Two deep sea samples from the Arabian Sea (AS-2) and GB were relatively far from the overall trend line with relatively high PLFA concentrations and low RQ concentrations (Fig. 1).

Relationship Between Organic Carbon Contents and Bacterial PLFAs and RQs

The sediment organic carbon (OC) content ranged from 0.4 to 60 mg gdw^{-1} (mean 8.8±11, mg gdw^{-1} $n=51$) over more than two orders of magnitude in all samples (Table 3). A positive power

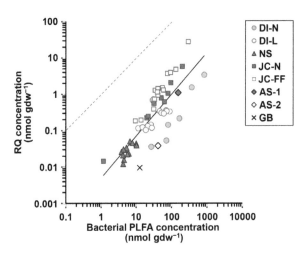

Figure 1. Comparison between bacterial PLFA and RQ concentration in the sediment with different sample sets. Line indicates trend for the full dataset. The dotted line indicates the 1:1 relationship.

correlation between the OC contents and the bacterial PLFA concentrations was observed (Fig. 2a and Table 4). This correlation was similar for all sample sets, except for the Dutch intertidal incubation and North Sea samples (Table 4). Given the correlation, it was not surprising to find that the correlation between the OC contents and the RQ concentrations was also positive (Fig. 2b and Table 4).

RQ/PLFA Ratios

We used a ratio based on mole concentration between total RQ and total bacterial PLFA (RQ/bacterial PLFA). The ratios of RQ/bacterial PLFA ranged from 0.0007 to 0.095 with lowest value for the deep sea sediment (GB) and highest values for Japanese fish farm (JC-FF-13) (Fig. 3a). Strong positive log-log correlations between the OC contents and the RQ/bacterial PLFA ratios of the Japanese fish farm samples were observed ($r = 0.888$, $P<0.01$) (Fig. 3a), while ratios for the other sample sets, except deep sea samples, showed no significant correlation with OC content.

Degradation Index Values for Japanese Sediments

DI values for all Japanese samples were ranged strongly from −1.1 to −0.2 (Fig. 3b) with more negative values indicating more

degraded (refractory) material. The DI value was positively correlated with the OC content ($r_s = 0.738$, $P<0.05$ for Japanese natural coast and $r_s = 0.702$, $P<0.01$ for Japanese fish farm), meaning the OC in the sediments with the highest OC content was relatively fresh (labile). A positive linear correlation between DI and the RQ/bacterial PLFA ratios only for the Japanese fish farm sediments was observed ($r = 0.751$, $P<0.01$) (Fig. 3b), whereas there was no significant correlation for Japanese natural coastal sediments ($r = 0.665$, $P = 0.072$). Note that the positive relationship for Japanese fish farm sediments was due to the sample from Stn. JC-FF-13 (without the plot of Stn. JC-FF-13, $r = 0.469$, $P = 0.106$).

Relative Composition of PLFA and Quinone Pools

The composition of the bacterial PLFA (general [SFAs ($\leq C_{19}$), 16:1ω7c, and 18:1ω9c] + specific) in the sediment showed less variation as compared to the composition of RQ (Fig. 4). The three dominant PLFAs, 16:0, 16:1ω7c and 18:1ω7c, were present generally in almost all the samples (Fig. 4a). SFAs ($\leq C_{19}$), 16:1ω7c, and 18:1ω9c as a *general* marker for bacteria accounted for 57 mol% of the total bacterial PLFA pool in all samples (range 47–71 mol%). Bacteria-*specific* PLFAs showed variation in the full dataset. Together, i15:0, a15:0, 10Me16:0, and 18:1ω7c as a *specific* marker for bacterial groups accounted for average 25 mol% (range 11–46 mol%) of the total bacterial PLFA pool in all samples (Fig. 4a).

In general, the relative composition of RQ varied more strongly (Fig. 4b). The most obvious difference is seen between the deep-sea and the other (coastal and estuarine) sediments. Almost all coastal sediments except Japanese fish farm sediments were dominated by PQ-9 and UQ-8, while two deep sea sediments (AS-2 and GB) were dominated by MK-8(H₂) and MK-8. Japanese fish farm sediments were dominated by UQ-10 and UQ-8. Together, UQ-8, -9, and -10 accounted for 45 mol% (range 9–83 mol%) of the total RQ pool in all samples. PQ-9 and K1, which are derived from photosynthetic organisms, were observed in not only coastal area, but also in the oxygen minimum deep-sea sediment (AS-1) (Fig. 4b).

Cluster Analysis of the Pattern of Differences Among Samples in Individual PLFAs and RQs

The differences in the bacterial PLFA and RQ profiles for the different sample sets (Fig. 4) were further clarified by two cluster analyses. The first analysis was performed to investigate the co-

Table 3. Concentration of bacterial PLFAs, respiratory quinones (RQ) and organic carbon (OC) in marine sediments in this study.

	Bacterial PLFA (nmol gdw⁻¹)		RQ (nmol gdw⁻¹)		OC (mg-C gdw⁻¹)	
	Range	Mean ± SD	Range	Mean ± SD	Range	Mean ± SD
All samples	1.17−834	67.0±122	0.01−28.0	1.22±3.75	0.37−60.4	8.8±11.2
Dutch intertidal (DI):						
Natural	27.2−834	202±280	0.03−3.4	0.84±1.14	-*	-
Lab incubations	11.5−89.3	43.0±26.2	0.10−0.36	0.23±0.11	3.6−16.3	9.8±6.2
North Sea (NS):	3.84−10.3	6.05±2.2	0.01−0.05	0.03±0.01	0.44−3.0	1.6±1.2
Japanese coast (JC):						
Natural coast	1.17−202	63.3±30.2	0.01−5.8	1.26±1.8	0.37−23.7	10.1±8.6
Fish farm	9.55−295	68.8±72.9	0.18−28.0	3.54±7.19	1.7−49.6	10.5±11.7

*Not determined.

Table 4. Log/log power regressions and Spearman's rank coefficients between the bacterial PLFA (nmol gdw^{-1}) and respiratory quinone (RQ) concentrations (nmol gdw^{-1}), and between organic carbon (mg-C gdw^{-1}) and the bacterial PLFA and quinone concentration of individual sample set.

	Bacterial PLFA (x)		OC (x)		OC (x)	
	versus RQ (y)		*versus* bacterial PLFA (y)		*versus* RQ (y)	
	Power regression	r_s	Power regression	r_s	Power regression	r_s
All samples	$y = 0.0049 \times^{1.151}$	0.823**	$y = 6.326 \times^{0.882}$	0.946**	$y = 0.039 \times^{1.148}$	0.809**
Dutch intertidal (DI):						
Natural	$y = 0.0096 \times^{0.774}$	0.643	–	–	–	–
Lab incubations	$y = 0.0156 \times^{0.716}$	0.825**	$y = 5.991 \times^{0.861}$	0.781**	$y = 0.045 \times^{0.720}$	0.982**
North Sea (NS):	$y = 0.0059 \times^{0.899}$	0.624*	$y = 5.650 \times^{0.138}$	0.253	$y = 0.028 \times^{0.115}$	0.263
Japanese coast (JC):						
Natural coast	$y = 0.0094 \times^{1.122}$	0.983**	$y = 6.027 \times^{1.011}$	1.000**	$y = 0.069 \times^{1.149}$	0.983**
Fish farm	$y = 0.0043 \times^{1.487}$	0.952**	$y = 5.962 \times^{1.022}$	0.880**	$y = 0.056 \times^{1.570}$	0.847**

Levels of significance are *$P<0.05$, **$P<0.01$.

variation in the relative abundance of the individual bacteria-*specific* PLFAs (sum of BFAs and MUFAs ($\leq C_{19}$) except 16:1ω7c and 18:1ω9c) and RQs (Fig. 5). When different compounds cluster closely, this indicates that these compounds are probably derived from the same bacterial groups. Two main clusters (cluster-1 and -2) were observed that were further divided into two sub-clusters (cluster-1a, -1b, 2a, and -2b) (Fig. 5). These five clusters were characterized by a relatively high mole fraction of group-specific bacterial PLFAs and RQs among all samples. It is noteworthy that UQs were present in cluster-1, whereas almost all partially saturated MKs were in cluster-2.

Cluster Analyses of Bacterial PLFA and RQ Profiles

The second cluster analysis was conducted to investigate the differences in bacterial community structure of the different sediments based on the bacterial PLFA and RQ profiles separately in order to compare the chemotaxonomic resolution of these two methods (Fig. 6). The bacterial PLFA profiles clearly separated into three main groups (group PI, PII, and PIII in Fig. 6a). Group PI comprised all Dutch intertidal natural sediments, while all other samples were included in group PII. The only exception here is a single Japanese natural coastal sediment sample (JC-N-9) that formed a separate cluster (PIII). Further differentiation involved division of group PII into six different groups (group PI-1~6 in

Fig. 6a) based on the threshold value of 0.13 (representing the observed level of dissimilarity between replicate samples, *see* Fig. S1). The RQ profiles were divided clearly into four main groups (group QI, QII, QIII, and QIV in Fig. 6b). Within these main groups, almost all sample sets were distinguished as separate groups based on the threshold value of 0.1 for sample discrimination of different RQ profiles [52] (Fig. 6b). The general sample classification of the different sediments between bacterial PLFAs versus RQs based on the dissimilarity index was significantly correlated (using 10,000 randomizations, Mantel's coefficient $r = 0.435$, $P = 0.0001$).

Discussion

Analysis of lipid biomarkers is a powerful tool for quantification of bacterial abundance and community structure. While PLFAs have been widely utilized as quantitative bacterial biomarkers in marine sediment [22,23], applications of the quinone profiling method to marine sediments are still very few [24,25]. In this study, we analyzed concentrations of PLFAs and RQs in a broad range of marine sediments to investigate and compare their application as indicators of bacterial biomass and community composition.

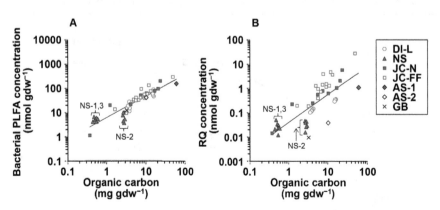

Figure 2. Comparisons between: a) organic carbon and bacterial PLFA concentration, b) organic carbon and RQ concentration.

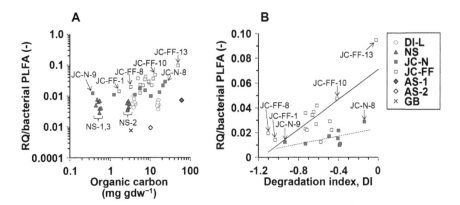

Figure 3. Relationships between: a) organic carbon and RQ/bacterial PLFA ratio in the sediment with different sample sets, b) degradation index and RQ/bacterial PLFA ratio in the Japanese coastal natural- and fish farm sediments.

PLFAs and RQs as Bacterial Biomass Indicators

We found a strong correlation between total concentrations of the bacterial biomarkers PLFAs and RQs across several orders of magnitude, both for the individual sample sets and for the whole dataset (Fig. 1). PLFA concentrations have been used frequently as a measure of bacterial biomass in seawater and marine sediments (e.g. [22,23,54]), because PLFA concentrations are relatively constant in bacterial biomass and PLFAs degrade rapidly upon death of the source organism, meaning that they are specific for *living* bacterial biomass [23]. The strong correlation across several orders of magnitude between PLFAs and RQs indicates that RQs also provide an estimate of *living* bacterial biomass in sediment. This allowed us to determine the conversion from RQ concentration (nmol gdw^{-1}) to bacterial biomass (mg C gdw^{-1}; biomass = 0.192 RQ$^{0.586}$, $r_s = 0.853$, $P < 0.001$, $n = 59$). The equation is based on the correlation between the RQ concentration and summed concentrations of four bacteria-specific PLFAs (i14:0, i15:0, a15:0 and i16:0) calculated by the equation detailed in [30] using the conversion factors from [55]. Interestingly, this relationship is not linear, which is probably due to metabolic variation in quinone concentrations in bacterial cells in different environments.

Previous studies have already demonstrated that total RQ concentrations correlated very well with microbial biomass carbon in soil (measured by a fumigation-extraction method, $r = 0.96$, [31]), with the total bacterial cell count in various environments ($r = 0.98$, [32]), and with bacterial cell volume in lake water ($r = 0.98$, [33]). Our study is the first to demonstrate the good correlation between concentrations of RQ versus PLFAs as a compound-specific biomarker for bacteria in marine sediment. Thus, these results indicate that RQ concentration can be utilized as a proxy for bacterial biomass in sediment samples.

Despite the overall strong correlation between PLFAs and RQs, a more detailed look at Fig. 1 reveals that there is residual variation to be explained. Firstly, the range in RQ concentrations was around one order of magnitude higher than that of the bacterial PLFA concentrations in the full sample set. Secondly, the slopes of the fits for the individual sample sets were different (Fig. 1 and Table 4), which implies that the different sediments contained bacterial communities with different RQ/PLFA ratios. We consider two possible explanations for this varying RQ/PLFA ratio. The first explanation is inherent group-specific differences in the RQ/PLFA ratios of the different groups of bacteria contributing to the overall bacterial community. This can be related, for example, to the type of energetic metabolism of the

bacteria (e.g. [56]). The second explanation concerns the activity of the bacteria. While bacterial PLFA concentrations (being a structural biomass component) are relatively stable under different conditions [57,58], concentrations of RQs in bacterial cells can also depend on the metabolic activity due to growth phase [59], substrate utilization [60], and redox state [61]. The strong PLFA versus RQ correlation over a broad range of sediments suggests that RQ concentration reflects mainly bacterial biomass but that may have an additional component related to the activity/metabolism of bacteria.

If RQ concentrations are also dependent on the activity of the bacteria, RQ concentrations relative to PLFA concentrations (the RQ/bacterial PLFA ratio) may also depend on the quality and quantity of the OM in the sediment as these two factors directly influence bacterial activity [36,62,63]. We investigated this relationship through assessment of the correlation between the RQ/bacterial PLFA ratio versus OM *quantity* (total OC content) for all samples and OM *quality* (i.e. DI, the amino acid-based degradation index) for the Japanese sediments (Fig. 3b). The absence of a clear correlation between OC content and the RQ/bacterial PLFA ratio for the full dataset (Fig. 3a) indicates that this ratio was not influenced by OM *quantity*. In addition, we also investigated the relationship between the RQ/bacterial PLFA ratio versus OM *quality* for the Japanese samples. The OM *quality* was determined by the degradation index (DI), which is based on the relative composition of hydrolysable amino acids in the sediment [47]. This index provides an indication of the quality (or 'freshness') of the organic matter in the sediment with most negative values indicating relatively low quality (or 'refractory') material. Despite the wide range of observed DI values (-1.10 to -0.02), which indicates substantial variation in OM *quality* between samples for both the natural and fish farm sediments (Fig. 3b), there was no correlation between DI values and the RQ/bacterial PLFA ratio for the natural sediments and only a weak correlation for the fish farm sediments (Fig. 3b). Overall, our results indicate that there was no strong control of bacterial activity on the RQ/PLFA ratio by both quantity and reactivity of the OC pool.

Still, the RQ/bacterial PLFA ratios for Japanese fish farm sediments were clearly higher than the natural sediments. According to previous studies, the ratio between total RQ and total PLFA concentration has been used to indicate mainly two aspects: a presence of aerobic bacteria and facultative heterotrophic bacteria and a respiratory activity in comparison with fermentation processes [38,40,41]. Further investigation of the

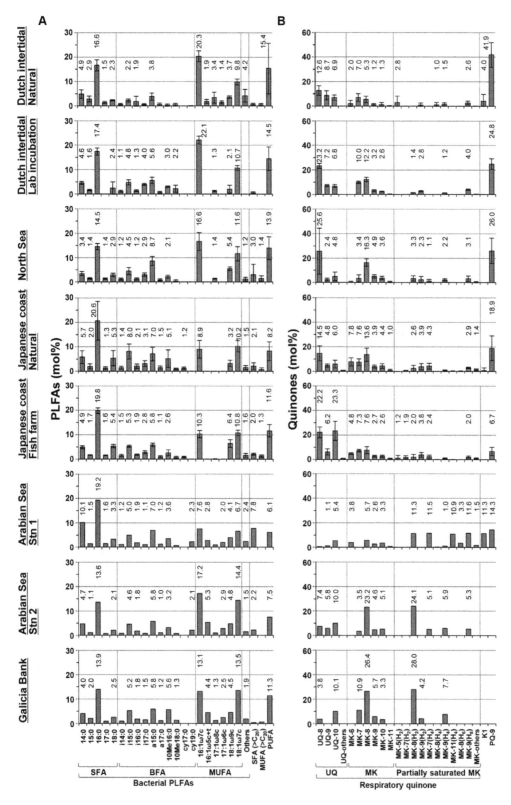

Figure 4. Summarized compositions of a) PLFAs and b) quinones with different sample sets. More than 3 mol% of components to total pool of each PLFAs and RQs were indicated as others. Note that the full range of PLFAs and quinones analyzed is shown here, meaning that this includes both bacteria-specific and non-specific compounds.

RQ/PLFA ratio, combined with a study on bacterial metabolism in marine sediments, is needed to explain the observed residual variation and the role of these two aspects.

Linking PLFA and Quinone Biomarkers

The cluster analysis as shown in Fig. 5 was conducted to investigate the co-variation between the bacterial PLFAs and RQs

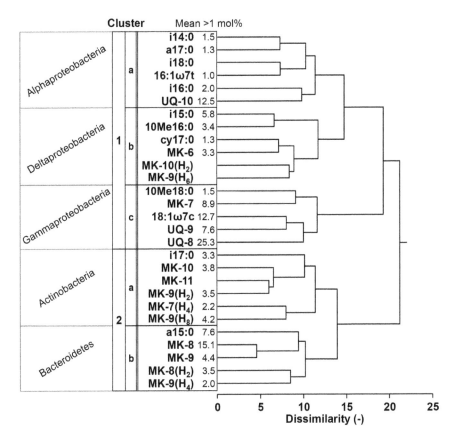

Figure 5. Cluster analysis of the pattern of differences among samples in the individual bacterial PLFAs and RQs. The mean mole percentage value indicates the mean mole fraction among all samples.

in the different sediments and their association with specific bacterial groups. The cluster analysis showed two main groups: cluster-1 comprised UQs, which are specific for Gram-negative Proteobacteria (*see* Table 2). Cluster-2 comprised almost all partially saturated MKs, which are predominantly present in Gram-positive bacteria, thereby indicating that cluster-2 was dominated by Gram-positive bacteria (Fig. 5 and Table 2). Based on the taxonomic assignment of PLFAs and RQs in Table 2, the subcluster can be analyzed in more detail. Subcluster-1a comprised UQ-10, indicating that this cluster was dominated by members of the class Alphaproteobacteria. Subcluster-1b was characterized by the presence of MK-6, i15:0, 10Me-16:0, and cy17:0, indicating that this cluster was predominance of the class Deltaproteobacteria. Subcluster-1c was characterized by UQ-8 and 18:1ω7c, indicating that it comprised mainly members of the class Gamma- and Beta-proteobacteria. Betaproteobacteria are well known to be a minor group in marine sediment [36], therefore, Subcluster-1c must have been dominated by mainly members of the class Gammaproteobacteria. Cluster-2a was characterized by MK-10, MK-9(H$_8$), and i17:0, indicating that this cluster is relatively rich in members of the Actinobacteria. Subcluster-2b was characterized by MK-8, MK-9, and a15:0, indicating that this cluster comprised members of the Bacteroidetes. Our study is the first to demonstrate a general agreement in the chemotaxonomic classification based on bacterial PLFAs versus RQs. This strengthens the use of these biomarkers for characterization of the sediment bacterial community. Although taxonomic resolution of both analyses is limited to identify phylogenetic groups of bacteria (low phylogenetic resolution), the value of this approach can be in combination with stable isotope

probing to allow researchers to trace the flow of elements within communities [23].

Bacterial Communities of the Different Sediments

The cluster analyses as shown in Fig. 6 were performed to investigate the resolution of the two types of bacterial biomarkers and their ability to distinguish between bacterial communities from different sediments. In general, the bacterial PLFA- and RQ profiles revealed a similar classification pattern in bacterial community differences in our wide range of marine sediments (Fig. 6). However, there is a clear difference in the resolution of both methods with sample classification based on the RQ profile distinguishing 37 groups, whereas classification based on of the bacterial PLFA profile distinguishes only 13 groups. In other words, the level of dissimilarity between RQ profiles was substantially higher than the level of dissimilarity between the PLFA profiles.

The higher sample discrimination in the RQ profile can be explained by the higher specificity of the RQs for specific bacterial groups (*see* Table 2) as well as the more pronounced differences between RQ profiles of different bacterial groups (that are typically dominated by one RQ while PLFA profiles typically comprised 5~17 PLFAs) [38,64,65]. In addition to this general trend, there were also some notable differences in the classification of the deep sea sediments (AS-1, AS-2, and GB), Japanese natural coastal sediment (JC-N-9), Japanese fish farm sediment (JC-FF-10 and JC-FF-13), North Sea sediment (NS-2), and Dutch natural intertidal sediment (DI-N-K) (Fig. 6).

Two groups of RQs that were particularly important for the higher level of dissimilarity in RQ profiles are UQ-*n* and the

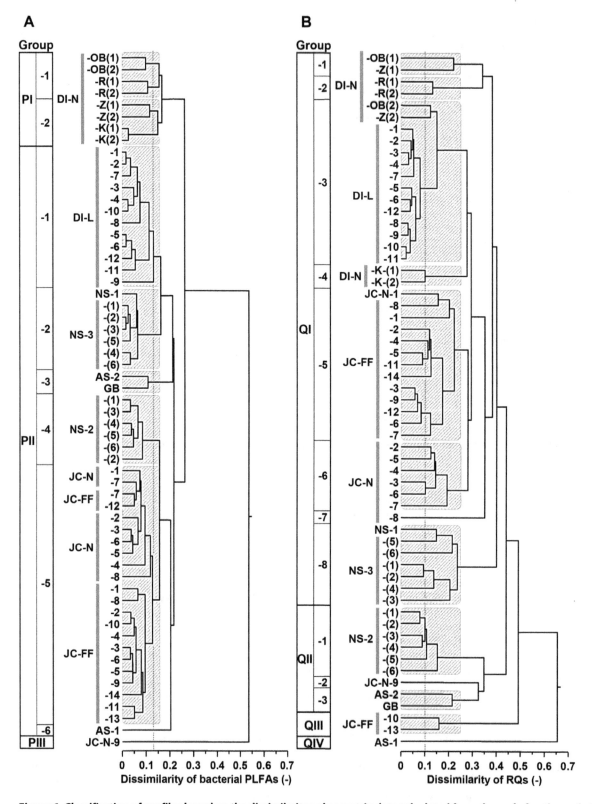

Figure 6. Classification of profiles based on the dissimilarity value matrix data calculated from the mole fractions of a) the bacterial PLFAs and b) the RQs of the sediments. Abbreviation of each sample indicates the system and other information of the sample (*see* Table 1). Parentheses in the abbreviation indicate the depth layer at the sampling station.

partially saturated MKs. UQs are good markers for Alpha-, Beta-, and Gamma-proteobacteria, whereas as for PLFAs, the only specific marker for proteobacteria is 18:1ω7c (see Table 2). Partially saturated MKs, which exist in Actinobacteria, Deltaproteobacteria, and Epsilonproteobacteria, showed most variation between sample sets, especially, MK-8(H$_2$), MK-9(H$_2$), MK-7(H$_4$), MK-9(H$_4$) and MK-9(H$_8$) which were present in more than 29 samples (15.4±9.2 mol% in the total RQ pool). Actinobacteria generally show a larger variation in MKs than PLFAs (e.g. [66]). Thus, higher sample discrimination in RQ profile could be due to presence of compounds originated from members of mainly Proteobacteria (UQ-n) and Actinobacteria (MK-n(Hx)).

Overall, we demonstrated that both concentration of bacterial PLFAs and RQs are good indicators for bacterial biomass and RQ profile can discriminate community difference more clearly than the bacterial PLFA profile. Thus, the combination between PLFA analysis (in combination with the stable isotope probing (SIP) technique) and quinone profiling method is a good strategy for studies on the role of bacteria in sediment biogeochemistry. These methods, and their applications, can be further expanded with the development of a method for stable isotope analysis of quinones so that quinones can also be applied in combination with stable isotope probing.

Acknowledgments

We are grateful to Dr. Laura Villanueva and Prof. Arata Katayama for their constructive comments on this work, to Prof. Akira Hiraishi for his critical comments on Table 2, and to Pieter van Rijswijk, Marco Houtekamer, Peter van Breugel, and Cobie van Zetten for their analytical support. We thank Dr. Katsutoshi Ito, Dr. Hideki Hamaoka, Hidejiro Onishi, Dr. Jun-ya Shibata, and Dr. Atsushi Sogabe for their assistance to collect samples in Japan, and Prof. Hiroaki Tsutsumi for the opportunity to use the facilities of the Prefectural University of Kumamoto. We also thank to Dr. Todd W. Miller for critically reading the manuscript.

Author Contributions

Conceived and designed the experiments: TK BV DvO. Performed the experiments: TK BV DVC LP MlG KM MK KO. Analyzed the data: TK BV DVC LP MlG. Wrote the paper: TK BV HTSB DvO.

References

1. Kallmeyer J, Pockalny R, Adhikari RR, Smith DC, D'Hondt S (2012) Global distribution of microbial abundance and biomass in subseafloor sediment. Proc Natl Acad Sci U S A 109: 16213–16216.
2. Orcutt BN, Sylvan JB, Knab NJ, Edwards KJ (2011) Microbial Ecology of the dark ocean above, at, and below the seafloor. Microbiol Mol Biol Rev 75: 361–422.
3. Arnosti C, Jørgensen B, Sagemann J, Thamdrup B (1998) Temperature dependence of microbial degradation of organic matter in marine sediments:polysaccharide hydrolysis, oxygen consumption, and sulfate reduction. Mar Ecol Prog Ser 165: 59–70.
4. Boetius A, Lochte K (1996) Effect of organic enrichments on hydrolytic potentials and growth of bacteria in deep-sea sediments. Mar Ecol Prog Ser 140: 239–250.
5. Alongi DM (1994) The role of bacteria in nutrient recycling in tropical mangrove and other coastal benthic ecosystems. Hydrobiologia 285: 19–32.
6. Cammen LM (1980) The significance of microbial carbon in the nutrition of the deposit feeding polychaete Nereis succinea. Mar Biol 61: 9–20.
7. van Oevelen D, Moodley L, Soetaert K, Middelburg JJ (2006) The fate of bacterial carbon in an intertidal sediment: Modeling an in situ isotope tracer experiment. Limnol Oceanogr 51: 1302–1314.
8. Lozupone CA, Knight R (2007) Global patterns in bacterial diversity. Proc Natl Acad Sci U S A 104: 11436–11440.
9. Zinger L, Amaral-Zettler LA, Fuhrman JA, Horner-Devine MC, Huse SM, et al. (2011) Global patterns of bacterial beta-diversity in seafloor and seawater ecosystems. PLoS One 6: e24570.
10. Nemergut DR, Costello EK, Hamady M, Lozupone C, Jiang L, et al. (2011) Global patterns in the biogeography of bacterial taxa. Environ Microbiol 13: 135–144.
11. Amann RI, Ludwig W, Schleifer KH (1995) Phylogenetic identification and in situ detection of individual microbial cells without cultivation. Microbiol Rev 59: 143–169.
12. Keller M, Zengler K (2004) Tapping into microbial diversity. Nat Rev Microbiol 2: 141–150.
13. Gontang E a, Fenical W, Jensen PR (2007) Phylogenetic diversity of gram-positive bacteria cultured from marine sediments. Appl Environ Microbiol 73: 3272–3282.
14. Logares R, Sunagawa S, Salazar G, Cornejo-Castillo FM, Ferrera I, et al. (2013) Metagenomic 16S rDNA Illumina tags are a powerful alternative to amplicon sequencing to explore diversity and structure of microbial communities. Environ Microbiol. doi:10.1111/1462-2920.12250.
15. Bolhuis H, Stal LJ (2011) Analysis of bacterial and archaeal diversity in coastal microbial mats using massive parallel 16S rRNA gene tag sequencing. ISME J 5: 1701–1712.
16. Inagaki F, Suzuki M, Takai K, Oida H, Sakamoto T, et al. (2003) Microbial communities associated with geological horizons in coastal subseafloor sediments from the sea of Okhotsk. Appl Environ Microbiol 69: 7224–7235.
17. Haynes K, Hofmann TA, Smith CJ, Ball AS, Underwood GJC, et al. (2007) Diatom-derived carbohydrates as factors affecting bacterial community composition in estuarine sediments. Appl Environ Microbiol 73: 6112–6124.

18. Smith CJ, Nedwell DB, Dong LF, Osborn AM (2006) Evaluation of quantitative polymerase chain reaction-based approaches for determining gene copy and gene transcript numbers in environmental samples. Environ Microbiol 8: 804–815.
19. Llobet-Brossa E, Rosselló-Mora R, Amann R, Ramon Rosselló-Mora (1998) Microbial community composition of Wadden Sea sediments as revealed by fluorescence in situ hybridization. Appl Environ Microbiol 64: 2691–2696.
20. Boetius A, Ravenschlag K, Schubert CJ, Rickert D, Widdel F, et al. (2000) A marine microbial consortium apparently mediating anaerobic oxidation of methane. Nature 407: 623–626.
21. Smith CJ, Osborn AM (2009) Advantages and limitations of quantitative PCR (Q-PCR)-based approaches in microbial ecology. FEMS Microbiol Ecol 67: 6–20.
22. Findlay R, Trexler M, Guckert J, White D (1990) Laboratory study of disturbance in marine sediments: response of a microbial community. Mar Ecol Prog Ser 62: 121–133.
23. Boschker HTS, Middelburg JJ (2002) Stable isotopes and biomarkers in microbial ecology. FEMS Microbiol Ecol 40: 85–95.
24. Urakawa H, Yoshida T, Nishimura M, Ohwada K (2001) Characterization of microbial communities in marine surface sediments by terminal-restriction fragment length polymorphism (T-RFLP) analysis and quinone profiling. Mar Ecol Prog Ser 220: 47–57.
25. Kunihiro T, Miyazaki T, Uramoto Y, Kinoshita K, Inoue A, et al. (2008) The succession of microbial community in the organic rich fish-farm sediment during bioremediation by introducing artificially mass-cultured colonies of a small polychaete, Capitella sp. I. Mar Pollut Bull 57: 68–77.
26. Kaur A, Chaudhary A, Kaur A, Choudhary R, Kaushik R (2005) Phospholipid fatty acid – A bioindicator of environment monitoring and assessment in soil ecosystem. Curr Sci 89: 1103–1112.
27. Collins MD, Jones D (1981) Distribution of isoprenoid quinone structural types in bacteria and their taxonomic implications. Microbiol Rev 45: 316–354.
28. Boschker HTS (2004) Linking microbial community structure and functioning: stable isotope (13C) labeling in combination with PLFA analysis. In: Kowalchuk GA, Bruijn FJ de, Head IM, Akkermans AD, Elsas JD van, editors. Molecular microbial ecology manual II. Kluwer. 1673–1688.
29. Hu HY, Fujie K, Urano K (1999) Development of a novel solid phase extraction method for the analysis of bacterial quinones in activated sludge with a higher reliability. J Biosci Bioeng 87: 378–382.
30. Middelburg JJ, Barranguet C, Boschker HTS, Herman PMJ, Moens T, et al. (2000) The fate of intertidal microphytobenthos carbon: an in situ ^{13}C-labeling study. Limnol Oceanogr 45: 1224–1234.
31. Saitou K, Nagasaki K, Yamakawa H, Hu H-Y, Fujie K, et al. (1999) Linear relation between the amount of respiratory quinones and the microbial biomass in soil. Soil Sci Plant Nutr 45: 775–778.
32. Hiraishi A, Iwasaki M, Kawagishi T, Yoshida N, Narihiro T, et al. (2003) Significance of lipoquinones as quantitative biomarkers of bacterial populations in the environment. Microbes Environ 18: 89–93.

33. Takasu H, Kunihiro T, Nakano S (2013) Estimation of carbon biomass and community structure of planktonic bacteria in Lake Biwa using respiratory quinone analysis. Limnology 14: 247–256.

34. Polymenakou PN, Bertilsson S, Tselepides A, Stephanou EG (2005) Links between geographic location, environmental factors, and microbial community composition in sediments of the Eastern Mediterranean Sea. Microb Ecol 49: 367–378.

35. Kunihiro T, Takasu H, Miyazaki T, Uramoto Y, Kinoshita K, et al. (2011) Increase in Alphaproteobacteria in association with a polychaete, *Capitella* sp. I, in the organically enriched sediment. ISME J 5: 1818–1831.

36. Polymenakou PN, Bertilsson S, Tselepides A, Stephanou EG (2005) Bacterial community composition in different sediments from the Eastern Mediterranean Sea: a comparison of four 16S ribosomal DNA clone libraries. Microb Ecol 50: 447–462.

37. Saitou K, Fujie K, Katayama A (1999) Detection of microbial groups metabolizing a substrate in soil based on the [^{14}C] quinone profile. Soil Sci Plant Nutr 45: 669–679.

38. Hedrick DB, White DC (1986) Microbial respiratory quinones in the environment. J Microbiol Methods 5: 243–254.

39. Nowicka B, Kruk J (2010) Occurrence, biosynthesis and function of isoprenoid quinones. Biochim Biophys Acta 1797: 1587–1605.

40. Villanueva L, Navarrete A, Urmeneta J, Geyer R, White DC, et al. (2007) Monitoring diel variations of physiological status and bacterial diversity in an estuarine microbial mat: an integrated biomarker analysis. Microb Ecol 54: 523–531.

41. Peacock a D, Chang YJ, Istok JD, Krumholz L, Geyer R, et al. (2004) Utilization of microbial biofilms as monitors of bioremediation. Microb Ecol 47: 284–292.

42. Veuger B, van Oevelen D (2011) Long-term pigment dynamics and diatom survival in dark sediment. Limnol Oceanogr 56: 1065–1074.

43. Pozzato L, Van Oevelen D, Moodley L, Soetaert K, Middelburg JJ (2013) Sink or link? The bacterial role in benthic carbon cycling in the Arabian sea oxygen minimum zone. Biogeosciences Discuss 10: 10399–10428.

44. Pozzato L (2012) Prokaryotic, protozoan and metazoan processing of organic matter in sediments: a tracer approach Utrecht University.

45. Taylor J, Parkes RJ (1983) The Cellular Fatty Acids of the Sulphate-reducing Bacteria, *Desulfobacter* sp., *Desulfobulbus* sp. and *Desulfovibrio desulfuricans*. Microbiology 129: 3303–3309.

46. Tamaoka J, Katayama-Fujimura Y, Kuraishi H (1983) Analysis of bacterial menaquinone mixtures by high performance liquid chromatography. J Appl Bacteriol 54: 31–36.

47. Dauwe B, Middelburg JJ (1998) Amino acids and hexosamines as indicators of organic matter degradation state in North Sea sediments. Limnol Oceanogr 43: 782–798.

48. Lindroth P, Mopper K (1979) High performance liquid chromatographic determination of subpicomole amounts of amino acids by precolumn fluorescence derivatization with o-phthaldialdehyde. Anal Chem 51: 1667–1674.

49. Vandewiele S, Cowie G, Soetaert K, Middelburg JJ (2009) Amino acid biogeochemistry and organic matter degradation state across the Pakistan margin oxygen minimum zone. Deep Sea Res Part II Top Stud Oceanogr 56: 376–392.

50. Kreyszig E (1979) Advanced Engineering Mathematics. Fourth edi. John Wiley & Sons, Inc.

51. Hiraishi A, Morishima Y, Takeuchi J-I (1991) Numerical analysis of lipoquinone patterns in monitoring bacterial community dynamics in wastewater treatment systems. J Gen Appl Microbiol 37: 57–70.

52. Hu H-Y, Lim B-R, Goto N, Fujie K (2001) Analytical precision and repeatability of respiratory quinones for quantitative study of microbial community structure in environmental samples. J Microbiol Methods 47: 17–24.

53. R Development Core Team (2013) R: a language and environment for statistical computing <http://www.R- project.org>.

54. Dijkman N, Kromkamp J (2006) Phospholipid-derived fatty acids as chemotaxonomic markers for phytoplankton: application for inferring phytoplankton composition. Mar Ecol Prog Ser 324: 113–125.

55. Evrard V, Huettel M, Cook P, Soetaert K, Heip C, et al. (2012) Importance of phytodetritus and microphytobenthos for heterotrophs in a shallow subtidal sandy sediment. Mar Ecol Prog Ser 455: 13–31.

56. Collins MD, Jones D (1981) Distribution of isoprenoid quinone structural types in bacteria and their taxonomic implications. Microbiol Rev 45: 316–354.

57. Guckert JB, Antworth CP, Nichols PD, White DC (1985) Phospholipid, ester-linked fatty acid profiles as reproducible assays for changes in prokaryotic community structure of estuarine sediments. FEMS Microbiol Lett 31: 147–158.

58. Brinch-Iversen J, King GM (1990) Effects of substrate concentration, growth state, and oxygen availability on relationships among bacterial carbon, nitrogen and phospholipid phosphorus content. FEMS Microbiol Lett 74: 345–355.

59. Polglase WJ, Pun WT, Withaar J (1966) Lipoquinones of *Escherichia coli*. Biochim Biophys Acta 115: 425–426.

60. Hu H-Y, Lim B-R, Goto N, Bhupathiraju VK, Fujie K (2001) Characterization of microbial community in an activated sludge process treating domestic wastewater using quinone profiles. Water Sci Technol 43: 99–106.

61. Mannheim W, Stieler W, Wolf G, Zabel R (1978) Taxonomic significance of respiratory quinones and fumarate respiration in *Actinobacillus* and *Pasteurella*. Int J Syst Bacteriol 28: 7–13.

62. Hoppe H-G, Arnosti C, Herndl GF (2002) Ecological significance of bacterial enzymes in the marine environment. Enzymes in the environment. Marcel Dekker, Inc. 73–107.

63. Mayor DJ, Thornton B, Zuur AF (2012) Resource quantity affects benthic microbial community structure and growth efficiency in a temperate intertidal mudflat. PLoS One 7: e38582.

64. Haack SK, Garchow H, Odelson DA, Forney LJ, Klug MJ (1994) Accuracy, reproducibility, and interpretation of fatty acid methyl ester profiles of model bacterial communities. Appl Environ Microbiol 60: 2483–2493.

65. Zelles L (1999) Fatty acid patterns of phospholipids and lipopolysaccharides in the characterisation of microbial communities in soil: a review. Biol Fertil Soils 29: 111–129.

66. Kroppenstedt RM (1985) Fatty acid and menaquinone analysis of Actinomycetes and related organisms. In: Goodfellow M, Minnikin DE, editors. Chemical methods in bacterial systematics. Academic Press. 173–199.

67. Martens T, Heidorn T, Pukall R, Simon M, Tindall BJ, et al. (2006) Reclassification of *Roseobacter gallaeciensis* Ruiz-Ponte, et al. 1998 as *Phaeobacter gallaeciensis* gen. nov., comb. nov., description of *Phaeobacter inhibens* sp. nov., reclassification of *Ruegeria algicola* (Lafay, et al. 1995) Uchino, et al. 1999 as Marinovu. Int J Syst Evol Microbiol 56: 1293–1304.

68. Hwang CY, Cho BC (2008) *Cohaesibacter gelatinilyticus* gen. nov., sp. nov., a marine bacterium that forms a distinct branch in the order Rhizobiales, and proposal of *Cohaesibacteraceae* fam. nov. Int J Syst Evol Microbiol 58: 267–277.

69. Romanenko LA, Tanaka N, Svetashev VI, Kalinovskaya NI (2011) *Pacificibacter maritimus* gen. nov., sp. nov., isolated from shallow marine sediment. Int J Syst Evol Microbiol 61: 1375–1381.

70. Oyaizu H, Komagata K (1981) Chemotaxonomic and phenotypic characterization of the strains of species in the Flavobacterium-Cytophaga complex. J Gen Appl Microbiol 27: 57–107.

71. Van Trappen S, Tan T-L, Samyn E, Vandamme P (2005) *Alcaligenes aquatilis* sp. nov., a novel bacterium from sediments of the Weser Estuary, Germany, and a salt marsh on Shem Creek in Charleston Harbor, USA. Int J Syst Evol Microbiol 55: 2571–2575.

72. Lim JH, Baek S-H, Lee S-T (2008) *Burkholderia sediminicola* sp. nov., isolated from freshwater sediment. Int J Syst Evol Microbiol 58: 565–569.

73. Vancanneyti M, Witt S, Abraham W-R, Kersters K, Fredrickson HL (1996) Fatty acid content in whole-cell hydrolysates and phospholipid and phospholipid fractions of Pseudomonads: a taxonomic evaluation. Syst Appl Microbiol 19: 528–540.

74. Jean WD, Huang S-P, Liu TY, Chen J-S, Shieh WY (2009) *Aliagarivorans marinus* gen. nov., sp. nov. and *Aliagarivorans taiwanensis* sp. nov., facultatively anaerobic marine bacteria capable of agar degradation. Int J Syst Evol Microbiol 59: 1880–1887.

75. Londry KL, Jahnke LL, Des Marais DJ (2004) Stable carbon isotope ratios of lipid biomarkers of sulfate-reducing bacteria. Appl Environ Microbiol 70: 745–751.

76. Smith JL, Campbell BJ, Hanson TE, Zhang CL, Cary SC (2008) *Nautilia profundicola* sp. nov., a thermophilic, sulfur-reducing epsilonproteobacterium from deep-sea hydrothermal vents. Int J Syst Evol Microbiol 58: 1598–1602.

77. Kim HM, Hwang CY, Cho BC (2010) *Arcobacter marinus* sp. nov. Int J Syst Evol Microbiol 60: 531–536.

78. O'Sullivan LA, Rinna J, Humphreys G, Weightman AJ, Fry JC (2006) Culturable phylogenetic diversity of the phylum "Bacteroidetes" from river epilithon and coastal water and description of novel members of the family Flavobacteriaceae: *Epilithonimonas tenax* gen. nov., sp. nov. and *Persicivirga xylanidelens* gen. nov., sp. Int J Syst Evol Microbiol 56: 169–180.

79. Kaur I, Kaur C, Khan F, Mayilraj S (2012) *Flavobacterium rakeshii* sp. nov., isolated from marine sediment, and emended description of *Flavobacterium beibuense* Fu, et al. 2011. Int J Syst Evol Microbiol 62: 2897–2902.

80. Kroppenstedt RM (2006) The Family Nocardiopsaceae. Prokaryotes. Springer New York. 754–795.

81. Knittel K, Kuever J, Meyerdierks A, Meinke R, Amann R, et al. (2005) *Thiomicrospira arctica* sp. nov. and *Thiomicrospira psychrophila* sp. nov., psychrophilic, obligately chemolithoautotrophic, sulfur-oxidizing bacteria isolated from marine Arctic sediments. Int J Syst Evol Microbiol 55: 781–786.

82. Akagawa-Matsushita M, Itoh T, Katayama Y, Kuraishi H, Yamasato K (1992) Isoprenoid quinone composition of some marine *Alteromonas*, *Marinomonas*, *Deleya*, *Pseudomonas* and *Shewanella* species. J Gen Microbiol 138: 2275–2281.

83. Shin N-R, Whon TW, Roh SW, Kim M-S, Kim Y-O, et al. (2012) *Oceanisphaera sedimini* sp. nov., isolated from marine sediment. Int J Syst Evol Microbiol 62: 1552–1557.

84. Collins MD, Weddel F (1986) Respiratory quinones of sulphate-reducing and sulphur-reducing bacteria: a systematic investigation. Syst Appl Microbiol 8: 8–18.

85. Devereux R, Delaney M, Widdel F, Stahl DA (1989) Natural relationships among sulfate-reducing eubacteria. J Bacteriol 171: 6689–6695.

86. Lancaster C (2002) Succinate:quinone oxidoreductases from ε-proteobacteria. Biochim Biophys Acta - Bioenerg 1553: 84–101.

87. Nakagawa Y, Yamasato K (1993) Phylogenetic diversity of the genus Cytophaga revealed by 16S rRNA sequencing and menaquinone analysis. J Gen Microbiol 139: 1155–1161.

88. Yamada Y, Inouye G, Kondo K (1976) The menaquinone system in the classification of coryneform and nocardioform bacteria and related organisms. J Gen Appl Microbiol 22: 203–214.

89. Athalye M, Goodfellow M, Minnikin DE (1984) Menaquinone composition in the classification of *Actinomadura* and related taxa. J Gen Microbiol 130: 817–823.

90. Boschker HTS, Kromkamp JC, Middelburg JJ (2005) Biomarker and carbon isotopic constraints on bacterial and algal community structure and functioning in a turbid, tidal estuary. Limnol Oceanogr 50: 70–80.

91. Ratledge C, Wilkinson S (1988) Microbial lipids volume 1. London: Academic Press.

92. Yokota A, Akagawa-Matsushita M, Hiraishi A, Katayama Y, Urakami T, et al. (1992) Distribution of quinone systems in microorganisms: Gram-negative eubacteria. BullJFCC 8: 136–171.

93. Hiraishi A (1999) Isoprenoid quinones as biomarkers of microbial populations in the environment. J Biosci Bioeng 88: 449–460.

94. Fujie K, Hu H-Y, Tanaka H, Urano K, Saitou K, et al. (1998) Analysis of respiratory quinones in soil for characterization of microbiota. Soil Sci Plant Nutr 44: 393–404.

16

Biodiversity of Prokaryotic Communities Associated with the Ectoderm of *Ectopleura crocea* (Cnidaria, Hydrozoa)

Cristina Gioia Di Camillo[1]*, Gian Marco Luna[2], Marzia Bo[3], Giuseppe Giordano[1], Cinzia Corinaldesi[1], Giorgio Bavestrello[3]

1 Department of Life and Environmental Sciences, Polytechnic University of Marche, Ancona, Italy, 2 Institute of Marine Sciences–National Research Council (CNR), Venice, Italy, 3 Department for the Study of the Land and its Resources, University of Genoa, Genoa, Italy

Abstract

The surface of many marine organisms is colonized by complex communities of microbes, yet our understanding of the diversity and role of host-associated microbes is still limited. We investigated the association between *Ectopleura crocea* (a colonial hydroid distributed worldwide in temperate waters) and prokaryotic assemblages colonizing the hydranth surface. We used, for the first time on a marine hydroid, a combination of electron and epifluorescence microscopy and 16S rDNA tag pyrosequencing to investigate the associated prokaryotic diversity. Dense assemblages of prokaryotes were associated with the hydrant surface. Two microbial morphotypes were observed: one horseshoe-shaped and one fusiform, worm-like. These prokaryotes were observed on the hydrozoan epidermis, but not in the portions covered by the perisarcal exoskeleton, and their abundance was higher in March while decreased in late spring. Molecular analyses showed that assemblages were dominated by Bacteria rather than Archaea. Bacterial assemblages were highly diversified, with up to 113 genera and 570 Operational Taxonomic Units (OTUs), many of which were rare and contributed to <0.4%. The two most abundant OTUs, likely corresponding to the two morphotypes present on the epidermis, were distantly related to Comamonadaceae (genus *Delftia*) and to Flavobacteriaceae (genus *Polaribacter*). Epibiontic bacteria were found on *E. crocea* from different geographic areas but not in other hydroid species in the same areas, suggesting that the host-microbe association is species-specific. This is the first detailed report of bacteria living on the hydrozoan epidermis, and indeed the first study reporting bacteria associated with the epithelium of *E. crocea*. Our results provide a starting point for future studies aiming at clarifying the role of this peculiar hydrozoan-bacterial association.

Editor: Jack Anthony Gilbert, Argonne National Laboratory, United States of America

Funding: This work was partly supported by the programme FIRB Futuro in ricerca 2008 funded by the MIUR (contract n.I31J10000060001). The funders had no role in study design, data collection and analysis, decision to publish, or preparation of the manuscript. No additional external funding was received for this study.

Competing Interests: The authors have declared that no competing interests exist.

* E-mail: c.dicamillo@univpm.it

Introduction

Anthozoan cnidarians are known to host rich populations of associated bacteria: the mucus layers of the hard coral *Porites astreoides* Lamarck, 1816 and those of the zoanthid *Palythoa* sp., for instance, host rich procariotic assemblages, whose abundance is regulated by the self-cleaning mechanisms of the cnidarian host [1,2]. Bacteria isolated from the ectodermal surface typically produce bioactive compounds, this is the case of the gorgonians *Subergorgia suberosa* (Pallas, 1766) and *Junceella juncea* (Pallas, 1766) [3] and of the soft coral *Dendronephthya* sp. that, by inhibiting the larval settlement, help the host in maintaining its surface clean [4]. Recently, bacteria were also found associated with black corals (such as the wire-coral *Stichopathes lutkeni* Brook, 1889 [5]). While the symbioses between microorganisms and anthozoans have been widely investigated, the interactions between bacteria and hydrozoans are still largely unknown, and this is especially evident for the bacterial assemblages colonizing the epithelia of their hosts.

Hydroids present an exoskeleton made up of polysaccharides and proteins (the perisarc), that can either envelop the zooids or can be limited to stolons and branches. This structure is frequently colonized by a complex assemblage of protists and prokaryotes [6–9]. Recently, the presence of chitinolytic bacteria belonging to the genus *Vibrio* has been reported in association with the perisarc of different species of hydroids [10,11]. Several epibiotic microorganisms were reported using electron microscopy and cultivation approaches on different hydroid species and within the same species in various parts of the colony (stem, branches, hydrothecae, tentacles etc) [12].

Very few cases of bacteria living on the hydrozoan epidermis were until now reported: gram-negative, rod-shaped bacteria were found on the epidermis of the green hydra along the margin of the ectodermal cells and becoming particularly abundant on the hypostome and the foot [13]. TEM analyses also revealed the presence of elongated bacterial cells on the epithelium of *Pennaria disticha* (Goldfuss, 1820) [14].

Several hydrozoan species are known to host bacteria inside their tissues, for example, spirochaetes in *Hydra circumcincta* Schulze, 1914 were localized i) extracellularly, – in the spaces between epitheliomuscular cells, ii) intracellularly – inside vacuoles of the epitheliomuscular cells and iii) through the mesoglea [15]. The gastroderm of some strains of *Hydra viridissima* (Pallas, 1766) contained bacterial vesicles associated with cells containing the symbiotic green alga *Chlorella* sp., that could promote metabolic exchanges between the host and the algal symbionts [16]. The recent discovery of nine species of bacteria living inside the epidermal cells of the hydrozoan *Tubularia indivisa* Linnaeus, 1758 has led to the hypothesis of their involvement in the production of cnidarians toxins [17].

The main caveats of the available studies on the associations between marine invertebrates (including hydrozoa) and bacteria, is that almost all of them have been based only on morphological descriptions based on microscopy or on classical cultivation-based techniques [18] that often led to erroneous identifications. The use of culture independent and highly sensitive techniques (e.g., tag-encoded amplicon pyrosequencing of hypervariable regions of the 16S rRNA gene) allows to overcome these problems and to identify also "rare" prokaryotic taxa, which could play a significant role in these associations [19].

Here we investigated the association between the Anthomedusan hydroid *Ectopleura crocea* (L. Agassiz, 1862) and the prokaryotic assemblages colonizing the hydranth surface and tested the hypothesis that the interaction and characteristics of the microbial assemblages change over time and according to the anatomical regions of the hydranths. We used a combination of electron and epifluorescence microscopy and molecular techniques (the tag-encoded pyrosequencing of the 16S rRNA gene). The application of the high-resolution pyrosequencing technique makes this study the most detailed investigation carried out so far to describe the biodiversity of prokaryotes associated with a marine hydroid.

Materials and Methods

Samplings

Samplings of *Ectopleura crocea* (Figure 1A, B) were carried out in 2009 and 2010 by SCUBA diving on the starboard bow of the wreck *Nicole* sunk at water depth of 10–12 m about 2 miles off the Conero Promontory (Numana, Italy) (43°30′076′′N – 13°40′191′′E). At each sampling time (April 2009, March 2010, April 2010 and May 2010), three hydroid colonies of about 100 hydranths each were detached from the wreck walls and, using previously sterilized steel forceps, immediately put into sterilized test-tubes in order to avoid microbial contamination. Once back in the laboratory, the samples for microbiological analyses were washed twice in autoclaved and 0.2-μm prefiltered seawater to remove seawater microbes which can influence the composition of the hydrant-associated bacterial assemblage. Hence, the samples were processed according to the protocols required for each specific analysis detailed here below.

Morphological description and distribution

In order to observe, by means of electron microscopy (SEM and FE-SEM), the epibionts' morphology and their distribution on the host, several hydroid polyps were cut from each collected colony with micro-forceps and then fixed for three hours in 2.5% glutaraldehyde buffered with filtered seawater (pH adjusted to 7.5–7.8, with 0.1N NaOH). Then, for the Scanning Electron Microscopy (SEM), part of the samples was washed with distilled water, dehydrated in a graded ethanol series and dried with the Critical Point Dryer. They were then coated with gold-palladium in a Balzer Union evaporator and examined with a Philips XL20 SEM and a FESEM Zeiss Supra 40. Each considered portion of the polyp (oral and aboral tentacles, gastric column, hydranth base, neck, gonophores and hydrocaulus) was analysed by SEM to investigate the bacterial distribution along the colony (Figure 2). To investigate the possible mechanism of transmission of bacteria, we collected and analysed the actinulae just released and the actinulae developed inside the female gonophores. For ultrastructural investigation by Transmission Electron Microscopy (TEM),

part of the fixed samples was placed in filtered seawater, then post-fixed in 2% osmium tetroxide for 30 min, dehydrated in a graded ethanol series and embedded in Epon 812. Sections were cut on LKB ultramicrotome and the thin sections (70 nm) were stained with uranyl acetate and lead citrate and observed in a TEM CM200 operating at 100KV.

Comparison with other geographic areas

During the same study period, three other populations of *E. crocea* collected from different geographic areas were examined under electron microscopy to establish whether the association of microbes was a constant feature in this hydroid species. We analyzed colonies found in the shallow waters of the Murano's channels (Venice, North Adriatic Sea, located approximately 100 nautical miles apart northward from the wreck *Nicole*) and collected from the ship keels. Additional colonies were collected from a floating buoy deployed 10 miles off Brindisi (South Adriatic Sea, located approximately 300 miles apart southward from the wreck *Nicole*). Finally, additional samples were collected from fish farm cages located in the Ligurian Sea (Western Mediterranean Sea, located approximately 400 miles apart from the wreck *Nicole* in a different biogeographic region of the Mediterranean Sea). Other athecate hydroid species, namely *Eudendrium glomeratum* (Picard, 1952) and *Eudendrium racemosum* (Cavolini, 1785), which are common on the wreck *Nicole*, were also examined to verify the presence of associated microbial assemblages using electron microscopy.

Variations of abundance of the epibiotic prokaryotes over time and along the hydranths using electron microscopy

In the samples collected in March 2010, April 2010 and May 2010, five of the hydranths prepared for SEM were randomly chosen to determine the total microbial abundance (expressed as number of cell mm^{-2}). Five pictures of each hydroid portion were taken under the electron microscope; in each photo, bacteria were counted in nine randomly chosen areas having dimension of 10×10 μm^2. The results were reported according to the examined portion of the hydroid (aboral tentacles, oral tentacles, gonophores, gastric column, base of the hydranth, neck region and hydrocaulus).

Variations of abundance of the epibiotic prokaryotes over time using epifluorescence microscopy

The abundance of prokaryotes associated with hydrozoan hydranths was determined in April 2009, March 2010 and May 2010 by using epifluorescence microscopy. Ten hydranths from each of the 3 freshly-collected *Ectopleura crocea* colonies were preserved in 1 ml of 0.2 μm pre-filtered formalin (2% concentration in sterile seawater). To check the detachment efficiency of the prokaryotic cells from the hydranths, aliquots from each suspension (i.e., the formalin solution containing the hydranths) were treated with ultrasounds for 0, 1, 3, 8, and 15 min by using a Branson Sonifier 2200 (60W) in an ice bath to prevent overheating. Each replicate was then homogenized, vigorously vortexed and treated with ultrasounds for three times for one minute each, with 30 seconds intervals within each cycle, as this procedure optimized the detachment of prokaryotes from the hydrants.

Aliquots from each of the obtained suspensions were filtered onto 0.2 μm pore-size aluminium oxide filters (Anodisc, Whatman). Filters were stained using SYBR Green I (10000× in anhydrous dimethyl sulfoxide, Molecular Probes) by adding 20 μl

Figure 1. Main characteristics of *Ectopleura crocea.* A) Underwater photograph of the hydroid. B) Scheme of the colony. 1. Feeding and reproductive polyps; 2. Stems (hydrocauli); 3. Tangled stolons anchoring the colony to the substrate (hydrorhizae).

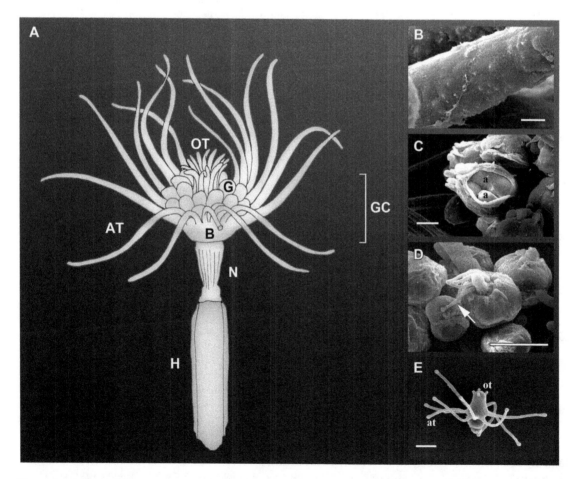

Figure 2. Portions of *E. crocea* **examined to find prokaryotes.** A) Scheme illustrating the main features of the polyp stage. GC gastric column, AT aboral tentacles, OT oral tentacles, G gonophores, B basal portion of hydranth, N neck zone, H hydrocaulus (portion covered with perisarc). B–E. Close-up view at scanning electron microscope. B). Aboral tentacle colonized by bacteria. C. Broken female gonophore containing immature actinulae round in shape (a). Bacteria were rarely found on actinulae at this stage. D. Mature female gonophore with actinula's tentacles protruding through the opening (white arrow). Bacteria were often observed on these tiny tentacles. E. Released and free-living actinula with developed aboral (at) and oral (ot) tentacles. Scale bars: B 20 μm; C 500 μm D, E 200 μm.

of the stock solution (previously diluted 1:20 with 0.2 μm filtered Milli-Q water). The filters were incubated 15 minutes in the dark, washed twice with 3 ml of sterilized Milli-Q water, then mounted onto microscope slides and added with 20 μl of antifade solution (50% phosphate buffer and 50% glycerol containing 0.5% ascorbic acid). Filters were analyzed by epifluorescence microscopy (Zeiss Axioskop 2, magnification ×1,000) under blue light excitation. For each filter, at least 20 microscope fields were observed. The data were expressed as number of cells per hydrant.

Metagenetic analysis of the epibiotic prokaryotes

To identify the epibiotic prokaryotes, we used the tag-encoded amplicon pyrosequencing of hypervariable regions (V5 and V6) of the prokaryotic 16S rRNA gene. The genomic DNA was extracted from ten hydrants from freshly-collected *Ectopleura crocea* samples in April 2009 and March 2010. These two samples were selected on the basis of the results of total abundance, which showed that these samples were characterized by the highest abundance of epibiotic prokaryotes. For DNA extraction, the hydrants were put into a sterile tube and the DNA extracted by using the UltraClean Soil DNA Isolation kit (MoBio Laboratoires) following the manufacturer protocol. The concentrations of extracted DNA were determined by using a NanoDrop^{TM} fluorospectrometer (Thermo Scientific) and SYBR Green I as a stain. Extracted DNA was stored at −80°C until further molecular analyses of prokaryotic diversity. Bacterial and archaeal 16S ribosomal DNA (rDNA) amplicons were generated using the universal primers 789F (5′-TAGATACCCSSGTAGTCC-3′) and 1046R (5′- CGACAGC-CATGCANCACCT-3′; [20,21]). All PCR reactions were performed in a volume of 50 μl in a thermalcycler (Biometra, Germany) using the MasterTaq® kit (Eppendorf AG, Germany), which reduces the effects of PCR-inhibiting contaminants. Thirty PCR-cycles were used, consisting of 94°C for 1 minute, 55°C for 1 minute and 72°C for 2 minute, preceded by 3 minutes of denaturation at 94°C and followed by a final extension of 10 minutes at 72°C. To check for eventual contamination of the PCR reagents, negative controls containing the PCR-reaction mixture but without the DNA template were run during each amplification. Positive controls, containing genomic DNA of *Escherichia coli*, were also used. PCR-products were checked on agarose-TBE gel (1%), containing ethidium bromide for DNA staining and visualization. Twelve different reactions were run for each sample and then combined together to reduce possible PCR biases and to reach the amount necessary for 454 analysis. The amplicons were purified using Amicon Ultra 50 k device (Millipore). The amplicon length and concentration were estimated using the BioAnalyzer microfluidics device (Agilent), and then each amplicon was sequenced via emulsion PCR (which was performed using the recommended kit and protocol from 454 Life Sciences) by using a Genome Sequencer FLX Titanium (Roche). The analysis of the 16S rDNA sequences obtained was performed using the RDP's Pyrosequencing Pipeline (http://pyro.cme.msu. edu/ [22]). Briefly, the sequences were firstly analysed using the "Pipeline Initial Process", in order to trim off the key tag and primers, and to remove sequences of low quality (using a minimum average exp. quality score of 20). Then, bacterial and archaeal sequences were downloaded from each original FASTA sequence file using the option "FASTA Sequence Selection". Sequences were aligned using the "Pyrosequencing Aligner" tool and clustered using the "Complete Linkage Clustering" tool. Rarefaction curves, the number of OTUs (Operational Taxonomic Units) and the non-parametric Chao1 estimator were determined using the "Analysis tools" available in the RDP pipeline. The taxonomical identification of bacterial and archaeal sequences

was performed using the RDP classifier [18], using the default bootstrap cutoff of 80%. The partial 16S rDNA sequences obtained in this study have been deposited in the NCBI Short Read Archive under the accession number SRA052825.3.

Ethics statement

No specific permits were required for the described field studies. The wreck Nicole is not private, the buoy and the fish farm cage are private, but no specific permissions were required for all the mentioned locations or activities and the sampling procedures employed did not cause damage to the privately-owned properties. The field studies did not involve endangered or protected species and the locations were not protected in any way. The D.P.R. n° 1639 of the 2th October 1968, articles 26 – 29 regards only fish samplings or researches inherent to fishing activities (art. 7). The collection of hydroids is excluded by this kind of regulation. No permits are necessary to collect hydroids.

Results

Ectopleura crocea is a colonial hydroid distributed worldwide in temperate waters. In the study area, this hydroid settles on the wreck *Nicole* showing a marked temporal variation [23]. The hydroid species appears in autumn (November or December, when the *in situ* temperature is ca. 15°C), reaches its highest abundance in April and then declines in late spring (at the end of May or early June, when the sea temperature ranges between 18 and 21°C). Colonies are formed by dense tufts of polyps, each one showing an erect stem – the hydrocaulus – surrounded by perisarc and a hydranth lacking of perisarc and bearing mouth and tentacles (Figure 2A–E).

Morphological description of prokaryotes and their distribution

The electron microscopy clearly revealed the presence of dense populations of prokaryotes living in association with *E. crocea* hydranths. Two microbial morphotypes were observed, both displaying an elongated shape. The first, here named *Type I*, is horseshoe-shaped, 2 to 3.6 μm long (3 μm ±0.17) and 0.2 to 0.3 μm wide (0.2 μm ±0.01) (Figure 3 A, C, E), while the second (named *Type II*) is fusiform, worm-like and measures 2.8 to 5.8 μm in length (4 μm ±0.34) and 0.2 to 0.3 μm in width (0.2 μm ±0.01) (Figure 3 A, B, D, F). SEM micrographs of tentacles' transversal sections showed that these prokaryotes were present only on the external surface, and not inside the epitheliomuscular cells (Figure 3B). Prokaryotes were present all around the epidermis but, particularly, on the aboral tentacles. The two morphotypes were often simultaneously present on the hydroid's surface, but the fusiform, worm-like *Type II* was most commonly observed. Both morphotypes were generally observed on the hydranths and mature gonophores, while they were rarely seen on the neck region and never seen along the hydrocaulus.

E. crocea shows a free-moving stage bearing tentacles called actinula (Figure 2E). The SEM analysis of just released actinulae revealed the presence of prokaryotes on their epidermis, but microbial cells were rare and scattered. The early developed actinulae, ovoid and lacking tentacles, found inside the female gonophores (Figure 2C), rarely hosted microbial cells.

The transmission electron microscopy (TEM) confirmed that prokaryotes were only on the external surface of the hydranths, directly lying on the hydroid mucoproteinic coating (periderm) (Figure 4A–F). The glycocalyx of the microorganisms kept contact with the hydroid periderm (Figure 4A–F). The hydranth surface

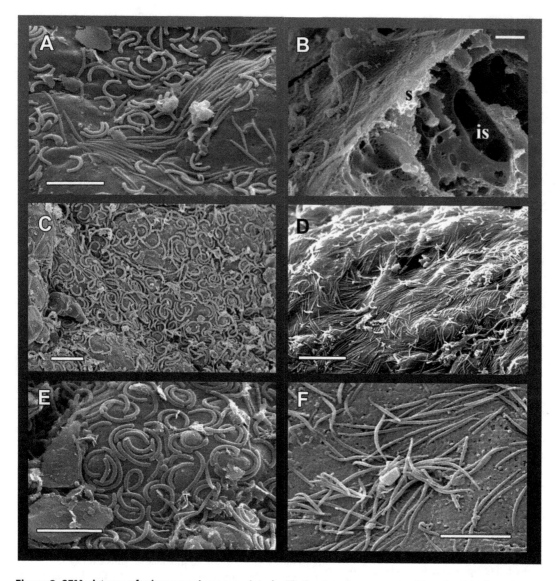

Figure 3. SEM pictures of microorganisms associated with the *Ectopleura crocea* epidermis. A) Surface of a tentacle densely covered with the two morphotypes of microorganisms living on the epidermis: one is horseshoe-shaped (green; named *Type I*) and the other is fusiform, worm-like (red; named *Type II*). B) Portion of a broken tentacle, bacteria were present on the surface (s) but not inside (is). C–D) Enlargements of tentacle portions which were densely covered by the two microorganisms. E) Particular of the horseshoe-shaped *Type I*, showing the peculiar arrangement in the tentacle grooves. F) Particular of a cluster of the worm-like *Type II*. Scale bars: A, C, E, F 5 μm; B 2 μm; D 10 μm.

observed by means of TEM and SEM did not show any signs of damage due to the presence of microorganisms.

Variations of abundance of the epibiotic prokaryotes over time and along the hydranths using electron microscopy

The analysis of SEM pictures showed that prokaryotic abundance varied considerably in the different anatomic regions of the hydranth and along the investigated period. The fusiform bacteria (Type II) were the most abundant. The highest density was observed on the aboral tentacles, with more than 500,000 cells mm^{-2}, while the lowest values were recorded on the neck region and the hydranth base. Concerning the temporal trend the highest values were observed in March 2010 (when the water temperature was 9.3°C) and the lowest in May 2010 (temperature = 16.6°C) (Figure 5A). Cell abundance on the aboral tentacles in the period

March-May was inversely correlated with temperature (r = 0,885; p<0.05).

Variations of abundance of epibiotic prokaryotes over time using epifluorescence microscopy

The abundance of epibionts on *E. crocea*, determined using the epifluorescence microscopy technique, was in the range 3,334±453 to 548,558±113,880 cells per hydrant, with a peak observed in March 2010 (Figure 5B).

Comparison with Ectopleura from other geographic areas

SEM analyses *of Ectopleura* specimens collected from the Murano's channel and the fish farm cages from the Ligurian Sea displayed the presence of prokaryotes of both types (*Type I* and *II*), despite with a lower density compared with those collected

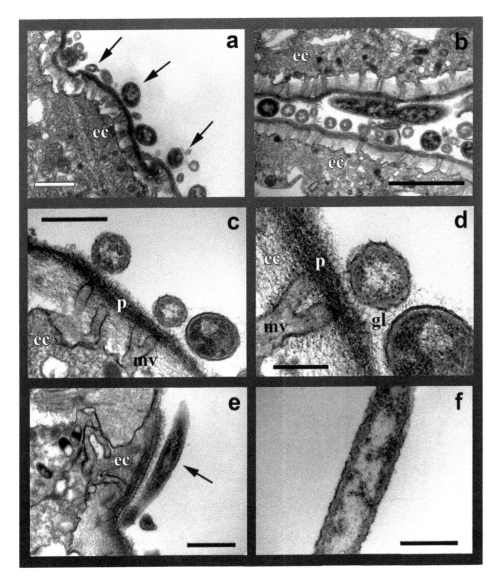

Figure 4. TEM pictures of *Type II* **bacteria associated with** *Ectopleura crocea.* A) Numerous bacteria in transversal sections (arrows) observed on the hydroid ectoderm (ec). B) Bacteria (longitudinal and tranversal sections) present in a groove of the hydroid ectoderm. C) Bacteria lying on the hydroid periderm (p) are often found in correspondence to the microvilles (mv) of the ectodermal cells. D) Close-up view showing the glycocalyx (gl) surrounded the microorganisms. e–f. Longitudinal section of a bacterium. Scale bars: a, c, e 1 μm; b 2 μm; d, f 0.5 μm.

from the wreck *Nicole*. Conversely, the specimens from the floating buoy from Brindisi did not show any associated prokaryote, as well as the other analyzed hydroid species (*E. glomeratum* and *E. racemosum*) from the wreck *Nicole*.

Metagenetic analysis of the epibiotic prokaryotes

The tag-encoded amplicon pyrosequencing analyses of the 16S rRNA gene from the *E. crocea* hydrants resulted in a total of 105,023 reads (32,882 in the samples collected in April 2009 and 72,141 in the samples collected in March 2010), with an average length of 238 and 233 nucleotides (April 2009 and March 2010, respectively) corresponding to a total of 16,324,661 bases sequenced. After filtering out the low-quality and short sequence reads, the number of high quality reads remained was 6,406 and 13,076 (April 2009 and March 2010, respectively). Of these, only 14 and 4 (respectively) matched with Archaea, indicating that the prokaryotic assemblages were largely dominated by Bacteria. In

either April 2009 and March 2010, the prokaryotic assemblages were dominated by bacteria belonging to the phyla Proteobacteria and Bacteroidetes (Figure 6A and B), which together accounted for 80–90% of the assemblage (April 2009 and March 2010, respectively). Actinobacteria was the third more represented phylum (15 and 2%, respectively), followed by Firmicutes (both ca 2%). Unclassified bacteria accounted for ca. 2 and 4%, respectively. Rarefaction analyses, based on OTUs definition at 97% similarity, indicated that rarefaction curves almost reached the plateau, and that the diversity was well described with the sequencing effort here utilized (data not shown). Further evaluation of the community structure at finer taxonomical level showed, using the 97% similarity cut-off, the presence of 112 bacterial genera (Figure 7A and B) and 570 bacterial OTUs in the sample collected in April 2009, and of 113 genera and 522 bacterial OTUs in the sample collected in March 2010. In both samples, the assemblage was dominated by two clusters of sequences matching with two genera (Figure 7A and B, reporting the 20 most abundant

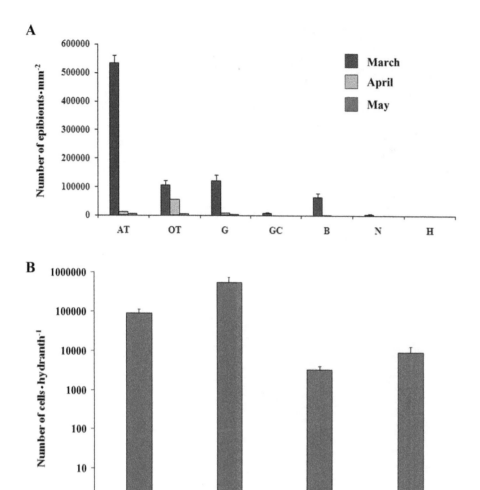

Figure 5. Total prokaryotic abundance, as determined by: A) analysis of the TEM pictures of prokaryotes present on several hydroid portions and B) epifluorescence microscopy. AT: aboral tentacles; OT: oral tentacles; G: gonophores; GC: gastric column; B: base of hydranth; N: neck region; H: hydrocaulus. Standard errors are shown.

bacterial genera). In April 2009, 23% of sequences clustered within the family Comamonadaceae, genus *Delftia* (average similarity 91%) while 26% of sequences clustered within the family Flavobacteriaceae (100% similarity) and showed the highest match with sequences of the genus *Polaribacter*. However, the similarity was very low (on average 50%), suggesting that it is probably a new genus/species. In March 2010, the *Polaribacter* cluster accounted for 76% of sequences and the *Delftia* cluster accounted for 5% of sequences. The remaining sequences belonged to many bacterial genera, most of which were "rare" and accounted for a minor percentage within the assemblage (from 0.2% – 0.4% down to 0.01 – 0.02%; see Table S1 for a complete list).

Discussion

The study of prokaryotic assemblages associated with cnidarians, which belong to one of the earliest branches in the animal tree of life, is potentially important to better understand the specificity of their interactions, their evolution and influence on the cnidarian health [24]. This study, based for the first time on the use of massive parallel sequencing, has enabled the accurate

identification of the bacterial assemblages associated with their cnidarian host. The hydroid *E. crocea* represents a mosaic of microenvironments that can be colonized by several organisms, both eukaryotic and prokaryotic, that generally settle only on those portions covered by the exoskeleton [6–11,25]. Several hydroid species have been shown to live with an associated microbial flora [26], but this is the first study dealing with the presence of bacteria on the hydroid's epidermis documented with a molecular approach combined with electron microscopy. Our molecular analyses revealed that most of the observed prokaryotic diversity was within the domain of Bacteria, whereas Archaea accounted for a negligible fraction. The bacterial assemblage was highly diversified, and comprised more than 100 bacterial genera and more than 500 taxa (OTUs) per single hydroid. This high bacterial biodiversity is unprecedented for hydrozoans, and is due to the application of the high resolution tag-encoded amplicon pyrosequencing technique, as already reported for other marine organisms [18,19,27]. This high diversity includes a large proportion of "rare" bacterial species, which may be transient microbes or may represent specifically associated microbes, which may provide benefits to their hydroid host (e.g. preventing fouling

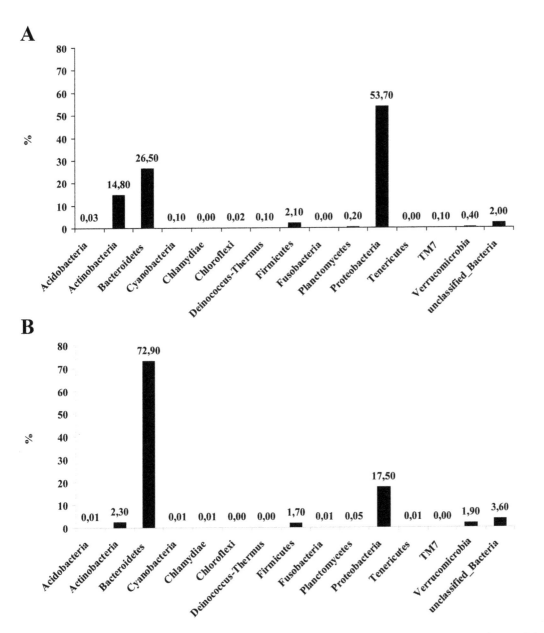

Figure 6. Community composition of bacterial assemblages associated with *E. crocea* **in April 2009 (A) and March 2010 (B), as revealed by tag-encoded amplicon pyrosequencing of the 16S rDNA gene.** Results are reported at the taxonomic level of phylum.

by other organisms, excluding potential pathogens and resource competitors, providing nutrients).

The high bacterial diversity observed through the metagenetic analysis was associated to two morphotypes revelaed by the electron microscopy analyses, which we named *Type I* and *Type II*. The elongated, worm-like morphotype (*Type II*) was most frequently observed and more abundant when compared to the other type. The two morphotypes were found exclusively on the epidermis of the hydranths and actinulae of *E. crocea*. Portions covered by the thick and stiff perisarcal exoskeleton, such as the stem and the hydrorhiza, were never colonized by these bacteria. The morphotypes *Type I* and *II* were rarely found on the neck, the transition region between the stem and the hydranth, covered with a very thin and soft skeletal layer, probably having a different protein/chitin ratio and therefore partially allowing the epibiosis. This non-random association between the host and the microbes

suggests that the two microbes may play a role in the hydroid life and survival.

Molecular analyses indicated the dominance of two OTUs, one clustering within the family Flavobacteriaceae, genus *Polaribacter*, and the other within the family Comamonadaceae, genus *Delftia*. The OTU related within the Flavobacteriaceae family dominated the two libraries, especially the one carried out on the samples collected in March 2010. The similarity with the genus *Polaribacter* was very low (on average 50%), suggesting that this OTU possibly represents a new genus within the Flavobacteriaceae family, or a new species within the *Polaribacter* genus, but further investigations are needed to confirm this hypothesis. The family Flavobacteriaceae is widely distributed in the marine environment, and several members of this family display a cellular worm-like morphology [28], similar to the one displayed by the morphotype *Type II*. In addition, some *Polaribacter* include psycro- or meso-

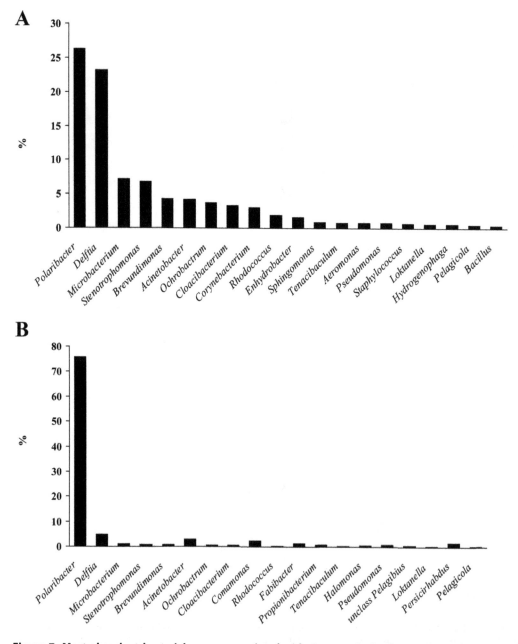

Figure 7. Most abundant bacterial genera associated with _E. crocea_ in April 2009 (A) and March 2010 (B), as revealed by tag-encoded amplicon pyrosequencing of the 16S rDNA gene. Reported are the 20 most abundant genera.

philic species, which do not tolerate high temperatures [29], and this could partly explain the negative relationship observed between sea water temperature and _Type II_ abundance on the hydroid surface. Nonetheless, the abundance of this bacterium varied during the study period, showing a peak in March and decreasing in late spring. This event anticipates the subsequent _E. crocea_ decrease which, in the North Adriatic Sea, typically occurs in June (till October when the resting stages of the quiescent hydrorhizas give rise to new colonies [23]). While it is possible that the hydroid regression can be triggered by environmental factors, such as the progressive temperature increase [23], the decline of _E. crocea_ following the disappearance of the _Polaribacter_ species lets to hypothesize that the hydroid might remain susceptible of infections when not protected by its own associated flora. However, further studies are needed to clarify this hypothesis.

The OTU belonging to the family Comamonadaceae with the highest similarity with the genus _Delftia_ could be associated to the _Type I_ morphotype. This bacterium was less frequently observed compared with the _Type II_ morphotype, and this was also reflected in the abundance of this OTU in the two pyrosequencing libraries. Members of this genus are typically marine bacteria and are associated to marine invertebrates, such as sponges or crustaceans. Several _Delfia_ spp. are reported to have a curved-rod, horseshoe morphologies [30]. Further cultivation analyses aimed at isolating _Delphia_ sp. and _Polaribacter_ (cfr) sp. will allow the identification at the species level and the investigation of their functional role within the host.

SEM analyses indicated that epibiontic bacteria were present on _E. crocea_ samples collected from different geographic areas, but were absent in other hydroid species of the same areas. This suggests the presence of a species-specific relationship, similar to the one described for *Hydra* [24], and supports the hypothesis that these two species are resident species. This is also confirmed by molecular analyses conducted in different periods (March and April), which indicate a potential selection operated by *E. crocea* on the associated bacterial assemblage. In this regard it has been recently shown that *Hydra* is able to distinguish between the epibionts and other potentially pathogenic prokaryotes colonizing its surface [31].

The morphotypes *Type I* and *Type II* were also observed on the newly released actinulae, suggesting the presence of a vertical transmission of the prokaryotes. While studying the reproductive structures of *E. crocea*, it is common to observe some of the actinula tentacles protruding from the gonophore before the release (Figure 2D). This lets to hypothesize that the bacterial colonization of the *E. crocea* hydrants can occur when prokaryotes, present on the gonophore surface, colonize the protruding tentacles of the actinulae or because of accidental contact between actinula tentacles and other portions of the hydranth. However, additional analyses are needed to test for the ability of the actinulae to collect these bacteria from the surrounding seawater.

It is likely that other similar associations between bacteria and hydroids occur, especially within the family Tubulariidae. However, the investigation of prokaryotes-invertebrate associations in the marine environments using modern molecular techniques, is still in its infancy, with most of the studies being referred to Leptomedusan species [6–11,32]. Since the three genera of hydrozoans showing association with bacteria – *Ectopleura*, *Hydra* and *Pennaria* – are all included in the order Capitata [33], we hypothesize that this group of marine invertebrates has the ability to establish specific microbe-host interactions.

Bacteria can interact with cnidarian epidermis in several ways. There is a mutualistic relationship between prokaryotes and *Hydra* polyps: while the prokaryotes obtain nutritional benefits from their host, the hydroid partially loses the ability to intake phosphate (used by symbiotic chlorellae) when bacteria are removed [13]. Non-symbiotic *Hydra* species were not able to bud under bacteria-free conditions, suggesting that associated prokaryotes could

provide a budding factor [34]. Many coral species are able to produce antifouling molecules, and several bacteria species are known to be directly involved in the biosynthesis of these secondary metabolites [35]. Several coral-associated bacteria grow in the coral mucus layer [36] and produce antimicrobial metabolites, that are believed to protect the coral host from pathogens [2]. It is also possible that the bacteria colonizing the surface of *Ectopleura* release substances that discourage predation. The nudibranch *Cuthona gymnota* (Couthouy, 1838), for instance, feeds on the hydroid piercing the stems while never eats the hydranths. However, hydroid cnidocysts can represent an important deterrent toward predators [37]. This study is one of the first reports of bacteria living on hydrozoan epidermis, and indeed the first reporting a stable association of bacteria with the epithelium of *E. crocea*. Our results represent a starting point for future studies, which may provide insights on the significance and the role of this specific association between bacteria and hydrozoa.

Acknowledgments

We are grateful to the staff of the Department of Biology of the University of Padova (Italy) that prepared samples for TEM analyses and to Prof. Cinti and Dr. Fernando of the Polytechnic University of Marche for the possibility to use the electron microscope.

Author Contributions

Conceived and designed the experiments: CGDC GML. Performed the experiments: CGDC GML GG. Analyzed the data: CGDC GML MB. Contributed reagents/materials/analysis tools: GB CC. Wrote the paper: CGDC GML GB CC. Species identification: CGDC GML.

References

1. Ducklow HW, Mitchell R (1979) Bacterial populations and adaptations in the mucus layers on living corals. Limnol Oceanogr 24: 715–725.
2. Rohwer F, Seguritan V, Azam F, Knowlton N (2002) Diversity and distribution of coral-associated bacteria. Mar Ecol Prog Ser 243: 1–10.
3. Gnanambal MEK, Chellaram C, Patterson J (2005) Isolation of antagonistic marine bacteria from the surface of the gorgonian corals at Tuticorin, Southeast coast of India. Indian J Mar Sci 34: 316–319.
4. Dobretsov S, Qian PY (2004) The role of epibotic bacteria from the surface of the soft coral *Dendronephthya* sp. in the inhibition of larval settlement. J Exp Mar Biol Ecol 299: 35–50.
5. Santiago-Vázquez LZ, Brück TB, Brück WM, Duque-Alarcón AP, McCarthy PJ, et al. (2007) The diversity of the bacterial communities associated with the azooxanthellate hexacoral *Cirrhipathes lutkeni*. Int Soc Micro Ecol J 1: 654–659.
6. Romagnoli T, Bavestrello G, Cucchiari E, De Stefano M, Di Camillo CG, et al. (2007) Microalgal communities epibiontic on the marine hydroid *Eudendrium racemosum* in the Ligurian Sea, during an annual cycle. Mar Biol 151:537–552.
7. Bavestrello G, Cerrano C, Di Camillo CG, Puce S, Romagnoli T, et al. (2008) The ecology of protists epibiontic of marine hydroids. J Mar Biol Ass UK 88: 1611–1617.
8. Di Camillo C, Puce S, Romagnoli T, Tazioli S, Totti C, et al. (2006) Coralline algae epibionthic on thecate hydrozoans (Cnidaria). J Mar Biol Ass UK 86: 1285–1289.
9. Di Camillo CG, Bo M, Lavorato A, Morigi C, Segre Reinach M, et al. (2008) Foraminifers epibiontic on *Eudendrium* (Cnidaria: Hydrozoa) from the Mediterranean Sea. J Mar Biol Ass UK 88: 485–489.
10. Stabili L, Gravili C, Piraino S, Boero F, Alifano P (2006) *Vibrio harveyi* associated with *Aglaophenia octodonta* (Hydrozoa, Cnidaria). Microb Ecol 52: 603–608.
11. Stabili L, Gravili C, Tredici SM, Piraino S, Talà A, et al. (2008) Epibiotic *Vibrio* luminous bacteria isolated from some Hydrozoa and Bryozoa species. Microb Ecol 56: 625–636.
12. Gorelova OA, Kosevich IA, Baulina OI, Fedorenko TA, Torshkhoeva AZ, et al. (2009) Associations between the White Sea Invertebrates and Oxygen-Evolving Phototrophic Microorganisms. Moscow Univ Biol Sci Bull 64: 16–22.
13. Wilkerson FP (1980) Bacterial symbionts on green hydra and their effect on phosphate uptake. Microb Ecol 6: 85–92.
14. Östman C (2000). A guideline to nematocyst nomenclature and classification, and some notes on the systematic value of nematocysts. Sci Mar 64: 31–46.
15. Hufnagel LA, Myhal ML (1977) Observations on a Spirochaete Symbiotic in *Hydra*. Trans Amer Micr Soc 96: 406–411.
16. Margulis L, Thorington G, Berger B, Stolz J (1978) Endosymbiotic bacteria associated with the intracellular green algae of *Hydra viridis*. Current Microbiol 1: 227–232.
17. Schuett C, Doepke H (2009) Endobiotic bacteria and their pathogenic potential in cnidarian tentacles. Helgol Mar Res 64: 205–212.
18. Lee OO, Wang Y, Yang J, Lafi FF, Al-Suwailem A, et al. (2011) Pyrosequencing reveals highly diverse and species-specific microbial communities in sponges from the Red Sea. The ISME Journal 5: 650–664.
19. Barott KL, Rodriguez-Brito B, Janouškovec J, Marhaver KL, Smith JE, et al. (2011) Microbial diversity associated with four functional groups of benthic reef algae and the reef-building coral *Montastraea annularis*. Environ Microbiol 13: 1192–1204.
20. Wang Y, Qian PY (2009) Conservative fragments in bacterial 16S rRNA genes and primer design for 16S ribosomal DNA amplicons in metagenomic studies. PLoS ONE 4: e7401.

21. Sogin ML, Morrison HG, Huber JA, Welch DM, Huse SM, et al. (2006) Microbial diversity in the deep sea and the underexplored "rare biosphere". PNAS 103: 12115–12120.

22. Cole JR, Wang Q, Cardenas E, Fish J, Chai B, et al. (2009) The Ribosomal Database Project: improved alignments and new tools for rRNA analysis. Nucleic Acids Res. 37 (Database issue): D141–D145.

23. Di Camillo CG, Giordano G, Bo M, Betti F (in press) Seasonal patterns in the abundance of *Ectopleura crocea* (Cnidaria: Hydrozoa) on a shipwreck in the Northern Adriatic. Mar Ecol.

24. Fraune S, Bosch CG (2007) Long-term maintenance of species-specific bacterial microbiota in the basal metazoan *Hydra*. PNAS 104: 13146–13151.

25. Di Camillo C, Puce S, Romagnoli T, Tazioli S, Totti C, et al. (2005) Relationships between benthic diatoms and hydrozoans (Cnidaria). J Mar Biol Ass UK 85: 1373–1380.

26. Gravili C, Boero F, Alifano P, Stabili L (2011) Association between luminous bacteria and Hydrozoa in the northern Ionian Sea. J Mar Biol Ass UK (in press).

27. Neave MJ, Streten-Joyce C, Glasby CJ, McGuinness KA, Parry DL, et al. (2011) The bacterial community associated with the marine polychaete *Ophelina* sp. 1 (Annelida: Opheliidae) is altered by copper and zinc contamination in sediments. Microb Ecol online early DOI 10.1007/s00248-011-9966-9.

28. Bowman JP, McCammon SA, Lewis T, Skerratt JH, Brown JL, et al. (1998) *Psychroflexus torquis* gen. nov., sp. nov., a psychrophilic species from Antarctic sea ice, and reclassification of *Flavobacterium gondwanense* (Dobson et al. 1993) as *Psychroflexus gondwanense* gen. nov., comb. nov. Microbiol 144: 1601–1609.

29. Nedashkovskaya OI, Kim SB, Lysenko AM, Kalinovskaya NI, Mikhailov V, et al. (2005) *Polaribacter butckevichii*, a novel marine mesophilic bacterium of the family Flavobacteriaceae. Current Microbiology 51: 408–412.

30. Wen A, Fegan M, Hayward C, Chakraborty S, Sly LI (1999) Phylogenetic relationships among members of the Comamonadaceae, and description of *Delftia acidovorans* (den Dooren de Jong 1926 and Tamaoka et al. 1987) gen. nov., comb. nov. Int J Syst Evol Microbiol 49: 567–576.

31. Altincicek B (2009) The innate immunity in the cnidarian *Hydra vulgaris*. Inv Surv J 6: 106–113.

32. Svoboda A, Cornelius PFS (1991) The European and Mediterranean species of *Aglaophenia* (Cnidaria: Hydrozoa). Zool Verh Leiden 274: 1–72.

33. Collins AG, Winkelmann S, Hadrys H, Schierwater B (2005) Phylogeny of Capitata and Corynidae (Cnidaria, Hydrozoa) in light of mitochondrial 16S rDNA data. Zool Scripta 34: 91–99.

34. Rahat M, Dimentman C (1982) Cultivation of bacteria-free *Hydra viridis*: missing budding factor in nonsymbiotic hydra. Science 216: 67–68.

35. Proksch P, Edrada RA, Ebel R (2002) Drugs from the seas – current status and microbiological Implications. Appl Microbiol Biotechnol 59: 125–134.

36. Sharon G, Rosenberg E (2008) Bacterial growth on coral mucus. Curr Microbiol 56: 481–488.

37. Stachowicz JJ, Lindquist N (2000) Hydroid defenses against predators: the importance of secondary metabolites versus nematocysts. Oecol 124: 280–288.

PERMISSIONS

LIST OF CONTRIBUTORS

Francisco Barros-Becker, Alvaro Pulgar and Carmen G. Feijóo
Departamento de Ciencias Biologicas, Facultad de Ciencias Biologicas, Universidad Andres Bello, Santiago, Chile

Jaime Romero
Instituto de Nutrición y Tecnología de los Alimentos, Universidad de Chile, Santiago, Chile

Sarah A. Castine
AIMS@JCU, School of Marine and Tropical Biology, Australian Institute of Marine Science, Centre for Sustainable Tropical Fisheries and Aquaculture, James Cook University, Townsville, Queensland, Australia

Dirk V. Erler and Bradley D. Eyre
School of Environmental Science and Management, Centre for Coastal Biogeochemistry, Southern Cross University, Lismore, New South Wales, Australia

Lindsay A. Trott
Australian Institute of Marine Science, Townsville, Queensland, Australia

Nicholas A. Paul and Rocky de Nys
School of Marine, Tropical Biology and Centre for Sustainable Tropical Fisheries and Aquaculture, James Cook University, Townsville, Queensland, Australia

Darren M. Green, David J. Penman, Herve Migaud, James E. Bron, John B. Taggart and Brendan J. McAndrew
Institute of Aquaculture, University of Stirling, Stirling, Stirlingshire, United Kingdom

Hsiao-Che Kuo, Young-Mao Chen and Tzong-Yueh Chen
Laboratory of Molecular Genetics, Institute of Biotechnology, National Cheng Kung University, Tainan, Taiwan
Translational Center for Marine Biotechnology, National Cheng Kung University, Tainan, Taiwan
Agriculture Biotechnology Research Center, National Cheng Kung University, Tainan, Taiwan
University Center for Bioscience and Biotechnology, National Cheng Kung University, Tainan, Taiwan
Research Center of Ocean Environment and Technology, National Cheng Kung University, Tainan, Taiwan

Ting-Yu Wang and Hao-Hsuan Hsu
Laboratory of Molecular Genetics, Institute of Biotechnology, National Cheng Kung University, Tainan, Taiwan
Translational Center for Marine Biotechnology, National Cheng Kung University, Tainan, Taiwan

Szu-Hsien Lee
Institute of Nanotechnology and Microsystems Engineering, National Cheng Kung University, Tainan, Taiwan
Department of Engineering Science, National Cheng Kung University, Tainan, Taiwan

Tieh-Jung Tsai and Ming-Chang Ou
Laboratory of Molecular Genetics, Institute of Biotechnology, National Cheng Kung University, Tainan, Taiwan

Hsiao-Tung Ku
Research Division I, Taiwan Institute of Economic Research, Taipei, Taiwan
Office for Energy Strategy Development, National Science Council, Taipei, Taiwan

Gwo-Bin Lee
Institute of Nanotechnology and Microsystems Engineering, National Cheng Kung University, Tainan, Taiwan
Department of Engineering Science, National Cheng Kung University, Tainan, Taiwan
Department of Power Mechanical Engineering, National Tsing Hua University, Hsinchu, Taiwan

Luke A. Rogers
Department of Zoology, University of Otago, Dunedin, Otago, New Zealand

Stephanie J. Peacock
Department of Biological Sciences, University of Alberta, Edmonton, Alberta, Canada

Peter McKenzie
Mainstream Canada, Campbell River, British Columbia, Canada

Sharon DeDominicis
Marine Harvest Canada, Campbell River, British Columbia, Canada

Simon R. M. Jones
Pacific Biological Station, Fisheries and Oceans Canada, Nanaimo, British Columbia, Canada

Peter Chandler and Michael G. G. Foreman
Institute of Ocean Sciences, Fisheries and Oceans Canada, Sidney, British Columbia, Canada

Crawford W. Revie
Atlantic Veterinary College, University of PEI, Charlottetown, Prince Edward Island, Canada

Martin Krkošek
Department of Zoology, University of Otago, Dunedin, Otago, New Zealand
Department of Ecology and Evolutionary Biology, University of Toronto, Toronto, Ontario, Canada

Yifei Dong, Hua Tian, Wei Wang, Xiaona Zhang, Jinxiang Liu and Shaoguo Ru
Marine Life Science College, Ocean University of China, Qingdao, Shandong Province, The People's Republic of China

Nina Sandlund
Research group Disease and Pathogen transmission, Institute of Marine Research, Bergen, Norway

Britt Gjerset, Renate Johansen and Ingebjørg Modahl
Section of Virology, National Veterinary Institute, Oslo, Norway

Øivind Bergh
Research group Oceanography and climate, Institute of Marine Research, Bergen, Norway

Niels Jørgen Olesen
Section of Virology, Technical University of Denmark, Frederiksberg C, Denmark

Mohd Ashraf Rather, Rupam Sharma, Subodh Gupta, S. Ferosekhan, V. L. Ramya and Sanjay B. Jadhao
Central Institute of Fisheries Education, Versova, Mumbai, India

Leydis Zamora and Lucas Domínguez
Centro de Vigilancia Sanitaria Veterinaria (VISAVET), Universidad Complutense, Madrid, Spain

José F. Fernández-Garayzábal and Ana I. Vela
Centro de Vigilancia Sanitaria Veterinaria (VISAVET), Universidad Complutense, Madrid, Spain
Departamento de Sanidad Animal, Facultad de Veterinaria, Universidad Complutense, Madrid, Spain

Cristina Sánchez-Porro and Antonio Ventosa
Departamento de Microbiología y Parasitología, Facultad de Farmacia, Universidad de Sevilla, Sevilla, Spain

Mari Angel Palacios
Piszolla, S.L., Alba de Tormes, Salamanca, Spain

Edward R. B. Moore
Culture Collection University of Gothenburg (CCUG) and Department of Infectious Disease, Sahlgrenska Academy of the University of Gothenburg, Göteborg, Sweden

Titus Chemandwa Ndiwa
Kenya Wetlands Biodiversity Research Group, Ichthyology Section, National Museums of Kenya, Nairobi, Kenya
Département Conservation et Domestication, UMR IRD 226 CNRS 5554, Institut des Science de l'Evolution, Universitéde Montpellier 2, Montpellier, France

Dorothy Wanja Nyingi
Kenya Wetlands Biodiversity Research Group, Ichthyology Section, National Museums of Kenya, Nairobi, Kenya

Jean-François Agnese
Département Conservation et Domestication, UMR IRD 226 CNRS 5554, Institut des Science de l'Evolution, Universitéde Montpellier 2, Montpellier, France

Kevin A. Glover, Vidar Wennevik, François Besnier, Anne G. E. Sørvik and Øystein Skaala
Section of Population Genetics and Ecology, Institute of Marine Research, Bergen, Norway

María Quintela
Dept of Animal Biology, Plant Biology and Ecology, University of A Coruña, Spain

Clara Boglione, Michele Scardi, Elisa Palamara, Tommaso Russo and Stefano Cataudella
Laboratory of Experimental Ecology and Aquaculture, Department of Biology, "Tor Vergata" University of Rome, Rome, Italy

Domitilla Pulcini
Laboratory of Experimental Ecology and Aquaculture, Department of Biology, "Tor Vergata" University of Rome, Rome, Italy
Council for Research in Agriculture – Animal Production Centre, Rome, Italy

Chunsheng Liu., Hongyan Xu., Siew Hong Lam, Zhiyuan Gong
Department of Biological Sciences, National University of Singapore, Singapore, Singapore

Zhi-Hua Li
Department of Biological Sciences, National University of Singapore, Singapore
Yangtze River Fisheries Research Institute, Chinese Academy of Fishery Sciences, Wuhan, China

Hongyan Xu, Weiling Zheng, Siew Hong Lam and Zhiyuan Gong
Department of Biological Sciences, National University of Singapore, Singapore

Tadao Kunihiró
Department of Ecosystem Studies, Royal Netherlands Institute of Sea Research (NIOZ), Yerseke, The Netherlands
Department of Marine Microbiology, Royal Netherlands Institute of Sea Research (NIOZ), Yerseke, The Netherlands

Bart Veuger, Lara Pozzato, Marie Le Guitton and Dick van Oevelen
Department of Ecosystem Studies, Royal Netherlands Institute of Sea Research (NIOZ), Yerseke, The Netherlands

Kazuyoshi Moriya
School of Natural Science & Technology, Kanazawa University, Kakuma-machi, Kanazawa, Japan

Michinobu Kuwae and Koji Omori
Center for Marine Environmental Studies (CMES), Ehime University, Matsuyama, Ehime, Japan

Henricus T. S. Boschker
Department of Marine Microbiology, Royal Netherlands Institute of Sea Research (NIOZ), Yerseke, The Netherlands

Cristina Gioia Di Camillo, Giuseppe Giordano and Cinzia Corinaldesi
Department of Life and Environmental Sciences, Polytechnic University of Marche, Ancona, Italy

Gian Marco Luna
Institute of Marine Sciences–National Research Council (CNR), Venice, Italy

Marzia Bo and Giorgio Bavestrello
Department for the Study of the Land and its Resources, University of Genoa, Genoa, Italy

Index

Printed in the USA
CPSIA information can be obtained
at www.ICGtesting.com
JSHW052005011124
72840JS00003B/29